AN INTRODUCTION TO MOLECULAR MEDICINE AND GENE THERAPY

Dedicated with Love to my wife,
Laura Williams Cheever,
And my children,
Rachel Ann
Jenny Lynn
Rebecca Marie

CONTENTS

The research field of gene therapy and the clinical practice of medicine evolving from the research are fast moving, ever-changing disciplines. On an almost daily basis, there appears in the media a "breaking story" of a gene-based research finding. An implication of the story is that this research breakthrough will speedily transform, in the next few years, into a marvelous new therapy in molecular genetic medicine. *An Introduction to Gene Therapy and Molecular Medicine* provides a basis to interpret new clinical and basic research findings in the areas of cloning, gene transfer and targeting, the application of genetic medicine to clinical conditions, ethics, government regulation, genomics, and biotechnology and bioinformatics. The text provides the reader with fundamental and comprehensive basic as well as clinical research observations and findings relative to gene therapy and molecular medicine.

An Introduction to Gene Therapy and Molecular Medicine can be divided into three sections: basic science introductory Chapters 1 to 5; clinical application Chapters 6 to 12; and Chapters 13 to 15, and Appendix, addressing evolving issues related to gene therapy and molecular medicine. Each chapter, as well as the appendix, contains key concepts that the authors wish to leave the readers and a specific itemized listing of suggested readings in the field. The reading lists comprise state-of-the-art reviews, salient research articles, and articles useful for lay readership as well.

Chapter 1 is truly an overview of the field and the contents of the book. It presents in broad general terms the diseases targeted by gene therapy and the tools researchers use and the future needs of the field. It address what may be the "holy grail" of medicine, possible approaches for reversing the process of aging. A second approach to the aging issue is presented in Chapter 2. The techniques and usage of cloning are addressed. Chapter 3 provides a fundamental need for basic research in addressing human disease and the generation and use of animal models of disease. Specifically useful for gene therapy and molecular medicine are transgenic mouse models of human pathogenesis. Chapter 4 provides what appears to be endless detail related to vectors and their use in gene transfer. Chapter 5 rounds out the section on basic research by providing useful approaches to target genes to produce a specific desired expression of the gene.

Chapters 6 to 12 provide the clinical use of gene therapy approaches. Chapter 6 presents the use of gene therapy approaches to hematology with a special discussion of application of gene therapy using hemopoietic stem cells. Chapter 7 presents gene therapy in liver diseases describing approaches for inherited metabolic diseases as well as acquired infectious diseases such as hepatitis C. Chapter 8 presents impressive, realistic approaches to use gene therapy for the therapy of "broken hearts" and cardiovascular diseases. Chapter 9 presents equally challenging and controversial molecular approaches for gene therapy of the brain and diseases of the central nervous system. Chapter 10 provides an overview of how gene therapy can be used in the treatment of cancer. Chapter 11 details gene therapy approaches for infectious diseases with a specific emphasis on utilization for the therapy of HIV infection. Chapter 12 provides a "disease contrast" in that it addresses an incredibly debilitating disease, rheumatoid arthritis, and how gene therapy can be used in amelioration of joint destruction.

The last section of the book provides individual presentations related to gene therapy and molecular medicine. Chapter 13 provides an update on the issue of federal regulation and oversight of gene therapy research. It is an issue that has been in the spotlight of late. Chapter 14 provides an ethical essay on gene therapy and the use of molecular medicine with an insight into health care rationing. Chapter 15 is a brief description of where the practice of molecular medicine is today. Finally, the appendix is an important presentation of some commercial aspects in molecular medicine and gene therapy. After all, if gene therapy is to reach the public in all walks of life a commercial venue is needed.

Thus, *An Introduction to Gene Therapy and Molecular Medicine* is a comprehensive manual that can be used as an aid through the rapidly moving field of gene therapy and its application in molecular medicine.

Thomas F. Kresina
Bethesda, Maryland

CONTRIBUTORS

Andrea D. Branch, Ph.D., Department of Medicine, Mount Sinai School of Medicine, One Gustave L. Levy Place, New York, NY 10029

Bruce Bunnell, M.D., Children's Hospital Research Foundation, 700 Children's Drive, Columbus, Ohio 43205-2696

Barbara A. Conley, M.D., Chief, Clinical Investigations Branch, Cancer Therapy Evaluation Program, Division of Cancer Therapy and Diagnosis, National Cancer Institute, Rockville, MD 20852

Laurie C. Doering, Ph.D., Department Pathology and Molecular Medicine, Health Science Center, McMaster University, 1200 Main Street West, Hamilton, Ontario, L8N 325, Canada

Cynthia E. Dunbar, M.D., Division of Intramural Research, National Heart, Lung, and Blood Institute, National Institutes of Health, Bethesda, MD 20892

Victor J. Dzau, M.D., Chairman of Medicine, Brigham and Women's Hospital, Harvard Medical School, Boston, MA 0215-6195

Afshin Ehsan, M.D., Research Institute and Department of Medicine, Harvard Medical School and Brigham and Women's Hospital, 75 Francis Street, Boston, MA 02115

Leonard M. Fleck, Ph.D., Center for Ethics and Humanities, C-201 East Fee Hall, Michigan State University, East Lansing, MI 48824-1316

Renate E. Gay, M.D., WHO Collaborating Center for Molecular Biology and Novel Therapeutic Strategies for Rheumatic Diseases, Department of Rheumatology, University Hospital, Gloriastrasse 25, CH-8091 Zurich

Steffen Gay, M.D., WHO Collaborating Center for Molecular Biology and Novel Therapeutic Strategies for Rheumatic Diseases, Department of Rheumatology, University Hospital, Gloriastrasse 25, CH-8091 Zurich

Simon J. Hall, M.D., Department of Urology, The Institute for Gene Therapy & Molecular Medicine, Mount Sinai School of Medicine, New York, NY 10029

D. Joseph Jerry, Ph.D., Department of Veterinary and Animal Sciences, University of Massachusetts, Amherst, Amherst, MA 01003

William C. Kisseberth, Department of Medicine, Hematology-Oncology Unit, University of Rochester Cancer Center, 601 Elmwood Avenue, Box 704, Rochester, NY 14642

Eric B. Kmiec, Ph.D., Deptartment of Biological Sciences, University of Delaware, Newark, DE 19716

Thomas F. Kresina, Ph.D., National Institute on Alcohol Abuse and Alcoholism, Room 602, 6000 Executive Blvd, NIH, Bethesda MD 20892-6600

Deborah Y. Kwoh, The Immune Response Corporation, Carlsbad, CA 92008

Charles Lollo, Ph.D., Associate Director, Gene Therapy & Chemistry, The Immune Response Corporation, Carlsbad, CA 92008

Michael J. Mann, M.D., Research Institute and Department of Medicine, Harvard Medical School and Brigham and Women's Hospital, 75 Francis Street, Boston, MA 02115

Roy Musil, The Immune Response Corporation, Carlsbad, CA 92008

Thomas Pap, M.D., WHO Collaborating Center for Molecular Biology and Novel Therapeutic Strategies for Rheumatic Diseases, Department of Rheumatology, University Hospital, Gloriastrasse 25, CH-8091 Zurich

Katherine Parker Ponder, M.D., Department of Biochemistry and Molecular Biophysics, Washington University School of Medicine, Box 8125, 660 South Euclid Avenue, St. Louis, MO 63110

James Robl, Ph.D., Department of Veterinary and Animal Sciences, University of Massachusetts, Amherst, Amherst, MA 01003

Eric Sandgren, V.M.D. Ph.D., Department of Pathobiological Sciences, School of Veterinary Medicine, University of Wisconsin-Madison, 2015 Linden Drive West, Madison, WI 53706-1102

Christy L. Schilling, Division of Gastroenterology-Hepatology, Department of Medicine, The University of Connecticut Health Center, Farmington, CT 06030

Martin J. Schuster, Division of Gastroenterology-Hepatology, Department of Medicine, The University of Connecticut Health Center, Farmington, CT 06030

Richard Trauger, Gene Therapy & Chemistry, The Immune Response Corporation, Carlsbad, CA 92008

George Wu, M.D., Ph.D., Division of Gastroenterology-Hepatology, Department of Medicine, The University of Connecticut Health Center, Farmington, CT 06030

Tong Wu, M.D., Hematology Branch, Building 10, Room 7C103, National Heart, Lung, and Blood Institute, Bethesda, MD 20892

AN INTRODUCTION TO MOLECULAR MEDICINE AND GENE THERAPY

CHAPTER 1

Molecular Medicine and Gene Therapy: An Introduction

THOMAS F. KRESINA, PH.D. and ANDREA D. BRANCH, PH.D.

INTRODUCTION

The use of genetics and genetic manipulation by humans for the therapy of human disease is a new and rapidly evolving field of both basic science and clinical medicine. The science of gene therapy is derived from significant research advances in the fields of genetics, molecular biology, clinical medicine, and human genomics. Thus, *gene therapy* can be defined as the use of genetic manipulation for treatment of disease. Experimental gene therapy research breakthroughs observed in model systems are modified for clinical or bedside use, forming the emerging practice of molecular medicine. Molecular medicine encompasses the elucidation of the genetic basis of disease, diagnosis of the disease, the design of an appropriate approach to disease management or therapy, the application of approved therapeutic protocols, and monitoring of clinical outcomes.

In the history of the practice of western medicine, initial concepts of disease were related to an imbalance in the persona or humus. Illness was treated on a whole-body or systemic level. As the practice of medicine advanced to and through the twentieth century, more information became available regarding the physiology of the body as well as its organ and tissue structure. Subsequently, advances were made into the cellular biology of health and disease. Most recently, research investigations opened insight into the genetic basis of inheritance and the biological processes at the molecular level. These were mainly in the genetics and molecular biology of selective breeding practices for plants and animals. The basic principles form a nidus for experimental treatments for human diseases.

The bases for this application to human disease are the successful development of the medical and surgical techniques in human organ transplantation, the western tradition of pharmacotherapy, and the continuing elucidation of the human genome and its regulatory elements. On what seems to be an almost daily basis, startling new molecular genetic discoveries are publicized. Some have profound moral

An Introduction to Molecular Medicine and Gene Therapy, Edited by Thomas F. Kresina
ISBN 0-471-39188-3 © 2001 Wiley-Liss

and ethical considerations, such as the cloning of sheep and primates. Others lead to a profound understanding of the pathogenesis of human disease, such as the identification of the mutation in the genes responsible for liver diseases, such as, hemochromatosis or, in pediatrics, Alagille syndrome. The cloning studies show us the new frontiers of genetic medicine and challenge us to use them wisely. The discoveries of mutant genes leading to disease pathology lend the promise of rapid diagnosis and potentially early clinical intervention allowing for better medical management. However, the discoveries of genes responsible for human pathology challenge us in the use of genetic population screening. The evolving field of genetic epidemiology can provide precise data on the incidence and prevalence of a specific inherited trait. The challenge here is to use this information ethically and in a medically beneficial manner (see Chapter 14).

GENETIC MANIFESTATIONS OF MOLECULAR MEDICINE

Gene therapy offers the potential of a one-time cure for devastating inherited disorders. It has application to many diseases for which current therapeutic approaches are ineffective or where the prospects for effective treatment are obscure. Current recombinant deoxyribonucleic acid (DNA) technologies allow for the rapid identification of genes and the facile manipulation of genetic material. This enables medical researchers to examine cellular physiology at a molecular level. Using these tools, scientists and clinicians can identify and determine a molecular basis of disease. There is a broad array of diseases in which specific protocols of gene therapy could provide novel therapeutic approaches. These are the "traditional genetic diseases" so called for their familiarity in clinical medicine (see Table 1.1). They consist of chromosomal disorders that are inherited as a single gene, Mendelian disorder (autosomal dominant, autosomal recessive, sex-linked recessive, or sex-linked dominant), and result from a mutation at a single locus. These compare to the multifactorially inherited disorders that involve multiple genes working in concert with known or enigmatic environmental factors.

Most diseases are complex and multifactorial. They result from a complex series of events involving changes in the level of expression of many genes and/or environmental factors and behavior. While many individual interventions may be partially effective at treating complex diseases, the greatest benefits are likely to be derived from combination therapies. Although complexity is the rule in human pathogenesis, many first-generation gene therapies are designed as a single intervention to correct a disease by adding a functional version of a single defective gene, as illustrated in Figure 1.1a. Such strategies, for example, have been used to introduce a specific gene into the liver cells of patients with familial hypercholesterolemia (see Chapters 6 and 7). But, it is estimated that only 2% of human diseases are thought to be caused by direct one-to-one Mendelian expression of a single gene. Even in these monogenetic diseases, clinical heterogeneity occurs, and it is often difficult to predict the progress of the clinical course of a patient. Patient-to-patient variation results from many factors, including differences in alleles, environment, and genetic background. While the precise cause of variable penetrance of a genetic lesion is usually not known, it likely reflects the genome's extensive series of "back-up" systems and feedback loops. For example, this premise has been

TABLE 1.1 Selected Inherited Disorders and Their Genetic Basis

Classification	Nomenclature	Characterization	Frequency
Autosomal aneuploidies newborns	Trisomy 13	Karyotype: 47,XX or XY +13 (extra copy)	1 per 12,000
	Trisomy 18	Karyotype: 47 XX or XY +18 (extra copy)	1 per 6000 newborns
	Trisomy 21 Down's syndrome	Karyotype: 47,XX or XY +21 (extra copy)	1 per 800 newborns ↑ incidence with age
Sex chromosome aneuploidies	Klinefelter's syndrome	Karyotype: 47, XXY plus variants	1 per 700 newborns males
	Triple X female	Karyotype: 47,XXX	1 per 1000 newborns
	Turner's syndrome	Karyotype: 45,X; 45X/46XX or 45X/46XY	1 per 1500 newborn females
	XYY male	Karyotype: 47,XXY	1 per 800 newborns
Autosomal dominant	Aniridia, type I	Chromosome 2 defect	1 per 80,000
	Aniridia, type II	Chromosome 11 defect	1 per 80,000
	Polycystic kidney disease	Chromosome 16 linkage	1 per 1250
	Charcot–Marie–Tooth	Two forms type I and II	1 per 2800
	Familial polyposis coli and Gardner's syndrome	Chromosome 5; adenomatous polyposis coli (APC) gene	1 per 8000
	Huntington's disease	Linked to chromosome 4p	1 per 3000
	Intrahepatic cholestasis	Vanishing bile ducts	
	Alagille syndrome	*Jagged 1* gene—20p12	1 per 70,000
	Byler's disease	18q21	familial
	Marfan's syndrome	Chromosome 15: FBN1 gene	1 per 20,000
	Myotonic dystrophy	19q13.2–q13.3	1 per 8000
	Neurofibromatosis		
	Type I	Chromosome 17: NF-1 gene 17q11.2	1 per 2000–5000
	Type II	Chromosome 22: NF-2 gene 22q12.2	
	Retinoblastoma	Deletion or rearrangement chromosome 13 RB-1 gene	1 per 20,000
	Pancreatitis hereditary	Chromosome 7 cationic trypsinogen gene PRSSI Two mutations: R117H & N21I	Familial

TABLE 1.1 (*Continued*)

Classification	Nomenclature	Characterization	Frequency
	Idiopathic	SPINKI-Chromosome 5 Missense mutation- N345	
Autosomal recessive	α_1-Antitrypsin deficiency	Chromosome 14 Multiple alleles based on phenotype M, S, Z, I	1 per 3500
	Cystic fibrosis	7q31–q32, CFTR gene Multiple alleles: Δ 508 ↑ Also R117H, R75Q, D1270N	1 per 2500 (Caucasians)
	Gaucher's disease Ashkenazic Jewish descent	N370S allele (nonneuropathic)	1 per 625
	Caucasian population	L444P allele neuropathic	
	Hemochromatosis	HFE gene C282Y and H63D mutations	1 per 300
	Thalassemia (α)	Globulin gene complex on chromosome 16 Two alles α-thal 1 α-thal 2	1 per 250–1000
	Thalassemia (β)	Chromosome 11 Two alleles β(+) IVS-I β(+) IVS-II	

shown in several lines of "knock-out" mice, which lack genes involved in key cellular processes (see Chapter 3). Such mice can be phenotypically normal. Thus, the genome has an impressive ability to compensate for a missing part. Because of this ability, the most effective treatments for single-gene diseases may not always be replacement of the single defective gene. Options may exist as illustrated in Figure 1.1*b*, where either a functional copy of a frankly defective gene could be added to correct a deficiency (yielding genotype 3) or expression of a compensatory gene could be enhanced (yielding genotype 4).

Monogenetic Disorders

Single-gene disorders are relatively infrequent in incidence but contribute significantly to the chronic disease burden. They include sickle cell anemia, the hemophilias, inherited immune deficiency disorders such as adenosine deaminase deficiency, hypercholesterolemia, severe combined immune deficiency syndrome, as well as the inherited disorders of cystic fibrosis, phenylkentouria, Duchenne's

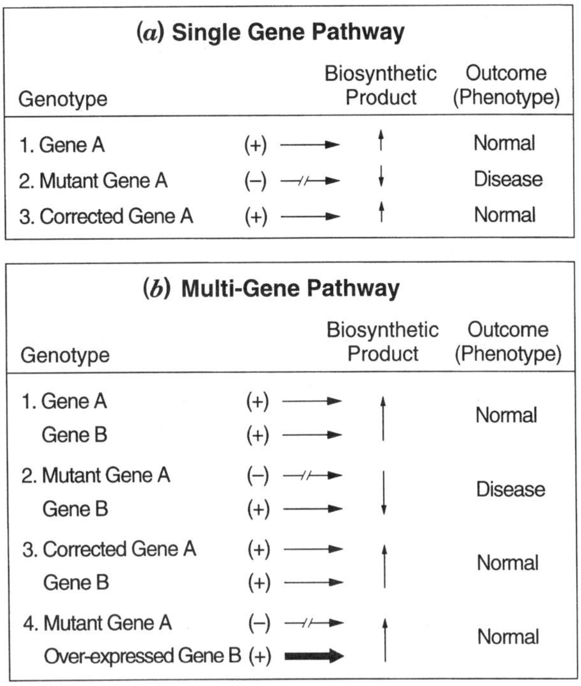

FIGURE 1.1 Pathology can result from a single gene defect, as illustrated in (*a*). More often, multiple genes are involved. In the latter case, a variety of gene therapy options may exist, as depicted in (*b*).

muscular dystrophy, emphysema, and fragile X syndrome. In deficiency disorders, pathology is a direct result of loss of function of the relevant protein. The straightforward application of gene therapy is replacement. Thus, the mutation needs to be identified and the normal gene isolated. In such situations, the transfer and (importantly) correct expression of the protein would benefit the patient, hopefully to the level of curative. In other dominantly inherited disorders where the presence of an abnormal protein interferes with the function and development of organ or tissue, only selective deletion of the mutant gene would be of benefit. Other diseases that are autosomal recessive (requiring two mutant alleles) manifest themselves in utero or at birth and thus require early diagnosis and intervention. Other difficulties in somatic gene therapy for monogenetic disorders are the necessity of direct therapy to a specific tissue or cell type, the number of cells or fraction of tissue needed to be transformed for therapy, and achievement of the therapeutic level of protein along with the long-term regulation of gene expression.

Mutifactorial Disorders

Multifactorial or polygenic disorders are well known because of their common occurrence in the population. In general, they involve several genes. An in-depth knowledge of the pathophysiology of the disease is required to discern the mechanism for therapy by gene-based therapeutic approaches. Examples of these disorders are coronary heart disease, diabetes mellitus, and essential hypertension.

Therefore, multifactorial disorders may not only have a complex genetic component but also be influenced by environmental factors. Elucidation of the pathophysiology of the disorder may suggest how the insertion of a specific gene may reverse or retard disease progression. For these diseases, it may be of most clinical importance to determine how a specific gene product influences tissue or cellular physiology. Currently, gene therapy for these disorders is in a relatively early stage of development.

When designing an appropriate approach to genetic disease management or gene therapy, it is important to ascertain the level of interactions between genes because the majority of diseases causing death in the United States result from processes influenced by many genes. These diseases are polygenic and/or epigenetic in origin. Epigenetic phenomena, such as imprinting, reflect the "state" of a gene and are influenced by environmental factors. Some measure of the magnitude of the gene expression changes that occur during a diseased state was provided by a recent comparison of gene expression profiles in normal and cancer cells (see Chapter 10). Using cellular DNAs (cDNA) as messenger ribonucleic acid (mRNA) surrogate markers of gene activation, it was found that almost 300 genes were expressed at significantly different levels in gastrointestinal tumors compared to normal tissue. The differential activation of such a large number of genes infers that all the genes will not be regulated through common mechanisms. Similar studies are now proceeding in the field of obesity research where the genetic basis of this disease is being elucidated. Thus, it is fundamental to the understanding of disease pathogenesis to identify all genes involved. Specific targeted interventions can then be aimed at the most accessible pathogenic targets. Since multiple experimental therapeutic approaches exist for treating even a "simple" monogenetic disorder, it will be most important to lay the groundwork for considering the potential numerous interventions for the multifactorial diseases that cause morbidity and mortality in the United States.

A specific example of the genetic manifestations of molecular medicine can be seen with the liver disease, α_1-antitrypsin deficiency (see Chapter 7). This liver disease results from a relatively common genetic lesion, in that, about 1 in 8000 infants born in the United States is homozygous for the most frequent mutant allele. Two entirely different organ-specific pathogenic processes can occur in these individuals. Liver injury can result from the accumulation of improperly folded α_1-antitrypsin protein in the endoplasmic reticulum of cells. Lung injury in the form of emphysema can result from the unrelenting proteolytic attack on lung elastin caused by the absence of α_1-antitrypsin. The severity of disease in individuals homozygous for the mutated gene is highly variable, indicating that the impact of the single-gene mutation depends on the "genetic background" of the individual. This example illustrates how the activity of compensatory genes can determine whether a genetic lesion becomes a genetic disease, suggesting that the up-regulation of compensatory genes might be an effective strategy for treating patients with certain genetic mutations.

For diseases that result in multiple organ-specific pathologies, one can question whether both organ pathologies can be cured by a gene therapy that merely adds a correct copy of the wild-type gene. In the case of the liver disease, α_1-antitrypsin deficiency, antisense strategies and ribozymes are being designed to destroy the mRNA of the mutant gene in an effort to eliminate the misfolded protein (see

Chapter 11). However, directed mutagenesis (induced by specialized oligonu-cleotides) is being explored as a way to repair the mutant gene and thereby "killing two birds with one stone" through the elimination of the aberrant protein as well as providing a source of functional polypeptide (gene product) at the same time.

GENE THERAPY AND PATTERNS OF GENE EXPRESSION

The clinical complexities of α_1-antitrypsin deficiency provide a window into the relationship between genotype and phenotype. The goal of somatic (nongermline) gene therapy is to achieve a healthy phenotype by manipulating gene expression. Gene therapy, thereby, corrects or compensates for genetic lesions or deficiencies whether inherited or acquired. Fully achieving this goal requires insight not only into the ways genes interact with each other, but also with the way genes interact with the environment. In biological systems, information flows in two directions—from the genome outward and from the extracellular milieu inward. Gene products perform important functions in this information transfer process. They serve as biosensors, forming a complex network that relays information about the intracel-lular and extracellular environment back to the genome. The genome can respond to the signals it receives in many ways, some of which are positive for the host and some of which could be detrimental to the host. For example, based on environ-mental stimuli the genome can up-regulate genes necessary for normal physiology, such as those encoding antiviral antibodies. Alternatively, the stimuli can up-regulate genes that accelerate a pathogenic process, such as those encoding auto-antibodies. The goal of innovative medical interventions, such as gene therapy, is to accentuate the positive potential of gene expression and eliminate or circumvent the negative.

Because genes are linked to each other through an information network, it is often possible to alter the expression of one gene by manipulating the products of another. As presented in Figure 1.2, manipulation leads to the up-regulation of one

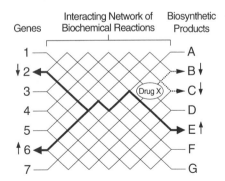

FIGURE 1.2 Schematic representation of a system in which genotype and phenotype are related by a complex network of interactions involving many proteins, RNAs, and reactants. Drug binding to a specific component leads to complex effects, lowering levels of some biosynthetic products, raising levels of others. Through a series of feedback loops, expression of some genes is up-regulated and of other genes down-regulated. (Adapted from Anderson and Anderson, Electrophoresis, 1996.)

gene and the down-regulation of another. Co-up-regulation and co-down-regulation can also take place. For example, the changes that occur in hypercholesterolemic patients (see Chapter 7) taking lovastatin provide an example of coordinately controlled gene expression. Mevacor (lovastatin) was developed to inhibit the enzyme, 3-hydroxy-3-methylglutaryl CoA reductase, and thereby lower plasma cholesterol levels. However, the biochemical reaction that has the greatest cholesterol-lowering effect occurs because lovastatin-induced enzyme inhibition produces a co-up-regulation of low-density lipoprotein receptor, which in turn removes low-density lipoprotein (LDL) cholesterol from plasma. Thus, a gene therapy protocol could follow this example and provide network effects or new interactions with environmental stimuli.

Infectious agents, such as human immunodeficiency virus (HIV) (see Chapter 11) and hepatitis C (HCV) (see Chapter 7), claim many lives in the United States. However, most death and disability in the United States is not caused by an infection but results from conditions causing chronic disabling diseases through an interplay of multiple genetic and environmental factors. These conditions include cardiovascular disease, malignant neoplasms, and cirrhosis. When the under (or over) expression of many different genes contributes to pathogenesis, it may be impossible to stop disease progression by replacing any single gene. However, it may be feasible to develop gene therapies to ameliorate these disease processes once they are fully understood at the molecular level.

Fortunately, knowledge of pathogenesis is taking a quantum leap forward because of several new techniques and technologies and the emergence of the field of "bioinformatics," which allow patterns of gene expression in diseased and healthy tissues to be determined (see the Appendix). As the molecular details of pathogenesis emerge and can be related to information about gene networks, the field of gene therapy may redefine its goals. Gene therapies may come to encompass all interventions specifically designed to promote health by altering patterns of gene transcription and translation.

Since patterns of gene expression vary from patient to patient, in part as a result of DNA polymorphisms, detailed information about the genotype of individual patients will be extremely important to consider when designing therapies. Advances in rapid DNA sequencing and gene expression analysis will soon reduce the cost of gathering data about a patient's genome and pattern of gene expression. This will pave the way for medical interventions tailor-made for an individual patient (see Chapter 15). Academic medical centers can contribute to the development of personalized medicine by providing high-quality specimen banks. They can establish interactive teams of scientists and physicians who are able to conduct the complex clinical trials needed to find the best matches between the expanding universe of therapeutic options and the genetic constitution of an individual patient.

GENE THERAPY AND MOLECULAR MEDICINE

A simple and concise definition of gene therapy (there are many) is the use of any of a collection of approaches for the treatment of human disease that rely on the transfer of DNA-based genetic material into an individual. Gene delivery can be

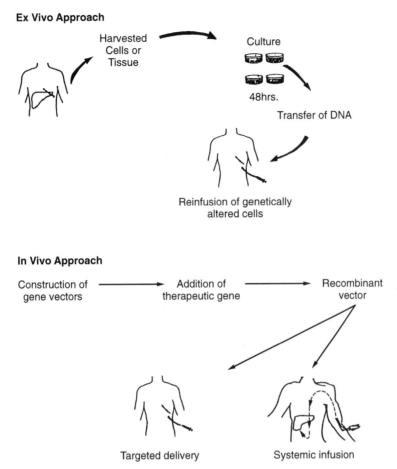

FIGURE 1.3 Two basic methods for delivery of genes. The upper panel shows the ex vivo approach. It requires removal of cells or tissue, culture of cells, and transfection. Successfully transformed cells are selected and returned to the patient where they home to the original location of removed cells or tissue. The lower panel shows the in vivo approach. A gene vector construct, suitable for the delivery of genes to the targeted cell or tissue, is generated. The therapeutic gene is incorporated onto the construct and the recombinant vector is delivered to the patient by any of a number of methods. The method of choice should be previously shown to provide the best level of transfection with minimal side effect.

performed in vivo through the direct administration of the packaged gene into the blood, tissue, or cell. Alternatively, the packaged DNA can be administered indirectly via ex vivo laboratory techniques (see Figure 1.3). Currently, somatic gene therapy, which targets nongermline cells (nonegg and nonsperm cells), is consistent with the extension of biomedical science and medical therapy in which treatment does not go beyond the individual. In altering the genetic material of somatic cells, gene therapy may correct the specific disease pathophysiology. Therapy to human germline cells, thereby modifying the genetic composition of an offspring, would represent a departure from current medical practices in addition to presenting specific ethical issues (see Chapter 14).

Cancer

Cancer is a genetic disease that is expressed at the cellular level (see Chapter 10). The generation of neoplasia is a multistage process driven by inheritance and relatively frequent somatic mutation of cellular genes. These genes include oncogenes, tumor suppressor genes, and DNA repair genes. In a minority of individuals with cancer and in pediatric cases, germline mutations of tumor suppressor or DNA repair genes are the primary neoplastic events. Germline mutations result in all cells of an individual becoming at risk for cancer development and thus are not suitable for somatic cell gene therapy. But in both somatic and germline mutations, clonal selection of variant cells results in a population of cells with increasingly aggressive growth properties.

In individuals with only somatic gene mutations, the insertion of a gene (such as a tumor suppressor gene) would alter the phenotype of a malignant cell only if the mutation is not dominant. Additionally, the level of corrective cellular therapy (possibly as high as 100% correction of all tumor cells) would need to be determined as well as the issue of gene therapy in distal metastasis. Thus, substantial biological obstacles remain to be overcome in the application of gene therapy in certain forms of cancer. Based on these formidable problems, indirect therapies have been proposed. These include: gene transfer of cytokines or other immune mediators to augment host immune responses, the genetic modification of neoplastic cells to promote immunogenicity, the treatment of localized cancers with genes encoding viral or bacterial enzymes that convert prodrugs into toxic metabolites, or the transfer of genes that provide enhanced resistance to conventional chemotherapy (see Chapter 10).

Infectious Diseases

Chronic infectious diseases are suitable targets for gene therapy. These include viral, bacterial, and parasitic infections such as the hepatitis, herpesvirus infection, HIV and its analogs, human papillomavirus infection, mycoplasma infection, Lyme disease, malaria, rabies, and *Listeria* infection. Gene therapy strategies for diseases caused by rapidly proliferating infectious pathogens include intracellular immunization and polynucleotide vaccines. Gene-therapy-induced vaccination for these pathogens may represent an effective strategy by acting classically to "prime" innate immunity prior to exposure to the pathogen. Intracellular immunization seeks to transform cells into cells that are refractory to infection. Protocols may include ribozymes, antisense RNA, RNA decoys, intracellular antibodies, or genetic suppressor elements (see Chapter 11).

Genetic Vaccination

Polynucleotide or genetic vaccination seeks to attenuate the host's immune response, thus having both prophylatic and therapeutic potential. The physiologic basis for polynucleotide vaccines, either RNA or DNA, is the direct inoculation and expression of specific pathogen gene(s) whose products are immunogenic and thus subsequently induce protective or neutralizing immunity. During the next decade, gene therapy may make its greatest contribution to medicine through the introduction of DNA vaccines. In part because DNA vaccines utilize simple vectors, they can be developed quicker than most other gene therapies. New and more effective vaccines are urgently needed in the United States and throughout the world to prevent infectious diseases. Furthermore, since they induce a broad range of immune responses, DNA vaccines may be useful in treating infectious diseases, such as chronic hepatitis B virus (HBV) infection, and it is hoped that they can be used to treat noncommunicable diseases, such as cancer and allergic reactions.

DNA vaccines have produced dramatic results in preclinical trials in many model systems, attesting to the simplicity and robustness of this technology. Immune responses have been generated against viral, bacterial, parasitic, allergy-inducing immunogens, and tumor-specific antigens. DNA vaccines are particularly useful for the induction of cytotoxic T cells. Furthermore, by varying the mode of delivery, it may be possible to select the type of immune response elicited by a DNA vaccine: intramuscular injection is associated with Th1-like helper cellular immune responses, while Th-2-like helper cellular immune responses are seen following progressive vaccinations in which DNA is literally "shot" into the epidermis with a gene gun.

Most DNA vaccines consist of a bacterial plasmid with a strong viral promoter, the gene of interest, and a polyadenylation/transcription termination sequence. The plasmid is grown in bacteria (*Escherichia coli*), purified and injected or blasted into target tissues of the recipient. The DNA is taken up, and its encoded protein is expressed. However, the plasmid does not replicate in mammalian cells, and it does not integrate into chromosomal DNA. This approach raises fewer concerns about mutagenesis and safety. The regulatory elements that have been used in DNA vaccines most frequently mediate high levels of gene expression in mammalian cell cultures or in transgenic mice. These include the human cytomegalovirus immediate/early promoter, the Rous sarcoma virus, and the SV40 virus early promoter, and the transcript termination/polyadenylation signal from either the SV40 virus or the bovine growth hormone 3′ untranslated region. Most vaccination vectors also contain an intron, which enhances expression of genes in mammalian cells. In some DNA vaccines, a cassette of CG dinucleotides is incorporated into the vector to boost immune responses, building on the discovery that DNA oligonucleotides containing centrally located CG dinucleotides stimulate B cells.

Rapid progress is being made toward the development of a DNA vaccine for HBV. It will be an interesting historical parallel if the first DNA vaccine for use in humans turns out to be for HBV. This is because the current HBV vaccine is the first vaccine produced from recombinant cells that is effective against a human virus. The yeast cells utilized for this vaccine were originally described in 1984 and contain an expression vector with an alcohol dehydrogenase I promoter with a segment encoding the HBV surface antigen of the adw subtype. Because the vaccine contains only a single viral protein, it is called a "subunit" vaccine, in contrast to vaccines comprised of attenuated live viruses or inactivated whole viruses, which contain many viral proteins. Unfortunately, the efficacy of the recombinant HBV vaccine has been difficult to duplicate in subunit vaccines for other infectious pathogens. Based on the ability to stimulate both T-cell and B-cell responses, it is hoped that DNA vaccines will be effective against a broad spectrum of agents. Thus, it is hoped that they will be effective not only as preventive modalities but also as therapeutic vaccines. Therapeutic vaccines would be given to infected patients to stimulate immune clearance of established pathogens.

Organ Transplantation and Cellular Engineering

Organ Transplantation   Organ and tissue transplantation are accepted treatments for end-stage organ damage. Current survival rates for major organ transplantation procedures range from 70 to 95% survival for 1 year to 30 to 75% for 5-year survival. These results indicate that the transplantation procedure itself is no longer a survival issue but that posttransplantation complications reduce long-term survival. Posttransplantation complications include acute and chronic allograft, rejection, infection, and the side effects of immunosuppresive treatments. Gene therapy approaches have been suggested as novel methods to control posttransplantation complications at the molecular level. Both ex vivo and in vivo approaches have been advanced.

For in vivo gene therapy, adenovirus vectors (see Chapter 4) have been used to obtain efficient gene transfer to the lung and heart in a posttransplantation setting. The efficacy of such procedures show the feasibility of genetic modification of the graft to reduce posttransplantation rejection, such as chronic graft vascular disease in cardiac allograft rejection, or other physiological processes. The graft rejection process could be modified by inserting specific genes of immunosuppressive molecules or by transfecting genes of antisense molecules to block expression of an important mediator of graft rejection. An example of a mediator to target would be an adhesion molecule. In addition to immune-mediated graft rejection, graft function is also important. Physiological processes could be modified for organ or tissue grafts that are malfunctioning. For instance, a liver allograft not producing therapeutic levels of factor VIII could be transfected with the gene for factor VIII.

The latter example has implications for ex vivo gene therapy approaches in organ transplantation. Organ, tissue, or cellular engineering could be performed on candidate grafts prior to transplantation during the cold storage time. This may be possible because recent studies have indicated that gene transfection may not be affected greatly by nonphysiological temperatures. Thus, organs or tissues may be transfected with genes of cytokines to reduce allorejection or other genes to suppress major histocompatibility (MHC) complex alloantigens or host MHC antigens. Studies, to date, have shown that transfection of immuno-modulating genes such as transforming growth factor beta (TGF-β) or interleukin 10 (IL-10) can induce local immunomodulation in transplanted vascularized organs or in cellular transplants such as pancreatic islet cells for diabetes.

Inherent in the ex vivo gene therapy technique is the opportunity to perform cellular engineering. Cells, tissues, or organs could be genetically modified or engineered to perform unique or specific functions. Host tolerance to a transplanted organ could be induced by the intrathymic administration of chimeric cells (part donor–part host phenotype; see Chapter 3). This would allow for a better "take" of the transplanted organ and less use of highly toxic immunosupressive regimens. Alternatively, the use of microencapsulated genetically engineered cells could be utilized. Microencapsulation is the procedure by which transduced cells secreting specific molecules are enclosed within microscopic, semipermeable containers. The encapsulated cell can be thought of as a naturally occurring microcapsule in which enzymes and organelles are contained within the plasma membrane. Current technology allows for the production of synthetic semipermeable microcapsules that are referred to as artificial cells. The permeable membrane allows for the diffusion or even active transport of specific molecules for therapeutic intervention. Thus, it could be possible to develop microcapsules with artificial chromosomes to be utilized for genetic therapy.

Cellular Transplantation   The unrelenting shortage of donor organs for whole organ transplantation has resulted in the use of every known method to promote successful transplantation outcomes. As inferred above, beyond organ and tissue transplantation is the experimental approach of cellular transplantation. For instance, hepatocyte cell transplantation has become an experimental treatment for individuals rejected as candidates for organ (liver) transplantation. Gene therapy's role in this avenue of research is multifold. Transfected cells may be suitable candidates to be grown in mass cell culture prior to transplantation. Alternatively, gene therapy approaches may genetically modify cells to become less immunogenic and thus less likely to undergo acute rejection. Insertion of genes that inhibit complement activation, insertion of genes that inhibit antibody reactivity or delete cellular MHC antigens in transplanted cells are protocols under investigation. The cells could be of allogenic origin (same species) or xenogenic origin (different species). In the case of human liver transplantation, current methods of organ transplantation could be augmented by the generation of human cloned cell lines with trangenes (see Chapter 3) expressing unique histocompatibility antigens to reduce allograft rejection. These cloned cells could be used in cases where cellular transplantation was feasible resulting in a benign and less costly procedure.

Alternatively, the use of cells or organs, cloned or produced in quality controlled herds of transgenic animals, is currently under investigation to augment human organ transplantation (see Chapters 2 and 3). Thus, it is likely that the field of organ transplantation, which is heavily comprised of aspects of molecular medicine, will be influenced by the emerging advances of gene therapy. To this point, gene-marking protocols have advanced aspects of autologous transplantation. The data generated from these clinical protocols may be utilized to advance new approaches to transplantation. The most important finding to date from these studies is the observation that genes may be transferred and expressed in vivo in hematopoietic progenitor cells (see Chapter 6). These cells are components of the current basic research efforts isolating and characterizing human stem cells. The effective transduction of stem cells would enable the maintenance of genetic expression in the human body for as long as the lifetime of the recipient.

Molecular Pathology and Laboratory Medicine

As gene therapies become more successful, they will be introduced into the mainstream of clinical medicine. As procedural therapies, they will consist of the collection and processing of cells from the patient, introduction of DNA into the patient's genome via some type of vector (see Chapter 4), the process of infusion of cells, and the monitoring of patient's status through the sophisticated techniques of molecular biology and genetics. As these procedures become routine, they enter the domain of the molecular pathologist and the realm of clinical laboratory medicine.

The molecular pathologist will have a role to play in genetic screening of individuals and in the assessment of efficacy of gene therapy. Issues of patient confidentiality and the resolution of ethical and procedural issues based on established guidelines will need to be addressed at a local level (see Chapter 13). As well, the introduction of molecular biological assays into the clinical pathology laboratory will need to be established. Laboratory medicine will need to develop gene therapy monitoring procedures as gene therapy protocols become pharmaceutical methodologies. Monitoring would likely follow therapeutic expression levels of the transgene as well as the nucleotides and vectors in serum and various tissues.

Aging

Although the proximal causes of the major diseases in the United States are diverse and include a multitude of both genetic and environmental factors, these diseases have one feature in common—their connection with the aging process. Although it is difficult to suggest that aging is a "disease," the aging process is a genetic-based scenario that results in degenerative biologic sequellae promoting pathogenesis. For instance, muscle atrophy occurs as part of the normal aging process. Muscle strength decreases up to one-third in humans between the ages of 30 and 80. The amelioration of such downstream consequences of aging, including heart disease, diabetes, and flabbiness as well as a quest for the human genes directly associated with the aging process itself, may ultimately be a target for gene therapies. Animal studies suggest that this approach is possible. Recent studies using an adeno-associated virus vector (see Chapter 4) and the gene for insulinlike growth factor 1 showed that injection of aged mice with this construct totally prevented the decrease in muscle mass seen in aging. Other studies show that genes strongly influence the rate of aging. For example, evolution has endowed individuals of certain species with the genes needed to sustain unusually long life spans. Thus, it is reasonable to envision that gene therapies could add decades to the human life span in the context of promoting a high quality of life in the extended years.

Gene Therapy Protocols

More than 310 clinical protocols have been submitted to the National Institutes of Health (NIH) Office of Recombinant DNA Activities, now part of the Office of Biotechnology Activities (see Chapter 13), for review, and at least 600 individuals have undergone gene transfer in protocols involving more than a dozen diseases. Currently, most gene therapy protocols are phase 1 clinical trials—small clinical experiments that test feasibility and safety. Thus, efficacy has not been established for any gene therapy protocol. The most significant outcome of the numerous cell marking and therapeutic trials appears to be a lack of observed toxicity due to gene transfer. However, a recent clinical trial has reported one death due to the approved experimental protocol (see Chapter 13). Additionally, it has come to light that other deaths have occurred in gene therapy clinical trials. However, it is unclear whether these deaths are related to the experimental therapy. The majority of human gene therapy protocols involve cancer, and the most common viral vector in use is the retrovirus. Most cancer studies are gene-marking studies where a cell is marked with a gene to elucidate metatasis or recurrence.

Adverse short-term effects of gene therapy protocols vary depending on the clinical condition and status of the patient. The limited clinical experience to date does not rule out long-term adverse effects from gene therapy protocols as noted in Chapter 13. Thus, the ability to bring recent laboratory-based advances to the bedside relies on the quantity and quality of the underlying science, the carefulness used in clinical protocol design and outcome measure, as well as a multidisciplinary approach to bridging basic science and medicine.

GENE THERAPY: CURRENT BASIC SCIENCE ISSUES

Two critical steps are required for gene therapy using gene transfer techniques: (1) the appropriate transfer of gene(s) or genetic material and (2) the continued gene expression at appropriate levels for therapy. Currently, numerous basic science issues need to be addressed in the development of human gene therapy protocols.

Gene Transfer

Gene transfer can be achieved by two methods: direct transfer (in vivo) or laboratory manipulation (ex vivo). Utilizing these methods, gene transfer should be administered to the patient without adverse side effects. The period between multiple transfers (if necessary) should be maximal. Targeted gene therapy to specific cells or tissue should not be required. Various gene transfer protocols (systems) are currently under development and should be tailored to the clinical condition. They each have specific advantages and disadvantages (see Table 1.2). They include naked or complexed DNA or RNA, retroviruses, adenoviruses, adenoassociated virus, hepesvirus, and poxvirus (see Chapter 4). In principle, studies in yeast have indicated that the development of artificial chromosome vectors may allow for the maintenance of transferred genes and obviating the problems of random insertion of viral constructs.

Gene Expression

Once a gene is transferred into a tissue or cell, expression of that gene is necessary for successful gene therapy. The knowledge base of DNA sequences and regulatory elements that direct tissue specificity and transgene expression is ever expanding. Currently, however, persistent high levels of gene expression are not consistently achieved in gene therapy protocols. It is unclear whether these experimental data reflect unknown cellular mechanisms needed for therapeutic gene expression, a selective disadvantage of the use of stem cells expressing transferred genes, or the failure to include appropriate regulatory elements in current gene constructs. What is clear from current human studies is that protocols that produce high levels of gene expression in mice do not reproduce similar gene expressions in clinical studies. Long-term expression of transferred genes and high levels of gene product have been reported in murine studies. But a deficiency arises when comparable protocols are employed in clinical studies. This is particularly notable in relative levels of gene transfer. The inefficiency seen in human studies reduces the potential benefit of the introduction of foreign genes and makes the measurement of gene product

TABLE 1.2   General Characteristics of Gene Therapy Vectors for Gene Transfer

Gene Transfer	Advantage	Disadvantage	Clinical Application
Viruses Retrovirus Adenovirus Adenoassociated herpes poxvirus HIV-1	Efficient entry into cells Stable integration Biology known	Need high titer Limited payload Immunogenetic Difficult to control and stabilize expression Can induce adverse events Random insertion	Suitable for permanent correction Extensive use in marking studies Specific virus for specific disease, e.g., herpes-neurology
Liposomes	Commercially available Easy to use Targetable Large payload	Entry into cells Integration rate	
Naked DNA	Ease in preparation; safe no size limitation; no moderate application extraneous genes	Inefficient entry into cells; not stable	Topical application
Complexed DNA	More efficient uptake than naked DNA; protected from degradation; targetable Unlimited construct size	Not stable Inefficient cell entry Limited tragetability	Limited clinical use; vaccination
Artificial chromosomes	Autonomous vectors No insertion required Regulatable tissue and temporally	Unpredictable chromosome formation Centromere formation	Experimental: only in human transformed cells
Artificial cells	Designer potential	Complexity	Conceptual

difficult. Studies have relied on molecular methods of detection of gene expression rather that direct protein assays. Thus, at the current stage the lack of expression of transferred genes compromises both the clinical benefit and scientific value of gene therapy.

Gene Targeting

Gene therapy approaches could be enhanced by directing gene transfer and expression to specific cells or tissues (see Chapter 5). The easiest approach would be ex vivo gene transfer where the transfer could be limited to specific isolated cells. Using such an approach would reduce the need for gene targeting required with in vivo transfer techniques. However, current ex vivo techniques could be enhanced by using targeting techniques such as that used in liver-cell-directed gene therapy (see Chapter 7). The use of ligands that bind to surface receptors could augment gene incorporation into the cell. Alternatively, genetic regulatory elements such as matrix attachment regions (MARs) or multiple enhancing genetic elements could be incorporated into viral constructs to augment gene expression levels.

Disease Pathology

The identification of a genetic mutation as a cause of disease pathology is an important step in gene therapy. However, equally important is the elucidation of the biological mechanisms through which the mutated polypeptide molecule induces pathogenesis. Mutations may cause loss of function so that gene therapy replaces the mutated gene product sufficiently for effective therapy. However, somatic mutation may also be dominant negative in the biological mechanism. Here, the mutated protein inhibits a cellular metabolic pathway and a therapeutic approach would be to delete expression of the mutated protein. Therefore, a detailed understanding of the pathophysiology of the disease is required for designing gene therapy protocols. Both the genes in question need to be revealed as well the cellular targets that could be utilized for therapy. For example, skin or muscle cells could be targeted for systemic diseases as opposed to liver cells. Regardless, the use of gene therapy to further understand disease pathophysiology could lead to the development of novel therapeutic approaches to disease remission.

Animal Models of Disease

As a correlate to the study of disease pathogenesis in the context of gene therapy, animal models of human disease provide the principles of disease pathogenesis (see Chapter 3). Specific hypotheses and experimental therapies can be tested in animal models. For gene therapy, the specific cells to be targeted for therapy as well as the number of cells needed for therapy can be elucidated. The following questions can be addressed by the use of experimental protocols in animals: Are transformed cells at a selective advantage or disadvantage? Are specific constructs immunogenic? Can a mutated human gene produce pathogenesis? What are the critical outcome measures? In addition, when the animal pathogenesis and human disease manifestations are dissimilar, important keys to the human pathogenesis can still be obtained. Thus, as the testing ground of advancing molecular techniques, animal models or even the generation of transgenic animals should not be undervalued (see Chapter 3).

HUMAN GENE THERAPY: CURRENT STATUS AND BASIC SCIENCE RESEARCH NEEDS

Currently, for the field of gene therapy, it is the worst of times and the best of times. As presented in Chapter 13, researchers conducting clinical trials using gene therapy protocols have not been forthcoming with the reporting of adverse events in patients in gene therapy clinical trials. With the report of the initial death of a patient in a gene therapy clinical trial, other issues have bubbled to the surface beyond adverse event reporting. These include patient safety and informed consent as well as federal oversight and coordination among agencies. Numerous investigations have led to some suggested recommendations for improvements in manufacturing and testing of gene transfer products and patient selection and monitoring. To instill public confidence in the research, adverse event data should be analyzed in a public forum. However, in the midst of this apparent disarray, the public has been emotionally stretched by the announcement and publication of the first success of gene therapy. In a recent clinical trial performed after successful preclinical studies, gene therapy was shown to provide full correction of a disease phenotype in two patients with severe combined immunodeficiency-X1 (SCID-X1). The data presented showed clinical benefit for a ten month follow-up period. For long-term data on clinical benefit, await further follow-up of this study.

Research efforts are needed to develop new vectors for gene transfer, to improve current viral and nonviral vectors, and to enhance genomic technology. Nonintegrating vectors such as artificial chromosomes need to be further developed, and techniques using antisense strategies and ribozymes need to be enhanced. Studies are needed detailing gene expression that encompass regulatory elements both upregulating and down-regulating gene expression. Optimal recipient cells for gene transfer and therapy need to be identified. Specific outcome measures need to be defined. For instance, are we interested in survival as the only endpoint or is quality of life important as well? Thus, the field of gene therapy is in a growing phase where further advances will have a profound effect on our current understanding of molecular medicine.

GENE THERAPIES: NEXT HORIZON

Advances in genomics, biotechnology (see the Appendix), and pharmaceutical drug development are generating a panoply of new therapeutic compounds. Phase I clinical trials will determine toxicity and efficacy in experimental systems. The expected advantages of gene therapies include their potential durability (if the therapeutic transgene inserts into chromosomal DNA), simple dosing schedule (a single treatment may suffice), minimal toxic metabolites (most therapies will involve molecules that naturally occur in the human body), and the potential for delivery to selected cells, tissues, and organs. Potential disadvantages include difficulties in modulating potency, the production of deleterious DNA mutations in bystander genes, and immune-mediated destruction of tissues expressing transgenes.

As it becomes technically feasible to perform human gene therapy, the medical conditions regarded as suitable targets for gene therapy will expand from those that are life threatening, such as acquired immunodeficiency syndrome (AIDS) and cancer, to those that have a much lower medical imperative, but a high commercial value and popular appeal. Thus, "gene therapy agents" that lower transcription of the gene for 5-α-reductase (which converts testosterone to dihydrotestosterone) might be developed to treat benign prostatic hyperplasia, for example, and later used to treat baldness. In each case, the clinical benefits will have to be carefully weighed against the risks. Specifically, for the case of "genetic enhancement" such as the case of baldness, ethical issues will be part of the equation in weighing risks vs. benefits (see Chapter 14).

The targets of many current phase I therapies are genetic lesions causing disease in children and young adults. Successful gene therapy in these cases will save lives but not necessarily increase life expectancy or longevity. If gene therapies are to produce major increases in longevity, they will have to target diseases of the elderly, but which diseases? The holy grail of gene therapy would be to identify a transgene that modifies the biological clock and the aging process. Aging causes an increase in vulnerability to many pathologies. The incidence of cardiovascular disease, diabetes, cancer, obesity, osteoporosis, dementia, and arthritis all increase with age. One approach to reducing the morbidity and mortality resulting from these conditions is to understand the biochemical pathways leading to each pathology in the context of aging and then develop interventions—using components of gene therapy. A second approach would be to modify the aging process itself. This latter, preventative, approach would be superior. Might this be possible? Evidence from animal studies suggests that it may be.

Genes have been identified that strongly influence the aging process. In addition, genetic manipulations can increase the life span of an organism. Longevity appears to be a polygenic characteristic to which individual genes make significant contributions. In a variety of biological systems, extended longevity is associated with enhanced ability to minimize oxidative stress. However, the first step in developing human gene therapies to delay aging will be to identify "longevity genes" in humans and other species.

This area of research is in an early stage but on the "fast tract." Just as molecular biology evolved from observation using the fruit fly, evidence of longevity genes has been noted in worms and fruit flies. Mapping of quantitative trait loci (QTLs) has revealed at least five genomic regions that may be associated with longevity in the nematode, *Caenorhabditis elegans*. Screening of mutants with long life spans allowed the first longevity gene in nematodes, *age-1*, to be identified. Subsequently, four "clock" genes, were discovered. Mutations in the clock genes lengthen the life of the worm from 9 days to almost 2 months. Clock genes are thought to set an internal pacemaker by regulating genes involved in metabolism. When clock gene mutations are combined with a mutation in *daf-2* (a member of a different set of genes, which also affects nematode life span) worms, living at a leisurely pace, survive more than five times longer than normal. The human homologs of *daf-2* are the insulin and insulinlike growth factor receptors, indicating that aspects of the regulatory system are evolutionarily conserved. Thus, there are candidate human genes to target for longevity studies.

In the fruit fly, the link between longevity and resistance to oxidative stress has been shown. Oxidative stress is considered to be a major cause of age-associated loss of function in many biological systems. Damage from reactive oxygen metabolites causes peroxidation of membrane polyunsaturated fatty acid chains, modification of DNA (including base alterations, breaks, sister chromatid exchanges, and DNA–protein crosslinks), and carbonylation and loss of sulfhydryls in proteins. The concept that oxidative damage normally reduces longevity in flies is supported by the finding that one group of long-lived flies is resistant to oxidative stress. Furthermore, the life span of control flies can be increased by adding transgenes for the antioxidants superoxide dismutase and catalase. Such transgenic *Drosophila* experience a 30% increase in mean and maximum life spans. Significantly, this increase occurs despite greater physical activity and oxygen consumption by the transgenic flies. These flies suffer measurably less oxidative damage to protein, DNA, and enzymes than controls.

Other *Drosophila* studies suggest that there are multiple mechanisms of aging and more than one route to extended longevity. Caloric restriction is the only widely validated method for extending the life span and postponing senescence in mammals. Caloric restriction apparently triggers responses that protect against stress, especially oxidative stress. While severe caloric restriction would not be palatable to most patients, studies of gene expression profiles in animals on very low calorie diets may identify pathways whose up- or down-regulation will enhance longevity. In addition to studies of food-deprived animals, studies of mice (and men) who out-live their brethren will help to identify genes associated with longevity.

Darwinian selection can also confer a long life span. Some of the strongest evidence that animal senescence can be modulated by the action of genes comes from studies of queen ants. In ant species with social structures that protect the queens from "external causes" of death, the queens live up to 30 years, while those of species that provide less protection have genetic constitutions that give them much shorter life spans. Since the queens in both groups are similar in overall physiology and metabolism, the difference appears to arise because the protected queens occupy a niche in which longevity confers a selective advantage. Short life is not a necessary consequence of ant physiology. These studies show that the "right" genes can make a major difference. The challenge now is to identify these genes and to find their human counterparts.

Finally, aging research has also focused on modifying the telomeric regions of chromosomes to add "time" to the cellular life span. In 1991, it was reported that the tips of chromosomes in cells shortened as a cell replicated. Thus, cells replicated approximately 50 times to the so-called Hayflick limit, which was established by the length of the telomeric region. Recent studies have reported the activation of the enzyme called telomerase, which extended telomeric regions and lengthened the life span of cells in vitro by at least 20 cell divisions beyond the Hayflict limit. Thus, it is conceiveable to suggest the successful transfection and expression of the telomerase gene may promote the life span of individual cells in gene therapy protocols. An alternative approach would be the reconstitution of the telomers of embryonic stem cells. This approach would suggest that target cells used in gene therapy could have extended life spans.

KEY CONCEPTS

- Molecular medicine is the application of molecular biological techniques to the treatment and diagnosis of disease. It is derived form the successful development of human organ transplantation, pharmacotherapy, and elucidation of the human genome.
- Gene therapy is the use of any of a collection of approaches to the treatment of human disease based on the transfer of DNA-based genetic material to an

individual. The successful application of gene therapy requires the achievement of therapeutic levels of protein along with the long-term regulation of gene expression. Somatic gene line therapy targets nongermline cells and is consistent with the extension of biomedical science into medical therapy.

· At the moment, the diseases most amenable to gene therapies are those requiring transient expression of an exogenous gene. These applications can make use of bacterial plasmid vectors. Plasmid vectors can be used to generate either a gene product, such as a growth factor, or, in the case of DNA vaccines, these vectors can be used to stimulate immune responses.

· For the immediate future, a major challenge is to develop vectors that can yield stable therapeutic concentrations of gene products in nondividing cells located deep within the body. A lingering concern is raised by the possibility that cosuppression occurs in humans and that the same biochemical machinery that carries out gene silencing may shut off high-level expression of therapeutic genes. If true, gene therapies face an unanticipated roadblock that may be difficult to circumvent.

· The key gap in the gene therapy field is our lack of knowledge of exactly what sets the stage for the serious diseases causing morbidity and mortality in the United States. At the molecular level, it is not clear what processes go awry. Therefore, it is not clear which gene products have the greatest potential to be curative.

· A group of promising new tools is emerging that will allow patterns of gene expression to be compared in healthy and diseased tissue. On the one hand, these gene-profiling techniques will detect gene therapy targets—genes whose products contribute to disease. On the other hand, they will identify genes whose products may be useful when delivered as replacement genes.

· In the future, it is likely that gene therapies will be defined more broadly than they are now and will evolve to include all types of drugs specifically designed to alter patterns of gene expression. Gene therapists will want to treat complex diseases, which cannot be cured, by adding or subtracting a single gene. Just as radiologists adapted their ability to read simple X-rays and became experts at interpreting computerized axial tomography (CAT) scans, gene therapists will be in a position to use their understanding of genetics and gene expression to develop medical interventions aimed at manipulating patterns of gene expression. In addition, pharmaceutical agents taking the form of conventional drugs may be found that are as effective at inducing "healthy" patterns of gene expression as transgenes. The small size of these pharmaceuticals will give them an advantage over gene therapy vectors.

· Long-term and complex clinical trials will be needed to optimize and deliver new therapies. The academic medical community can prepare for future manpower needs by training more clinical investigators, genetic counselors, and statisticians. High-throughput screens of pharmaceutical libraries may soon be used to identify compounds worthy of further development based on the gene expression profiles they induce in treated cells.

· Gene therapies to prevent aging await a fuller understanding of biological clocks and the aging process.

SUGGESTED READINGS

Gene Therapy

Anderson WF. Human gene therapy. Science 256:808–813, 1992.

Dickson G (Ed.). Human Gene Therapeutics. Chapman and Hall, London, 1995, pp. 195–236.

Francisco M. Gene therapy: Better vectors, less hype. Nat Biotech 15:815, 1997.

Morgan RA, Blaese RM. Gene therapy: Lessons learnt from the past decade. Br Med J 319:1310, 1999.

Mulligan RC. The basic science of gene therapy. Science 260:926–932, 1993.

Schwertz DW, McCormick KM. The molecular basis of genetics and inheritance. J Cardiovasc Nurs 13:1–18, 1999.

Stephenson J. Gene therapy trial show clinical efficacy. JAMA 283:589–590, 2000.

Touchette N. Gene therapy: Not ready for prime time. Nat Med 2:7–8, 1996.

Verma IM, Somia N. Gene therapy—promises, problems and prospects. Nature 389:239–242, 1997.

Wadman M. NIH panel to limit secrecy on gene therapy. Nature 402:6, 1999.

DNA Vaccines

Arntzen CJ. High-tech herbal medicine: Plant-based vaccines. Nat Biotech 15:221–222, 1997.

Donnelly JJ, Ulmer JB, Shiver JW, Liu MA. DNA vaccines. Annu Rev Immunol 15:617–648, 1997.

Mancini M, Davis H, Tiollais P, Michel ML. DNA-based immunization against the envelope proteins of the hepatitis B virus. J Biotech 44:47–57, 1996.

Robinson HL. DNA vaccines: Basic mechanism and immune responses. Int J Mol Med 4:549–555, 1999.

Tang DC, DeVit M, Johnston SA. Genetic immunization is a simple method for eliciting an immune response. Nature 356:152–154, 1992.

Genomics

Benson DA, Boguski M, Lipman DJ, Ostell J. GenBank. Nucleic Acids Res 24:1–5, 1996.

Cargill M, Altshuler D, Ireland J, Sklar P, Ardie K, et al. Characterization of single-nucleotide polymorphisms in coding regions of human genes. Nat Genet 22:231–238, 1999.

DeRisi JL, Iyer VR, Brown PO. Exploring the metabolic and genetic control of gene expression on a genomic scale. Science 278:680–686, 1997.

Stephenson J. Human genome studies expected to revolutionize cancer classification. JAMA 282:927–928, 1999.

Wright AF, Carothers AD, Piurastu M. Population choice in mapping genes for complex diseases. Nat Genet 23:397–404, 1999.

Gene Therapy and Aging

Apfeld J, Kenyon C. Regulation of lifespan by sensory perception in *Caenorhabditis elegans*. Nature 402:804–809, 1999.

Crozier RH. Be social, live longer. Nature 389:906–907, 1997.

Jazwinski SM. Longevity, genes, and aging. Science 273:54–59, 1996.

Keller L, Genoud M. Extraordinary life spans in ants: A test of evolutionary theories of aging. Nature 387:958–960, 1997.

Kim S, Kaminker P, Campisi J. TIN2, a new regulator of telomere length in human cells. Nat Genet 23:405–412, 1999.

Kuro-o M, Matsumura Y, Aizawa H, Kawaguchi H, Suga T, Utsugi T, Ohyama Y, Kurabayashi M, Kaname T, Kume E, Iwasaki H, Iida A, Shiraki-Iida T, Nishikawa S, Nagai R, Nabeshima Y. Mutation of the mouse klotho gene leads to a syndrome resembling aging. Nature 390:45–51, 1997.

Lakowski B, Hekimi S. Determination of life-span in *Caenorhabditis elegans* by four clock genes. Science 272:1010–1013, 1996.

Pennisi E. Worm genes imply a master clock. Science 272:949–950, 1996.

Rattan SIS. Is gene therapy for aging possible. Ind J Exp Biol 36:233–236, 1998.

Shay JW. At the end of the millennium, a view of the end. Nat Genet 23:382–383, 1999.

Sohal RS, Weindruch R. Oxidative stress, caloric restriction, and aging. Science 273:59–63, 1996.

Wyllie FS, Jones CJ, Skinner JW, Haughton MF, Wallis C, Wynford-Thomas D, Faragher RGA, Kipling D. Telomerase prevents the accelerated cell ageing of Werner syndrome fibroblasts. Nat Genet 24:16–17, 2000.

Gene Therapy, Tissue Engineering, and Laboratory Medicine

Knop AE, Arndt AJ, Raponi M, Boyd MP, Ely JA, Symonds G. Artificial capillary culture: Expansion and retroviral transduction of CD4+ T-lymphocytes for clinical application. Gene Ther 6:373–384, 1999.

Lysaght MJ, Aebischer P. Encapsulated cells as therapy. Sci Am 280:76–82, 1999.

Pilling AM. The role of the toxicologic pathologist in the preclinical safety evaluation of biotechnology-driven pharmaceuticals. Toxicol Pathol 27:678–688, 1999.

Powell C, Shansky J, Del Tatto M, Forman DE, Hennessey J, Sullivan K, Zielinski BA, Vandenburgh HH. Tissue-engineered human bioartificial muscles expressing a foreign recombinant protein for gene therapy. Hum Gene Therapy 10:565–577, 1999.

Salapongse AN, Billiar TR, Edington H. Gene therapy and tissue engineering. Clin Plast Surg 26:663–676, 1999.

Serabian MA, Pilaro AM. Safety assessment of biotechnology-driven pharmaceuticals: ICH and beyond. Toxicol Pathol 27:27–31, 1999.

Terrell TG, Green JD. Issues with biotechnology products in toxicologic pathology. Toxicol Pathol 22:187–193, 1994.

Fetal Gene Therapy

Yang EY, Flake AW, Adzick NS. Prospects for fetal gene therapy. Semin Perinatal 23:524–534, 1999.

Zanjani ED, Anderson WF. Prospects for in utero human gene therapy. Science 285:2084–2088, 1999.

Gene Therapy, Disease Pathogenesis, and Transplantation

Bingham PM. Cosuppression comes to the animals. Cell 90:385–387, 1997.

Cavazzana-Calvo M, Hacein-Bey S, de Saint Basile G, Gross F, Yvon E, Nosbaum P, Selz F,

Hue C, Certain S, Casanova J-L, Bousso P, Le Deist F, Fischer A. Gene therapy of human severe combined immunodeficiency (SCID)-X1 disease. Science 288:669–672, 2000.

Goldfine ID, German MS, Tseng H-C, Wang J, Bolaffi JL, Chen J-W, Olsen DC, Rothman SS. The endocrine secretion of human insulin and growth hormone by exocrine glands of the gastrointestinal tract. Nat Biotech 15:1378–1382, 1779.

Golub TR, Slonim DK, Tamayo P, Huard C, Gaasenbeek M, Mesirov JP, Coller H, Loh ML, Downing JR, Caligiuri MA, Bloomfiled CD, Lander ES. Molecular classifiaction of cancer: Class discovery and class prediction by gene expression monitoring. Science 286:531–537, 1999.

Handyside AH, Lesko JG, Tarin JJ, Winston RMI, Hughes MR. Birth of a normal girl after in vitro fertilization and preimplantation diagnostic testing for cystic fibrosis. N Engl J Med 327:905–909, 1992.

Hennighausen L. Transgenic factor VIII: The milky way and beyond. Nat Biotech 15:945–946, 1997.

Paleyanda RK, Velander WH, Lee TK, Scandella DH, Gwazdauskas FC, Knight JW, Hoyer LW, Drohan WN, Lubon H. Transgenic pigs produce functional human factor VIII in milk. Nat Biotech 15:971–975, 1997.

Zhang L, Zhou W, Velculescu VE, Kern SE, Hruban RH, Hamilton SR, Vogelstein B, Kinzler KW. Gene expression profiles in normal and cancer cells. Science 276:1268–1272, 1997.

CHAPTER 2

Nuclear Transplantation and New Frontiers in Genetic Molecular Medicine

D. JOSEPH JERRY, PH.D. and JAMES M. ROBL, PH.D. with Ethics Note by
LEONARD M. FLECK, PH.D.

BACKGROUND

Nuclear transplantation made its debut as a novel tool for defining the genetic basis for differentiation and probing the extent to which these mechanisms may be reversible. From these beginnings, a mature technology has emerged with applications ranging from animal agriculture to clinical medicine. Thus, nuclear transplantation can be used to generate identical animals and transgenic livestock. Cloned livestock can be used to intensify genetic selection for improved productivity and have also been proposed as a reliable source of tissues and cells for xenotransplantation in humans. At a more fundamental level, these cloning experiments demonstrate that somatic cells retain developmental plasticity such that the nucleus of a single cell, when placed within an oocyte, can direct development of a complete organism.

INTRODUCTION

Nuclear transplantation is the process by which the nucleus of a donor cell is used to replace the nucleus of a recipient cell (Fig. 2.1). Somatic cells are most often used as the nuclear donors and are transferred, using micromanipulation, to enucleated oocytes. The factors contained within the cytoplasm of oocytes appear responsible for reprogramming somatic cell nuclei and are essential for the success of nuclear transplantation. Genetic reprogramming may be harnessed to alter the developmental potential of cells to allow regeneration of tissues or provide cellular therapies. Conversely, it is possible that illegitimate activation of these factors/mechanisms may lead to deleterious genetic reprogramming resulting in the development of cancer. Reprogramming mechanisms may also provide novel targets for cancer therapy and other diseases that involve genetically programmed differenti-

An Introduction to Molecular Medicine and Gene Therapy, Edited by Thomas F. Kresina
ISBN 0-471-39188-3 © 2001 Wiley-Liss

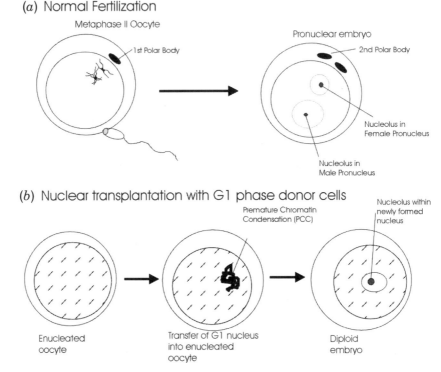

FIGURE 2.1   Comparison of embryos resulting from normal fertilization and nuclear transplantation. (*a*) Metaphase II oocytes (2N) come in contact with sperm (1N) causing the extrusion of the second polar body. This leaves a haploid (1N) complement of maternal chromosomes. The resulting pronuclear embryo is diploid containing equal genetic material from both parents. (*b*) In nuclear transplantation, the first polar body and metaphase chromosomes are removed leaving a cytoplast. The diploid donor cell is then introduced. The interphase chromatin undergoes premature chromatin condensation followed by reentry into S phase of the mitotic cell cycle resulting in a diploid embryo following division.

ation or developmental changes in cells. Along with these possibilities, great ethical questions surround the potential use of this technology for creating cloned humans.

In an effort to clarify these difficult questions, a historical perspective of the use of nuclear transplantation to define the molecular and cellular basis for differentiation is presented. The technical challenges and variations among organisms are considered in an effort to explore how these advances may be applied both in the laboratory and in the clinic. However, these accomplishments must also be considered within the framework of the limitations imposed by ethical concerns and technical challenges that remain.

NUCLEAR TRANSPLANTATION: A TOOL IN DEVELOPMENTAL BIOLOGY

In 1938, the door to human cloning was opened when it was proposed by Speman (1938) that the potency of a cell could be tested by transfer of nuclei from differentiated cells to unfertilized eggs. However, this experiment had to await the development of early nuclear transplantation techniques. By 1952, it could be shown that nuclei from blastula-stage *Rana pipiens* embryos could be transferred to enucleated frog oocytes and that these embryos could develop to blastocyst-stage embryos.

Blastomeres as Nuclear Donors

Early successes spawned a flurry of experiments demonstrating that blastomeres from early cleavage embryos could direct embryonic development when transferred to enucleated oocytes, and therefore, retained pleuripotency. Efforts using blastomeres as donor nuclei were soon followed by experiments using cells in more extreme states of differentiation.

Somatic Cells as Nuclear Donors

Initially, intestinal cells from *Xenopus laevis* feeding tadpoles were used as donor nuclei. A small fraction of the nuclear transplantation embryos developed to the swimming tadpole stage. In these experiments, seven embryos completed metamorphosis to produce normal adult males and females. The adult clones were fertile, demonstrating the completeness of the nuclear reprogramming. Other studies used renal adenocarcinoma cells from *R. pipiens* as donor nuclei to produce normal swimming tadpoles. Therefore, not only could differentiated somatic cell nuclei undergo reprogramming, but tumor cells could also be recruited to participate in normal embryonic development following nuclear transplantation.

Nonetheless, rates of development of embryos were reduced greatly when "differentiated" cells were used as nuclear donors compared to blastula or gastrula endodermal cells. Restrictions in the extent of development was most apparent in the Mexican axolotl. Only 0.6% of nuclear transplantation embryos from neurula-stage notochord cells formed swimming tadpoles, whereas 33% of nuclear transplantation embryos from blastulae cells reached this stage. Therefore, the vast majority of notochord nuclei were severely restricted in their developmental capacity. Failure of development following nuclear transplantation was associated with the presence of chromosomal abnormalities, which included ring chromosomes, anaphase bridges, chromosome fragments, and variable numbers of chromosomes. From these results, Briggs and co-workers (1964) concluded: "The central question therefore concerns the origin of these chromosomal abnormalities. Are they to be regarded as artifacts, or do they indicate a genuine restriction in the capacity of the somatic nuclei to function normally following transfer into egg cytoplasm?" Alternatively, others suggested that these differences may reflect the relative proportions actively dividing cells within the tissues. These questions remain to be settled despite the passage of three decades.

In subsequent experiments, primary cultures were used as a source of nuclei in an effort to provide more uniform populations. Also, "serial nuclear transplantation" gained favor to improve rates of development beyond the blastocyst stage. Serial nuclear transplantation involved a first round of nuclear transfer to produce partially cleaved blastocysts. Although the vast majority (<0.1%) of first-transfer embryos failed to develop beyond the blastocyst stage, they apparently contained a higher proportion of cells with nuclei that were capable of undergoing nuclear reprogramming following a secondary nuclear transplantation, also referred to as "recloning." Selection of the most well-developed embryos from initial nuclear transfers allowed enrichment for embryos that contained minimal genetic damage resulting from the manipulations. The positive effects of serial nuclear transplantation were not improved by additional rounds of nuclear transplantation, suggesting that sequential nuclear reprogramming was not taking place.

Using serial nuclear transplantation, partial or complete blastulae were obtained at rates of 22 to 31% using cultures of kidney, lung, heart, testis, and skin from adult frogs as donor nuclei for serial nuclear transplantation. Swimming tadpoles developed when nuclei for the initial transfers were from adult kidney, lung, and skin but not heart. Based on these results, it would appear that <10% of cells from the primary nuclear transplant embryos were able to undergo successful genetic reprogramming and direct successful development of tadpoles. It is also important to note that some developmental abnormalities were evident in tadpoles derived from nuclear transplantation. The descriptions were not extensive, but anal and cardiac edema were reported and resulted in subsequent death.

Since a relatively small proportion of donor nuclei were able to form even blastocysts following nuclear transplantation, it remained possible that embryos resulted only from a subpopulation of cells that retained stem cell-like characteristics. To rule out this possibility, primary cell cultures were established from foot-web explants and were shown to be differentiated by the expression of keratin in >99.9% of cells. Although no first-transfer embryos developed beyond early cleavage embryos, serial transplantation resulted in swimming tadpoles with well-differentiated organs.

Attempts to confirm these results in *Drosophila* yielded development of larvae but no adults. This result was extended by Schubiger and Schneiderman (1971) when it was shown that preblastoderm nuclei could be transplanted into oocytes, then develop 8 to 10 days when placed in a mature female. These implants were retrieved, then dissociated, and the nuclei were again used for serial nuclear transplantation. The serial nuclear transplant embryos were transferred into developing larvae where they underwent metamorphosis along with their hosts to form adult tissues. Therefore, extensive genetic reprogramming of donor nuclei was possible but required serial nuclear transplantation similar to that used in amphibia. Nonetheless, reprogramming was not sufficient to allow development of normal flies.

Conclusions

The work with amphibia clearly demonstrated that nuclear transplantation could be used to efficiently generate multiple cloned individuals using blastomeres from early cleavage embryos. Although rates of development were diminished when more highly differentiated cell types were used as donors for nuclear transplantation, it was possible to generate live offspring. Therefore, differentiation was reversible and developmental fates were subject to reprogramming under appropriate conditions. The extent of development following nuclear transplantation also varied considerably among tissues. Gurdon (1970) voiced caution that "nuclear transplantation experiments can only provide a minimum estimate of developmental capacity of a nucleus or a population of nuclei." It was a concern that the variable rates of success using cells from various tissues reflected technical challenges due to isolation or culture of specific cell types. Mechanical damage to cells during isolation may vary among tissues. The proportion of nuclear transfer embryos that result in live births may reflect the relative infrequency of specific stem cells that may be more amenable to nuclear reprogramming. The more limited reprogramming observed with Mexican axolotl and *Drosophila* may indicate that some changes are irreversible. If true, then some organisms or cell types may have biological barriers preventing nuclear reprogramming. At this point, the molecular basis for nuclear reprogramming was left to conjecture.

TECHNICAL DEVELOPMENTS IN NUCLEAR TRANSPLANTATION

The ability to create cloned frogs fueled hopes that mammalian nuclei might also be subject to nuclear reprogramming by the oocyte cytoplasm. The value of being able to make multiple clones of genetically superior livestock for the purpose of intensifying genetic selection was not lost on agricultural scientists. As a result, efforts to apply nuclear transplantation to create cloned livestock were undertaken by several groups. This required modifications of nuclear transplantation procedures.

Overview of the Procedures

The nuclear transplantation procedures were pioneered in 1952 in *R. pipiens* where it was possible to physically enucleate oocytes. However, the membranes surrounding the oocyte in *X. laevis* precluded this. Therefore, ultraviolet (UV) irradiation was used to destroy the nucleus. The donor cells were most conveniently handled in suspension following trypsinization. The donor cells were drawn into a glass micropipet, then inserted into the enucleated egg between the center and the animal pole. The intact donor cell, with its nucleus, cytoplasm, and membranes, was expelled into the recipient egg. The membranes surrounding the recipient cell should heal spontaneously as the pipet is withdrawn. The eggs were then transferred to buffered media and cleavage proceeded as manipulation of the oocyte was sufficient activation stimulus in amphibians.

Nuclear transplantation procedures in mammals involve four specific steps: (1) enucleation, (2) transfer of a donor nucleus along with its associated cytoplasm, (3) fusion of the donor nucleus and recipient cytoplasm, and (4) activation of cleavage (Fig. 2.2). Oocytes arrested in metaphase II of meiosis are most often used to prepare recipient cytoplasts because they are large cells that can be easily enucleated. Enucleation is accomplished by inserting a glass micropipet through the zona pelucida and withdrawing the polar body and metaphase chromosomes. Rather than direct injection, the intact donor cell (nucleus, cytoplasm, and membranes) is expelled into the perivitelline space adjacent to the enucleated oocyte with the aid of a micropipet. The enucleated oocyte and intact donor cell are then fused and treated to initiate the cell cycle, which is referred to as activation. Embryos resulting from this process would be genetically identical to the donor at the level of their genomic deoxyribonucleic acid (DNA) but are chimeric with respect to organelles. Therefore, animals prepared by nuclear transplantation are not true clones.

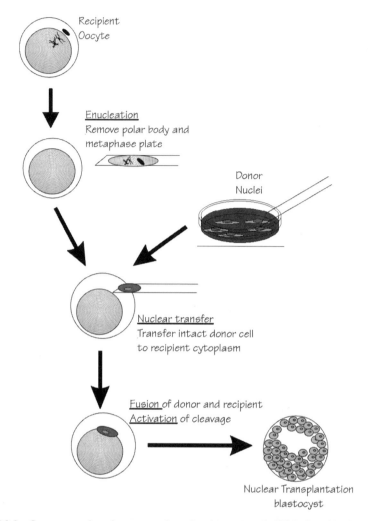

FIGURE 2.2   Summary of nuclear transplantaion in mammals. With the aid of a micropipet, the metaphase plate and first polar body are removed from an oocyte arrested in metaphase II to generate a recipient cytoplast. A donor cell (nucleus and cytoplasm) is transferred to the perivitelline space using a micropipet. Electrical pulses are used to stimulate fusion of the plasma membranes of the donor cell and recipient cytoplast causing the donor nucleus to enter recipient cytoplasm and initiation of cell division. During the first cell cycle, cytoplasm of the oocyte causes condensation of the chromatin followed by replication of the DNA. If successful, the embryo will continue to undergo cleavage to form a normal blastocyst.

Fusion

The first challenge was to develop more versatile methods for fusion of the donor and enucleated recipient cells. The use of Sendai virus to mediate fusion of the recipient oocyte and donor cells was ineffective in a number of species. The advent of electrical fusion of cell membranes provided a flexible and efficient method to stimulate fusion of the donor and recipient cells in a broad range of species.

Enucleation

Complete removal of chromosomes was also more challenging in mammalian oocytes. Treatment with cytoskeletal inhibitors, cytochalasin B, and colcemid stabilized the plasma membrane and prevented rupturing. This allowed a large pipet to be inserted through the zona pellucida and adjacent to the pronuclei without penetrating the membrane. The pronucleus can then be removed in a membrane-bound cytoplast along with the polar body as shown in Figure 2.2. Fluorescent vital dyes are now used to visualize chromatin to ensure complete removal of the metaphase II chromosomes.

Activation

Resumption of the cell cycle in metaphase II oocytes is referred to as activation and results in cleavage of the cell. Activation following nuclear transplantation also proved to be a formidable problem and variable among species. This may belie the lower efficiencies associated with nuclear transplantation in rodents. In cattle, fertilization of oocytes by sperm was shown to initiate changes in calcium concentrations in the oocyte cytoplasm. The electrical pulses used to induce fusion were also shown to cause calcium increases but were minimally effective in activating the oocyte following nuclear transplantation. Procedures to elevate calcium followed by the extended inhibition of MPF activity, using the kinase inhibitor 6-dimethylaminopurine, have been shown to support rates of development to the blastocyst stage that are equivalent to that of in vitro fertilized oocytes.

Cell Cycle Synchronization Between Nuclear Donor and Recipient Oocyte

Synchrony of the cell cycle between recipient oocyte and donor nucleus was also subject to refinements. Nuclear transplantation between metaphase donors and metaphase II recipient oocytes would appear to be the ideal match. Although modest success has been achieved, this approach remains technically challenging. The difficulty in using G2 or M-phase donor cells is that the cells are tetraploid at this stage of the cell cycle. Therefore, cell division must occur following nuclear transfer to produce a diploid two-cell embryo. The difficulty lies with the fact that premature chromatin condensation (PCC) occurs following nuclear transplantation followed by reentry into S phase leading to tetraploid embryos. Nuclei from cells that are in G1 also undergo PCC following nuclear transfer and proceed to S phase resulting in diploid embryos. To successfully utilize recipient oocytes in metaphase II with donor nuclei that are most likely in the G1 or S phases of the cell cycle, it is necessary that the oocyte be given an activation stimulus following fusion with the donor cell. The metaphase II oocyte cytoplasm has been shown to initiate immediate breakdown of the nuclear envelope of the donor cell, condensation of the chromosomes followed by reformation of the nuclear envelope and dramatic swelling of the nucleus as activation progresses. This sequence of events may be crucial for nuclear proteins of the donor cell to be lost and replaced by the oocyte nuclear proteins with nuclear reformation allowing reprogramming of the chromatin.

DEFINING THE LIMITS OF NUCLEAR REPROGRAMMING IN MAMMALS

With technical hurdles addressed, further investigations undertook the task of determining the point during development when cells lost their pluripotency and, therefore, had become differentiated. An initial report of successful nuclear transplantation in mice offered promise but was unable to be confirmed by other investigators.

Blastomeres as Nuclear Donors

In sheep, blastomeres from 8-cell and 16-cell embryos were shown to develop to blastocysts following nuclear transplantation and form viable embryos after transfer to the oviduct of recipient ewes. This was the first reproducible evidence that mammals could be cloned by nuclear transplantation as reported in *Nature* in 1986. Cattle (1987) and rabbits (1988) were soon added to the growing list of mammals that had been cloned with the assistance of nuclear transplantation. Full-term development of mice from nuclear transfer of blastomeres was eventually demonstrated in 1987. However, the rates were low compared to sheep and cattle, possibly due to differences in the requirements for activation following nuclear transfer. Cloning in pigs was also reported in 1989, but was limited to one live pig. These results emphasize the considerable variation in the success in cloning mammals using blastomeres as donor cells. Unlike earlier results using nonmammalian species, serial nuclear transplantation did not offer any substantial improvement in developmental potential.

Inner Cell Mass as Nuclear Donors

Efforts to obtain cloned animals using cells derived from the inner cell mass (ICM) were initially unsuccessful in mice. However, live births were reported in cattle using nuclear donors from the ICM. These data supported the concept that the ICM cells retained their primitive state and remain able to be reprogrammed by nuclear transplantation. Nonetheless, results from mice, rabbits, and cattle all suggest that reprogramming of cellular fates is dramatically restricted in eight-cell embryos and beyond.

Embryonic Stem Cells as Nuclear Donors

The more limited ability of ICM cells to participate in embryonic development following nuclear transplantation appeared to contradict results emerging from experiments with embryonic stem (ES) cells. ES cells had been derived from the ICM and maintained in vitro under conditions to prevent differentiation and were shown to contribute to many different tissues in aggregation chimeras. The most stringent verification of the totipotency of the ES cells was that they contributed to the germline, but this has been accomplished only in mice. Therefore, it appeared that ES cells retained totipotency.

An obvious extension of these experiments was to use ES cells as donors for nuclear transplantation. However, establishment of ES cell lines from species other than mice proved to be more difficult. Even in mice, success in establishing and maintaining totipotent ES cell lines has been largely limited to the 129 strain. Selection methods to eliminate differentiated cells have been developed recently to prepare ES cells from nonpermissive strains of mice. Use of the epiblast for deriving ES cell lines also appears promising. In spite of the challenges, ES-like cells have been produced from cattle, rabbits, pigs, and sheep.

Initial work using "short-term" cultures of bovine ES-like cells for nuclear transplantation resulted in live births. However, when bovine ES cell lines that had been in culture for extended periods were used as nuclear donors, the results were less promising. Normal fetal development was achieved following nuclear transplantation of bovine ES-like cells, but pregnancies failed due to improper development of the extraembryonic membranes of the fetal placenta. This occurred in spite of the fact that similarly derived ES cells were shown to contribute to a variety of tissues in aggregation chimeras. Rabbit ES cells were also used for nuclear transplantation. Fetal development of nuclear transplantation embryos derived from rabbit ES cells appeared to be normal, but no live births were reported. These data suggest that the ability of ES cells to form chimeras and their success in nuclear transplantation may be distinct features.

Somatic Cells as Nuclear Donors

Although nuclear transplantation was shown to be successful using blastomeres in a variety of species, the dramatic decreases in rates of success using ICM and ES cells had diminished the enthusiasm among developmental biologists for cloning mammals from somatic cells. The prevailing wisdom was thoroughly shaken by the reports of Dolly—a normal sheep that developed to term following nuclear transplantation of a donor nucleus from a single mammary epithelial cell. Not only was Dolly cloned from somatic cells but it was from adult cells providing a dramatic confirmation of the earlier work of Gurdon (1970). This was followed by nuclear transplantation of embryonic fibroblasts to clone cattle, sheep, and goats. Cumulus cells from adult animals have also been used as donor cells to clone mice and cattle.

The results from animals cloned using somatic cells from mammals substantiate much of the work performed in amphibians; however, the data are far from complete (summarized in Table 2.1). It is clear that a variety of somatic cell types are capable of undergoing nuclear reprogramming following nuclear transplantation and yield live offspring. However, efficiency of nuclear reprogramming is very dependent on the donor cells. Cumulus cells and fetal fibroblasts have proven to be competent donors in two species, whereas trophectodermal cells were consistently negative in two studies. Under different conditions, trophectodermal cells were used to produce cloned mice. These differences arise from differences in the techniques used, suggesting that procedures may be optimized further. The differences among cell types may also reflect incompatibilities in the cell cycle between donor and recipient cells. Some cell types may contain irreversible genetic blocks due to differentiation. Irreversible gene silencing can result from multiple G:C to A:T transition mutations, termed "repeat-induced point mutations," induced by methylation. The proportions of stem cells, which may be more amenable to undergoing nuclear reprogramming, are also likely to vary among tissues as well.

TABLE 2.1   Relative Success of Nuclear Transplantation Using Different Donor Nuclei

	Species	Blastocysts (%)	Live Births (% of transfers)	References
Tissue				
Embryonic or	Sheep (fibroblasts)	37.9	7.5	Wilmut et al., 1997
fetal cells	Sheep (fibroblasts)	6–20	5–20	Schnieke et al., 1997
	Cattle (fibroblasts)	12	14	Cibelli et al., 1998
	Goats (fibroblasts)	34–49	3.5	Baguisi et al., 1999
Adult somatic cells				
Fibroblasts	Sheep	11	3.4	Wilmut et al., 1997
Cumulus cells	Mice	—	2.3	Wakayama et al., 1998
	Cattle	49	83	Kato et al., 1998
Neuronal cells	Mice	22	—	Wakayama et al., 1998
Sertoli cells	Mice	40	—	Wakayama et al., 1998
Oviductal cells	Cattle	23	75	Kato et al., 1998
Granulosa cells	Cattle	69	10	Wells et al., 1999
Trophectoderm	Mice	—	0	Tsunoda et al., 1998
		—	0	Collas and Barnes, 1994
		32–64	8	
	Rabbit	0	—	

TOWARD AN UNDERSTANDING OF THE MECHANISMS OF GENETIC REPROGRAMMING

Cloning animals has been the focus of the efforts in nuclear transplantation to date because this provides the most stringent test of the underlying phenomenon of genetic reprogramming. Cloning has been, in some ways, an unfortunate endpoint because of the ethical dilemmas that arise from the potential application of this technology to humans. The prospect of human cloning and its moral and ethical implications has diverted both public and political attention away from the fundamental goal of identifying the molecular basis for reprogramming the DNA to allow cells to regain developmental plasticity. Once these mechanisms are understood, they may be harnessed to interconvert cell types. The implications and medical therapeutic applications of cellular interconversion are staggering (summarized in Table 2.2). For example, skin cells from a leukemia patient could be converted to hematopoietic stem cells for reconstituting the hematopoietic system following chemotherapy without risk of "residual disease" from the transplanted cells, a major reason for failure of autologous bone marrow transfers. Alternatively, new approaches toward disease etiology may be explored. Cancer could be viewed as the converse situation where a cell acquires new phenotypes as the result of inappropriate genetic reprogramming. Cancer cells harbor many genetic changes (see Chapter 11), but the phenotype is, in part, reversible. Thus the question arises: How to reverse the cancer phenotype through genetic reprogramming? The most dramatic example of such "reprogramming" of cancer cells is the ability of embryonal carcinoma cells to participate in normal development to produce chimeric mice. Adenocarcinoma cells have also been shown to produce normal offspring after nuclear transplantation. Additionally, the cellular microenvironment has been shown to "reprogram" globin

TABLE 2.2 Comparison of Nuclear Transplantation Donors and Applications to Therapies

Nuclear Donor	Blastocyst Development[a]	Benefits/Drawbacks	Applications
Blastomeres	10–40%	*Benefits* Generally result in higher rates of development following nuclear transplantation *Disadvantages* Negligible G1 phase due to rapid division Random genetic qualities of the cells as the donor cannot be predetermined Limited numbers of cells can be obtained Difficult to introduce transgenes or genetic modifications	Limited to experimental uses
Inner cell mass	10–40%	*Benefits* Easily prepared Relatively undifferentiated cells *Disadvantages* Random genetic qualities of the cells as the donor cannot be predetermined Limited numbers of cells can be obtained Difficult to introduce transgenes or genetic modifications	Limited to experimental uses

TABLE 2.2 (*Continued*)

Nuclear Donor	Blastocyst Development[a]	Benefits/Drawbacks	Applications
ES Cells	10–40%	*Benefits* Potentially unlimited source of cells because of indefinite life span Relatively undifferentiated cells Well suited to introducing genetic modifications *Disadvantages* Random genetic qualities of the cells as the donor cannot be predetermined Slow growing and cannot be clonally propagated in most species	Preparation of transgenic livestock for cell-based therapies and organs for xenotransplantation Cells for allotransplantation Precise genetic modifications can be introduced by homologous recombination
Adult somatic cells	10–70%	*Benefits* Numbers of cells from a primary culture are large Donor of cells can be genetically defined Suitable for genetic modifications *Disadvantages* Growth in vitro is limited by senescence in some Cell types with greatest success can only be obtained from females	Preparation of transgenic livestock for cell-based therapies and organs for xenotransplantation Cells for allotransplantation Cells for autologous transplantation Precise genetic modifications can be introduced by homologous recombination

[a]Overall efficiencies vary greatly among species.

gene expression in chimeric mice. Thus, one could envision new fields of investigation detailing cellular reprogramming mechanisms determining cell type and function based on the local tissue or organ microenvironment.

Methylation and Acetylation

Methylation and acetylation appear as prominent candidates in mediating nuclear reprogramming. These biochemical activities alter gene transcription not only during development but also function in oncogenic transformation as well. Methylation of DNA is a critical factor for remodeling the genome both during normal embryonic development and tumorigenesis (see Chapter 10). The hypomethylated genomic DNA of primordial germ cells undergoes extensive methylation during gametogenesis. The heavily methylated state of the genome, along with extensive deacetylation of chromatin-bound histones, is maintained in the newly formed zygote after fertilization and is associated with the transcriptionally inactive state of the embryonic genome. This is followed by a wave of demethylation during the eight-cell to blastocyst stages. A surge in methylation affecting the entire genome is observed on or about implantation. Global methylation of the embryonic genome coincides with lengthening of the G1 stage of the cell cycle and continues in a tissue-specific fashion in the developing embryo. These alterations in methylation appear essential for proper programming of developmental fates as targeted disruption of DNA methyltransferase (*Dmnt1*) resulted in embryonic lethality.

Imprinting

The uniparental expression of genes is referred to as imprinting. This involves the transcriptional silencing of specific genes during gametogenesis so that only the maternal or paternal allele is expressed in the embryo. The imprinting mechanism is fully reversible during gametogenesis in the next generation, and therefore, represents an epigenetic process that is subject to reprogramming by nuclear transplantation.

Expression of imprinted genes is often correlated with the methylation status of specific CpG islands within promoter regions. The allele-specific methylation at these sites must be preserved during the genomewide wave of demethylation occurring in the preimplantation embryo. Two themes emerge from the list of genes known to be imprinted. The *Igf2*, *Igf2r*, and *H19* genes affect the rate of fetal growth, while *Ins2*, *p57^{Kip2}*, and *Mash2* genes appear to participate in regulation of the cell cycle during fetal growth. Failure to maintain proper genetic imprinting may be responsible for the embryonic death in a large proportion of nuclear transplantation embryos beyond the blastocyst stage. Other genes undergo transcriptional silencing during embryonic development resulting in variagated gene expression, but patterns of gene silencing have not been studied in cloned animals.

Changes in methylation status of the genome following nuclear transplantation may also affect genes subject to parental imprinting. Regional methylation has been associated with X inactivation. Both male and female donor cells have been cloned by nuclear trasplantation, but whether X inactivation is properly coordinated in females remains to be considered. Several other loci need to be properly imprinted to allow normal development. Failure to maintain proper genetic imprinting may be responsible for the embryonic death that occurs beyond the blastocyst stage.

Cellular Senescence

Many questions as to the "age" of cells following nuclear transplantation persist. Erosion of telomeric repeats has been associated with aging and cellular senescence in vitro. It is clear from mice lacking telomerase ribonucleic acid (RNA) that telomerase activity is not essential for first-generation progeny. However, subsequent generations show shortened telomeres with karyotypic instability developing as a consequence. Most recently, cloned sheep were shown to have somewhat shortened telomeres indicating premature aging of their genomes. However, others found that the in vitro life span was regenerated by nuclear transplantation. Therefore, it will be important to resolve the effects of nuclear transplantation on telomerase activity, telomere stability, and senescence.

APPLICATIONS OF GENETIC REPROGRAMMING

Production of Human ES Cells

The ability to produce pluripotent cells from an adult donor would appear to be within our grasp. Procedures for establishing ES cell cultures from the ICM of in vitro fertilized embryos were reported recently in *Science* (Thomson et al 1998). Successful early embryonic development following nuclear transplantation has been reported in primate, for rhesus monkeys. As an extension of these studies, fibroblasts could be obtained from a patient and used as donor cells for nuclear transplantation to yield blastocysts. ES cells could then be prepared from the ICM of the nuclear transplantation blastocysts. This approach, precluded from federally funded research laboratories by Congress, would undoubtably cause rancorous debate based on ethical issues. It would require a blastocyst that has the potential for creating a human being to be sacrificed as a "tissue" donor for the patient. With the present-day technology, this experiment would require many attempts because of the inefficiency of obtaining blastocysts following nuclear transplantation and the challenges in creating ES cell cultures. Therefore, such an approach would be unsatisfactory because of both ethical reasons and technical obstacles. Based on these limitations, the major application of nuclear transplantation may be to identify the molecular mechanisms required for genetic reprogramming. Once the essential complex of factors is established, these may be provided in a purified form to reprogram cells obviating the need to create blastocysts.

Interconversion Between Cell Types for Cell Therapies

A more modest reprogramming might be envisioned in which cells need only be interconverted between lineages. Interconversion of brain stem cells to hematopoietic lineages was reported recently in *Science* (Bjornson et al 1999). In addition, it has been reported that mesenchymal stem cells can differentiate into adipose (fat), chondrocytes (cartilage), and osteocytes (bone). Nuclear transplantation may provide a means to generate various multipotent cell types in humans more efficiently than by purification used in these instances. Multipotent stem cells could be used as cellular therapies to replace cells lost in patients suffering from various maladies including diabetes or Parkinson is disease (see Chapter 9).

Genetically Modified Livestock for Xenotransplantation

However promising these methods may be, the most pressing limitation is the insufficient supply of human organs and tissues for transplantation. Use of genetically modified livestock is one source that has been considered. The most immediate drawback involves rejection of transplants. Efforts to induce immunological tolerance using gene therapy approaches have spawned increasing hopes for xenotransplantation. Alternatively, the antigenicity of the tissues may be reduced by targeted disruption of histocompatibility genes in donor animals. Nuclear transplantation provides the technical means to introduce the necessary genetic modifications to minimize rejection and produce genetically uniform donors.

Providing Cells for Tissue Engineering

Ideally, rejection of transplants could be eliminated by using cells from individual patients to regenerate their organs through tissue engineering. Great strides have been made in creating skin equivalents for treating chronic wounds. More complex tissues such as arteries and corneal tissues have also been successfully grown in vitro. The most stunning example was the growth of a functional bladder from precursor cells. Advances in biocompatible polymers will continue to extend the range of organs that can be grown in vitro. Presently, the amount that primary cell cultures can be expanded in vitro is limited by cellular senescence. Nuclear transplantation appears to reset the mechanisms regulating in vitro life spans. Tissue engineering approaches would also benefit from the ability to generate multipotent stem cell populations from individual patients to provide autologous transplants.

HUMAN EMBRYONIC STEM CELL RESEARCH: AN ETHICS NOTE

A frequent theme in the writings of medical ethicists is that ethical analyses need to be done before research is undertaken. The goal, here, is to prevent unethical research from being done in the first place. However, this directive may not be as feasible in practice as it appears in theory. Sometimes actual facts and consequences of research are not what was hypothetically predicted; and when such consequences are morally relevant, our moral judgments might be altered. This will likely be the case with regard to much of human embryonic stem cell research. For example, if it turns out to be the case that the few embryos used so far to generate several lines of stem cells are sufficient to satisfy research needs far into the future (because of the seemingly unlimited regenerative capacity of these cells), then it seems we have a minor ethical problem at best. On the other hand, if future research requires tens of thousands of embryos to be created and destroyed in order to meet highly specialized future research needs, then we have a more serious ethical problem. At present we have no way of knowing which state of affairs will prove true unless we allow the research to go forward.

The central moral conflict enunciated, to date, regarding this research is between its enormous therapeutic potential and the need to destroy human embryos (potential human persons) to realize that therapeutic outcome. Alzheimer's disease, Parkinson's, diabetes, heart disease, to name just a few, are seen as being substantially ameliorable (if not curable) if stem cell research realizes its potential (Wright, 1999), as the Patients' Coalition for Urgent Research hopes (a coalition of two dozen national organizations). Of course, that is a large "if." There are no guarantees that this line of research will achieve such success. Further, advocates of a "sanctity of life" ethic claim that even if such therapeutic gains were certain of realization, the means by which they were achieved would be evil, the destruction of embryos, which they regard as being persons with the same moral rights as you and I. One line of response is that the embryos so used have been discarded as "excess embryos" by couples who sought out in vitro fertilization (IVF) and have now achieved their goals. The moral argument is that the destruction of the embryos will happen anyway, and this research permits some substantial therapeutic good to be realized. No one is being paid to create embryos for research purposes (though that is another ethics issue that ought to be addressed). Further, the argument goes, the moral status of embryos is at least controversial. A more impartial description of their moral status would say they are "potential" persons whereas the individuals who must endure Alzheimer's or heart disease are clearly actual persons with actual moral rights and compelling health needs. (Is this latter point of sufficient moral weight that it would justify paying couples to create embryos that might be needed by a successful research effort and/or subsequent therapeutic deployment?)

There are alternate ways of satisfying the ethical concerns of a sanctity of life advocate. McGee and Caplan (1999) have pointed out that, strictly speaking, this research might not require the destruction of embryos. It may be medically possible to harvest a very few stem cells from an embryo that is then implanted in the womb for normal development. Further, it is imaginable that this would become a standard reproductive option that would allow those stem cells to be saved for possible future use by that person, thereby avoiding potential tissue rejection problems should future medical need require a transplant. It is not obvious that a defender of a sanctity of life ethic would be able to raise a strong moral objection to that procedure.

What Richard Doerflinger (1999) of the National Conference of Catholic Bishops has suggested is that there is an alternate line of research that ought to be pursued that might achieve the same therapeutic goals without having to destroy embryos, namely research that would begin with adult stem cells. The practical problem, however, is that this line of research has not thus far been established as viable. It may well be viable, sometime in the future. In the meantime proponents of the current research argue that it would be unconscionable to delay for years achieving the therapeutic promise of what we have now, especially if a substantial majority of Americans reject the view that embryos have the moral status of persons from the moment of conception.

Finally, there are two other moral concerns that need to be noted. First, there is no doubt that it would be a "very good thing," ethically speaking, if this stem cell research achieved its therapeutic objectives, such as curing diabetes, or substantially ameliorating Parkinson's or Alzheimer's. However, it is important to note that there is no ethical obligation to achieve that good. No one's moral rights would be violated if those research dollars were redirected to some other worthy medical or social use. There is no ethical imperative that this research be done. Second, there are other moral risks associated with this research that are rarely noted, namely risks to social justice. It is reasonable to ask who the expected beneficiaries of this research are likely to be. The short answer is that the beneficiaries will most likely be individuals who are well insured or otherwise financially well off, who have already benefited substantially from our health care system, and now may benefit even more. If a successful outcome to this research means more health care dollars for implementing the research, thereby increasing pressures for health care cost containment, thereby diverting resources from "charity care" for the poor and uninsured, then this is an ethically objectionable outcome. Further, currently, we are in the middle of the Medicare reform debates. An important moral issue will be this: Are we morally obligated, as a just and caring society, to provide access to the fruits of stem cell research as a Medicare covered benefit (especially if we continue to have 43 million or more uninsured)?

SUMMARY

Nuclear transplantation has been used to demonstrate that many differentiated cells can undergo nuclear reprogramming such that multipotency can be regained (summarized in Table 2.2). This has been demonstrated in diverse organisms ranging from amphibia to mammals, and therefore, appears to be a phenomenon that has been evolutionarily conserved. Development following nuclear transplantation was consistently lower in differentiated cells compared to blastomeres from early cleavage embryos in all organisms examined. Despite these similarities, the extent to which the developmental plasticity of cells can be reprogrammed appears extremely variable both among species and cell types. Discovery of the molecular basis for reprogramming will provide insights into the cause of the variation.

The ability to produce cloned livestock or reprogram somatic cells may have many applications. However, it should be remembered that organisms created by this process are not true clones because they are chimeric with respect to their mitochondrial genomes. Nuclear transplantation provides a means to efficiently introduce precise modifications into the genomes of domestic livestock, opening the possibility of engineering tissues for xenotransplantation. The ability to produce multipotent cells from an individual patient offers hope of providing cellular therapies without the risk of immunologic rejection.

KEY CONCEPTS

- Nuclear transplantation is the process by which the nucleus of a somatic cell is transferred to an enucleated oocyte. Fundamental to the success of nuclear transplantation is the identification of critical factors contained within the cytoplasm of oocytes, which allows reprogramming of somatic cell nuclei.

• Nuclear transplantation can be used to generate identical animals and transgenic livestock. Cloned livestock can be used to intensify genetic selection for improved productivity and have also been proposed as a reliable source of tissues and cells for xenotransplantation in humans. At a more fundamental level, these cloning experiments demonstrate that somatic cells retain developmental plasticity such that the nucleus of a single cell, when placed within an oocyte, can direct development of a complete organism.

• Nuclear donors define the limits of genetic programming in mammalian nuclear transfer and can be performed with blastomers, inner cell mass, embryonic stem cells, and somatic cells. A variety of somatic cell types are capable of undergoing nuclear reprogramming following nuclear transplantation to yield live offspring. However, efficiency of nuclear reprogramming is very dependent on the donor cells.

• A by-product of nuclear transplantation technology may be the ability to interconvert cell types for use in cell therapies. In addition, nuclear transplantation technology may allow for additional resources for organ transplantation through the generation of syngeneic cells, tissues, or organs (made from one's own cells) or xenogeneic cells, tissues or organ for use in xentotransplanatation.

SUGGESTED READINGS

Ethics

Doerflinger R. The threat of science without humanity. Catholic Standard (7/29), 1999, p. 7.

Geva E, Amit A, Lerner-Geva L, Lessing JB. Embryo transfer and multiple gestation. How many transfers are too many. Human Reprod 13:2988–2989, 1998.

McGee G, Caplan A. What's in the dish? Hastings Center Report 29(2):36–38, 1999.

Wright S. Human embryonic stem-cell research: Science and ethics. Am Sci 87(4):352–361, 1999.

Nuclear Transfer and Cloning

Baguisi A, Behboodi E, Melican DT, Pollock JS, Destrempes MM, Cammuso C, Williams JL, Nims SD, Porter CA, Midura P, Palacios MJ, Ayres SL, Denniston RS, Hayes ML, Ziomek CA, Meade HM, Godke RA, Gavin WG, Overstrom EW, Echelard Y. Production of goats by somatic cell nuclear transfer. Nat Biotechnol 17:456–461, 1999.

Berg H. Biological implications of electric field effects. V: Fusion of blastomeres and blastocysts of mouse embryos. Bioelectrochem Bioenerg 9:223, 1982.

Betteridge KJ, Rieger D. Embryo transfer and related techniques in domestic animals, and their implications for human medicine. Hum Reprod Update 8:147–167, 1993.

Briggs R, King TJ. Transplantation of living nuclei from blastula cells into enucleated frogs' eggs. Proc Natl Acad Sci USA 38:455, 1952.

Briggs R, Signoret J, Humphrey RR. Transplantation of nuclei of various cell types from neurulae of the Mexican Axolotl (*Ambystoma mexicana*). Dev Biol 10:233–246, 1964.

Burgess AM. The developmental potentialities of regeneration blastema cell nuclei as determined by nuclear transplantation. J Embryol Exp Morph 18:27–41, 1967.

Campbell KH, McWhir J, Ritchie WA, Wilmut I. Sheep cloned by nuclear transfer from a cultured cell line. Nature 380:64–66, 1996.

Cibelli JB, Stice SL, Golueke P, Kane JJ, Jerry J, Blackwell C, Ponce De Leon FA, Robl JM. Cloned transgenic calved produced from nonquiescent fetal fibroblasts. Science 280:1256–1258, 1998a.

Cibelli JB, Stice SL, Golueke PJ, Kane JJ, Jerry J, Blackwell C, Ponce de Leon FA, Robl JM. Transgenic bovine chimeric offspring produced from somatic cell-derived stem-like cells. Nat Biotechnol 16:642–646, 1998b.

Collas P, Barnes FL. Nuclear transplantation by microinjection of inner cell mass and granulosa cell nuclei. Mol Reprod Dev 38:264–267, 1994.

Kato Y, Tani T, Sotomaru Y, Kurokawa K, Kato J, Doguchi H, Yasue H, Tsunoda Y. Eight calves cloned from somatic cells of a single adult. Science 282:2095–2098, 1998.

Kubota C, Yamakuchi H, Todoroki J, Mizoshita K, Tabara N, Barber M, Yang X. Six cloned calves produced form adult fibroblast cells after long term culture. Proc Natl Acad Sci USA 97:990–995, 2000.

Mirsky S. What cloning means to gene therapy. Sci Am June 276:122–123, 1997.

Prather RS, Sims MM, First NL. Nuclear transplantation in early pig embryos. Biol Reprod 41:414–418, 1989.

Prather RS, Barnes FL, Sims MM, Robl JM, Eyestone WH, First NL. Nuclear transplantation in the bovine embryo: Assessment of donor nuclei and recipient oocyte. Biol Reprod 37:859–866, 1987.

Schnieke AE, Kind AJ, Ritchie WA, Mycock K, Scott AR, Ritchie M, Wilmut I, Colman A, Campbell KHS. Human factor IX transgenic sheep produced by transfer of nuclei from transfected fetal fibroblasts. Science 278:2130–2133, 1997.

Schubiger M, Schneiderman HA. Nuclear transplantation in *Drosophila melanogaster*. Nature 230:185–186, 1971.

Tsunoda Y, Yasui T, Shioda Y, Nakamura K, Uchida T, Sugie T. Full-term development of mouse blastomere nuclei transplanted into enucleated two-cell embryos. J Exp Zool 242:147–151, 1987.

Wakayama T, Perry ACF, Zuccotti M, Johnson KR, Yanagimachi R. Full-term development of mice from enucleated oocytes injected with cumulus cell nuclei. Nature 394:369–374, 1998.

Willadsen SM. Nuclear transplantation in sheep embryos. Nature 320:63–65, 1986.

Wilmut I, Schnieke AE, McWhir J, Kind AJ, Campbell KHS. Viable offspring derived from fetal and adult mammalian cells. Nature 385:810–813, 1997.

Wolf DP, Meng L, Ouhibi N, Zelinski-Wooten M. Nuclear transfer in the rhesus monkey: Practical and basic implications. Biol Reprod 60:199–204, 1999.

Zawada WM, Cibelli JB, Choi PK, Clarkson ED, Golueke PJ, Witta SE, Bell KP, Kane J, Ponce De Leon FA, Jerry DJ, Robl JM, Freed CR, Stice SL. Somatic cell cloned transgenic bovine neurons for transplantation in parkinsonian rats. Nat Med 4:569–574, 1998.

Developmental Biology, Nuclear Remodeling, Tissue Engineering and Xenotransplantation

Blasco MA, Lee H-W, Hande MP, Samper E, Lansdorp PM, DePinho RA, Greider CW. Telomere shortening and tumor formation by mouse cells lacking telomerase RNA. Cell 91:25–34, 1997.

Bjornson CR, Rietze RL, Reynolds BA, Magli MC, Vescovi AL. Turning brain into blood: A hematopoietic fate adopted by adult neural stem cells in vivo. Science 283:534–537, 1999.

Bracy JL, Sachs DH, Iacomini J. Inhibition of xenoreactive natural antibody production by retroviral gene therapy. Science 281:1845–1847, 1998.

Brook FA, Gardner RL. The origin and efficient derivation of embryonic stem cells in the mouse. Proc Natl Acad Sci USA 94:5709–5712, 1997.

Buhler L, Friedman T, Iacomini J, Cooper DK. Xenotransplantation—state of the art—update 1999. Front Biosci 4:D416–D432, 1999.

Collas P, Robl JM. Relationship between nuclear remodeling and development in nuclear transplant rabbit embryos. Biol Reprod 45:455–465, 1991.

Ferber D. Lab-grown organs begin to take shape. Science 284:423–425, 1999.

Germain L, Auger FA, Grandbois E, Guignard R, Giasson M, Boisjoly H, Gu. Reconstructed human cornea produced in vitro by tissue engineering. Pathobiology 67:140–147, 1999.

Gurdon JB. Nuclear transplantation and the control of gene activity in animal development. Proc R Soc Lond B 176:303–314, 1970.

Gurdon JB, Laskey RA. The transplantation of nuclei from single cultured cells into enucleate frogs' eggs. J Embryol Exp Morph 24:227–248, 1970.

Gurdon JB, Laskey RA, Reeves OR. The developmental capacity of nuclei transplanted from keratinized skin cells of adult frogs. J Embryol Exp Morph 34:93–112, 1975.

John R, Surani MA. Imprinted genes and regulation of gene expression by epigenetic inheritance. Curr Opin Cell Biol 8:348–353, 1996.

Langer RS, Vacanti JP. Tissue engineering: The challenges ahead. Sci Am 280:86–89, 1999.

Mackay AM, Beck SC, Murphy JM, Barry FP, Chichester CO, Pittenger MF. Chondrogenic differentiation of cultured human mesenchymal stem cells from marrow. Tissue Eng 4:415–428, 1998.

Niklason LE, Gao J, Abbott WM, Hirschi KK, Houser S, Marini R, Langer R. Functional arteries grown in vitro. Science 284:489–493, 1999.

Oberpenning F, Meng J, Yoo JJ, Atala A. De novo reconstitution of a functional mammalian urinary bladder by tissue engineering. Nat Biotechnol 17:149–155, 1999.

Piedrahita JA, Moore K, Oetama B, Lee CK, Scales N, Ramsoondar J, Bazer FW, Ott T. Generation of transgenic porcine chimeras using primordial germ cell-derived colonies. Biol Reprod 58:1321–1329, 1998.

Pittenger MF, Mackay AM, Beck SC, Jaiswal RK, Douglas R, Mosca JD, Moorman MA, Simonetti DW, Craig S, Marshak DR. Multilineage potential of adult human mesenchymal stem cells. Science 284:143–147, 1999.

Rossant J, Papaioannou VE. The relationship between embryonic, embryonal carcinoma and embryo-derived stem cells. Cell Differ 15:155–161, 1984.

Selker EU. Gene silencing: Repeats that count. Cell 97:157–160, 1999.

Shiels PG, Kind AJ, Campbell KHS, Waddington D, Wilmut I, Colman A, Schnieke AE. Analysis of telomere lengths in cloned sheep. Nature 399:317–318, 1999.

Smith LC, Wilmut I. Influence of nuclear and cytoplasmic activity on the development in vivo of sheep embryos after nuclear transplantation. Biol Reprod 40:1027–1035, 1989.

Speman H. Embryonic development and induction. Hafner, New York, 1938.

Stice SL, Robl JM. Nuclear reprogramming in nuclear transplant rabbit embryos. Biol Reprod 39:657–664, 1988.

Thompson EM. Chromatin structure and gene expression in the preimplantation mammalian embryo. Reprod Nutr Dev 36:619–635, 1996.

Thomson JA, Itskovitz-Eldor J, Shapiro SS, Waknitz MA, Swiergiel JJ, Marshall VS, Jones JM. Embryonic stem cell lines derived from human blastocysts. Science 282:1145–1147, 1998.

Tilghman SM. The sins of the fathers and mothers: Genomic imprinting in mammalian development. Cell 96:185–193, 1999.

Tsunoda Y, Kato Y. Not only inner cell mass cell nuclei but also trophectoderm nuclei of mouse blastocysts have developmental totipotency. J Reprod Fertil 113:181–184, 1998.

CHAPTER 3

Building a Better Mouse: Genetically Altered Mice as Models for Gene Therapy

William C. Kisseberth, D.V.M., M.S. and ERIC SANDGREN, V.M.D., PH.D.

BACKGROUND

Mice have been used in biomedical research for many years: their small body size, efficient reproductive characteristics, and well-defined genetics make mice an ideal experimental subject for many applications. In particular, the use of mutant mice as models of human disease, and more recently their use to explore somatic gene therapy, has been expanding. Multiple genetic assets of the mouse make the development of new models of human disease relatively straightforward in the mouse as compared to other species. These include the existence of inbred strains of mice, each with a unique but uniform genetic background, an increasingly dense map of the murine genome, and defined experimental methods for manipulating the mouse genome.

INTRODUCTION

In mice, genetic mutations may occur spontaneously or they can be induced by experimental manipulation of the mouse genome via high-efficiency germline mutagenesis, via transgenesis, or via targeted gene replacement in embryonic stem (ES) cells. Although each of these methods has potential advantages and disadvantages, all have been successful in generating models of human disease for use in developing gene therapy technology. Reviewing several common methods of manipulating the mouse genome, addressing questions relating to genetic disease that can be asked (and answered) using mouse models, describing how mouse models can be used to evaluate somatic gene transfer, and finally speculating on what experimental approaches to model development might be used in the future are the scope of this chapter.

An Introduction to Molecular Medicine and Gene Therapy, Edited by Thomas F. Kresina
ISBN 0-471-39188-3   © 2001 Wiley-Liss

PRODUCING MOUSE MODELS OF HUMAN DISEASE

Spontaneous Mutations

Spontaneous mutations occurring in existing mouse colonies are a historical and continuing source for models of genetic disease. In the past, pet mice were selected and propagated based on the presence of an unusual phenotype. Phenotypes such as coat color alterations or neurological disorders were chosen because of their striking visual impact. For phenotypes with a heritable basis, subsequent mating of affected animals produced "lines" of mice displaying the genetic-based phenotype. More recently, with the establishment of large scientific and commercial breeding facilities along with careful programs of animal monitoring, many additional lines of spontaneous mutants have been established. In some cases, an observed phenotype may be caused by mutation of a gene that is responsible, in humans, for a specific genetic disease. These models are usually identified based on phenotypic similarities between the mouse and human diseases. The mutated gene needs to be identified if these models are to assist in the research or testing of somatic gene therapies. Identification will require genetic mapping and positional cloning of the mutated gene, made easier in mouse by the availability of well-established gene mapping reagents. When the gene causing or associated with the human disease has been identified in the mouse, the mouse homolog of the human gene (a "candidate gene") can be screened for the presence of a mutation. A partial list of prominent spontaneous genetic disease mouse models is presented in Table 3.1.

High-Efficiency Germline Mutagenesis

As with selection of spontaneous mutations, high-efficiency germline mutagnesis using ethylnitrosourea (ENU) is phenotype driven. Young, sexually mature male mice are treated with the alkylating agent ENU, which introduces random base changes (mutations) into spermatogonial stem cells. Treated males are mated approximately 100 days later following recovery from a period of ENU-induced sterility. The resulting mutations can be transmitted to progeny, which are screened for the disease phenotype of interest (Fig. 3.1). In principle, ENU-induced muta-

TABLE 3.1   Selected Spontaneous Mouse Genetic Disease Models

Mouse Gene Symbol	Deficiency Produced	Disease Modeled
Btk^{xid}	Btk, Bruton's tyrosine kinase	X-linked agammaglobulinemia
Dmd^{mdx}	Duchenne muscular dystrophy protein (dystrophin)	Duchenne muscular dystrophy
$Hfh11^{nu}$	$Hfh11$, HNF	T-cell immunodeficiency
$Lepr^{db}$	$Lepr$, leptin receptor	Diabetes mellitus
$Lyst^{bg}$	Lysosomal trafficking disorder	Chediak–Higashi syndrome
NOD	Polygenic	Diabetes mellitus
$Pdeb^{rdl}$	$Pdeb$, phosphodiesterase, cGMP (rod receptor), betapolypeptide	Retinal degeneration
$Prkdc^{scid}$	$Prkdc$, protein kinase, DNA-activated catalytic peptide	Severe combined immunodeficiency
$Prph2^{Rd2}$	$Prph2$, peripherin	Retinal degeneration

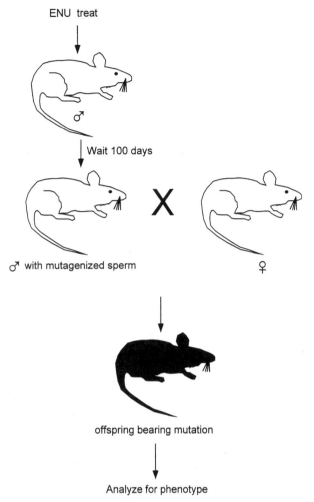

FIGURE 3.1 High-efficiency ENU-induced germline mutagenesis. Young, sexually mature male mice are treated with the mutagen ethylnitrosourea (ENU). After recovery from ENU-induced infertility, treated males with mutagenized sperm are mated with normal females. Offspring bearing the mutation are analyzed for the phenotype of interest.

tions are sufficiently frequent so that only 500–1000 offspring of treated males need to be screened to recover one animal with a mutation at a given genetic locus. Because of the number of animals to be screened, it is important for the phenotype to be well defined, easily and inexpensively identifiable, as well as expressed in young mice. Thus, large numbers of animals need not be maintained for an extended period of time prior to screening. Strategies for detecting phenotypes are quite variable. For example, dominant mutations may be based on an obviously visible phenotype, or altered electrophoretic mobility of a protein in a gel, or a change in behavior. Detection of recessive mutations generally requires (1) producing offspring from mice derived from mutagenized sperm, (2) interbreeding brothers with sisters from these litters, and (3) determining the phenotype of resulting offspring. If the original parent carried one mutant allele, half of its offspring also should be

TABLE 3.2   Selected ENU-Induced Mouse Genetic Disease Models

Gene Symbol	Deficiency Produced	Disease Modeled
Apc	Adenomatous polyposis coli (APC) protein	Adenomatous intestinal polyposis
Car2	Carbonic anhydrase II (CAII)	CA-II deficiency syndrome
Dma^{mdx}	Dystrophin dehydrogenase (G6PD)	Muscular dystrophy syndrome
G6pt	Glucose-6-phosphate	G6PD deficiency syndrome
	GTP-cyclohydrolase I	Tetrahydrobiopterin-deficient hyperphenylalaninemia
Hba	Hemoglobin, α-chain	α-Thalassemia
Hbb	Hemoglobin, β-chain	β-Thalassemia
Pah	Phenylalanine hydroxylase	Phenylketonuria
Sar	Sarcosine dehydrogenase	Hypersarcosinemia
Tpi	Triosephosphate isomerase (TPI)	TPI deficiency

carriers. A mating between two carrier offspring would produce progeny with a 25% chance of carrying two mutant alleles, thereby displaying a recessive phenotye. A strength of ENU mutagenesis is that disease models can be generated even though mutations are at unidentified loci. These new mutations can be mapped in the mouse genome and perhaps the human gene location inferred through synteny homologies. A partial list of ENU-induced animal models is presented in Table 3.2.

Transgenic Mice

Whereas the previous methods are phenotype driven, the following methods are genotype driven. Here a known genetic alteration is introduced into the germline and the phenotypic consequences are observed. As noted earlier, classical mutagenesis does have inherent limitations. Because mutations are produced randomly, extensive screening may be necessary to identify carriers of a mutation at the locus of interest. Second, induced mutations are not "tagged" in any way to facilitate identification of the mutant gene: ENU-induced deoxyribonucleic acid (DNA) lesions are typically single nucleotide changes. Transgenic animals circumvent some of these problems by allowing introduction of a precisely designed genetic locus of known sequence into the genome. Foreign DNA, or *transgenes*, can be introduced into the mammalian genome by several different methods, including retroviral infection or microinjection of ES cells, as well as by microinjection of fertilized mouse eggs (see Chapter 2).

The first step in the creation of transgenic mice is construction of the transgene (*trans* refers to the fact that, historically, the introduced DNA was not from the mouse; thus DNA was being transferred trans species). Most transgenes contain three basic components: the gene regulatory elements (enhancer/promoter), messenger ribonucleic acid (mRNA) encoding sequence, and polyadenylation signal (Fig. 3.2). The enhancer/promoter regulates transgene expression in an either/or developmental and tissue-specific manner. For example, gene regulatory elements from the albumin gene will be expressed in fetal hepatocytes beginning shortly after midgestation. Expression will reach a maximal (and steady state) level in young adult hepatocytes. The coding sequence may be in the form of genomic DNA or a

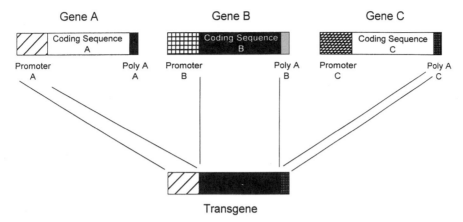

FIGURE 3.2   Transgene construction. Constituent parts of a simple transgene may come for one or more sources. Gene regulatory elements (promoters/enhancers) from gene A may be fused to the mRNA coding sequence from gene B and the polyadenylation signal of gene C. Transgene expression is directed in a developmental- and tissue-specific pattern specified by regulatory elements from gene A. The stability of transgene mRNA is modified by the polyadenylation signal from gene C.

cDNA and generally can be transcribed into an mRNA capable of being translated into a protein. Genomic DNA is preferred for transgene construction since it is more reliably expressed, possibly because of the presence of gene expression regulatory elements within introns. In practice, complementary DNA (cDNA) are commonly used because of their smaller size and ready availability. The use of cDNA's necessitates the use of special transgene construction techniques to enhance expression. Finally, for many applications transgene mRNA stability is an important issue. Message stability often can be improved by replacement of the gene's endogenous polyadenylation sequence with a heterologous polyadenylation sequence taken from a gene that produces a very stable message, such as the human growth hormone gene or the simian virus 40 (SV40) T antigens gene. The end result of the joining of these pieces of DNA is a transgene that will target stable expression of a selected coding sequence to specific tissue(s) during selected stage(s) of life. Most commonly, these elements of the transgene are assembled in plasmid vectors. Transgenes are then excised from the vector, isolated, and purified prior to injection into fertilized mouse eggs. More recently, transgenes have been created using large DNA fragments, including yeast and bacterial artificial chromosomes and P1 phage.

Once microinjected into the pronucleus of fertilized mouse eggs, as shown in Figure 3.3, transgenes can become integrated into chromosomal DNA in an apparently random manner and through an unknown mechanism. However, integration may be favored at sites of DNA double-strand breaks. In most instances, multiple copies of the DNA fragment will integrate in a head-to-tail tandem array at a single genomic locus. Microinjected eggs are then surgically transferred into the oviducts of pseudopregnant recipients and develop to term. Pseudopregnant females have been bred by vasectomized males, so that a state of "physiological pregnancy" is

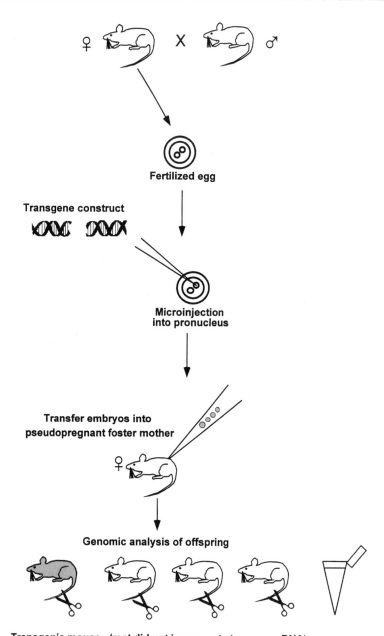

FIGURE 3.3   Transgenic mouse production. A dilute DNA solution containing the transgene construct is microinjected into the pronucleus of fertilized mouse eggs. The microinjected embryos are transferred to the oviduct of pseudopregnant foster mothers in which they develop until birth. Tissue samples (generally tail) are analyzed by Southern blotting or PCR for presence of the transgene. Mice that have incorporated the transgene into their genome and pass the transgene to their offspring are referred to as "founders" of a lineage.

induced by cervical stimulation during copulation. However, these females have no naturally occurring fertilized eggs. Offspring of implanted eggs with incorporated transgene in genomic DNA (termed *founder mice*) can be identified by PCR or Southern blotting of tissues using transgene-specific DNA as a probe. Presence of the transgene in the germline results in passage to progeny. A single founder mouse and its transgene-bearing offspring constitute a *lineage*.

Although present in every cell of the body, transgene expression is regulated as specified by its gene regulatory elements. In practice, transgene expression often is highly variable and dependent on the genomic site of integration. Thus, in a "typical" injection experiment in which nine lineages are generated that carry a particular transgene, mice in three lineages will not express the transgene. This may be a result of transgene integration into untranscribed or silent regions of the genome. Mice in another three lineages will express the transgene but in an unexpected tissue- or development-specific pattern. This outcome may be a consequence of transgene integration near powerful endogenous enhancer or promoter elements. These would overtly influence expression of the integrated DNA. Finally, mice in the final three lineages will express the transgene as expected based upon the transgene's regulatory elements. However, the level of expression may vary among lineages. One important aspect of transgenic animals is that transgenes permit assessment of the phenotypic consequences only of dominant acting genes. The transgenic mouse retains normal copies of all endogenous genes. Selected models created by the transgenic approach are listed in Table 3.3.

TABLE 3.3   Representative Transgenic Mice as Models of Human Disease

Transgene	Human Disease
F_c γ-RIII	Autoimmune hemolytic anemia (AIHA)
Bone morphogenic protein Protein-4	Inherited photoreceptor degeneration (Retina)
Epidermal growth factor receptor (EGFR)	Glioblastoma multiforme
Stromelysin-1 Matrix metalloproteinase 3 (MMP-3)	Atheroma
SAD mouse	Sickle cell disease
Db/Db	Type 2 diabetes
Ob/Ob	Obesity
EL/EL	Epilepsy
Juvenile cystic kidney *jck* mutation	Polycystic kidney disease (PKD)
Troponin I (TNI 1-193)	Reversible contractile heart failure (stunned myocardium)
Amyloid precursor protein or Presenilin-1	Early-onset familiar Alzheimer's disease
Connexins –CX43 or Cx40	Arrhythmias/sudden cardiac death
Neurotrophins and receptors	Nociceptive or analgesic pain
Lipoprotein	Atherosclerosis

Targeted Mutagenesis

The ideal model for the study of somatic gene therapy should exhibit the same genetic deficiency as the human disease. In general, the greater the similarity between the mouse mutation and the mutation as it occurs in humans, the greater the likelihood that the mouse will produce a reliable model of the human disease, that is, mimic human pathogenesis. Through the selective replacement of normal mouse genes with mutated genes, one can attempt to reproduce the molecular basis of human genetic diseases. A powerful method to accomplish this, developed in the 1990s involves inserting a mutant copy of the desired gene into a targeting vector and then introducing this vector into ES cells. ES cells are derived from cells of the inner cell mass of a blastocyst (see Chapter 2). They have retained an ability to differentiate into all cell types in the body. Thus, ES cells are "totipotent" and now can be maintained and manipulated in cell culture for animal model and gene therapy purposes. Most DNA targeting vectors that integrate into ES cell chromosomes do so randomly. However, with a low frequency, the construct will be "targeted " to the gene of interest in some cells and replaced by homologous recombination (Fig. 3.4; also see Chapter 5). ES cell colonies that have undergone homologous recombination carry the mutation in one allele of the targeted gene. They can be identified by PCR or Southern blotting. Individual cells from these colonies are microinjected into mouse embryos at the blastocyst stage of development (Fig. 3.5). Injected blastocysts develop into chimeric animals, whose tissues comprise a mixture of mutant ES cell-derived and blastocyst-derived (normal) cells. If mutant cells are incorporated into the germline, the mutation can be passed on to progeny heterozygous for the mutant allele. Matings between heterozygotes produce offspring, one-fourth of which carry two mutant alleles (homozygotes). This approach of targeted mutagenesis can identify the phenotypic consequences of deleting or modifying endogenous mouse DNA. Several models generated via targeted mutagenesis are listed in Table 3.4.

Analysis of Phenotype

Techniques for altering the mouse genome to create models of human disease depend upon a systematic and thorough evaluation of phenotype. Without a careful analysis of the consequences to the host of altered gene expression, the relevance of the model to the study of human disease is limited. Analysis of phenotype must take into account that the genetic change is expressed within a complex context: the living organism. Thus, the phenotype will be determined by specific molecular consequences of the mutation such as loss of gene expression, increased gene expression, and production of a mutant protein. In addition phenotypic expression is influenced by cellular biochemistry, tissue- and organ-specific physiology, as well as the environment, the organism-wide homeostatic mechanisms that regulate adaptation of an individual to its surroundings. The analysis of phenotype presupposes an understanding of normal anatomy and complex processes of physiology. However, for studies of gene therapy, these requirements represent an advantage because human disease does not exist in a test tube but within an environmental construct. Thus, it is within this organismal context that any therapy must be effective.

SEQUENCE REPLACEMENT VECTOR

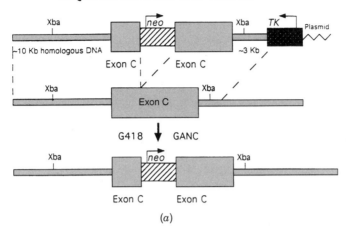

G418 ⟶ GANC

(a)

SEQUENCE INSERTION VECTOR

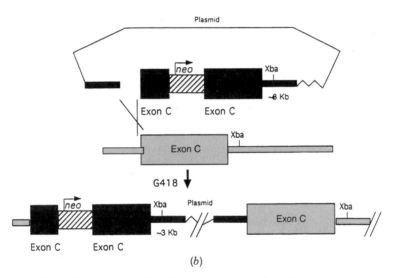

G418

(b)

FIGURE 3.4 Homologous recombination in the generation of gene-targeted animals. (a) Use of a replacement vector having a 10-kb homology with the endogenous locus and 3 kb of *neo* insertion splitting exon C. (*Top*) Arrows indicate transcriptional orientation of promoters and dotted lines indicate regions of homology where recombination may occur. (*Middle*) Wild-type locus. (*Bottom*) Predicted structure of locus after undergoing homologous recombination. (b) Homologous recombination using an insertional targeting vector. (*Top*) Plasmid with sequence insertion vector containing recombinant DNA homologous to the endogenous wild-type locus presented in the middle frame. Prior to electroporation the vector is linearized within the region of homology (5' and 3' ends lie adjacent to each other). (*Bottom*) Structure of altered gene locus based on homologous recombination.

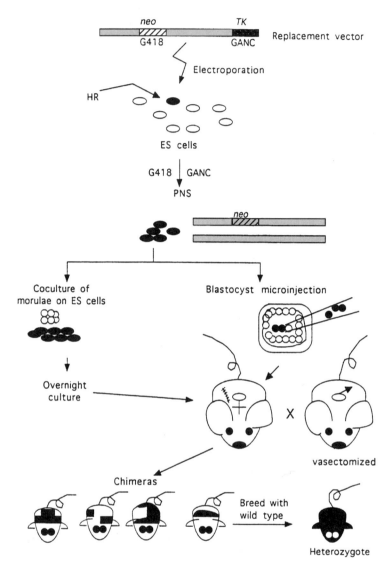

FIGURE 3.5 Microinjection of blastocysts, embryo-derived totipotent stem cells to generate germline chimeric animals. (From Jacenko, 1997.) HR, homologous recombination; PNS, positive–negative selection.

Phenotypic analysis usually involves examination of animal behavior, longevity, and cause of death, as well as gross and microscopic examination of animal tissues. Specialized physiological and behavioral tests also may be performed as a means to determine the cause of the observed abnormalities or because the induced mutation failed to alter the desired biological processes. In general, the analysis of phenotype focuses on detecting abnormalities that are expected from the specific mutation produced. However, these abnormalities would be correlative to those determined by the assessment of physiological and behavioral changes of human disease. Unanticipated phenotypic consequences should not be ignored. Results of

TABLE 3.4 Selected Animal Models Produced by Targeted Mutagenesis

Mouse Homolog	Human Disease	Candidate Human Gene	Genetic Characteristic
Magel 2–7C region	Prader–Willi syndrome (neurodegenerative disorder)	MAGEL-2 15q11–q13	MAGE proteins paternally expressed
Murine ceruloplasmin Gene (Cp)	Aceruloplasminemia (iron metabolism)	Ceruloplasmin gene	Autosomal recessive
β-adducin null mice or knock-out	Hereditary spherocytes (hemolytic disease)	β-Adducin delete exons 9–13	Loss of protein makes RBC fragile
OA-1 knock-out	Ocular albinism (loss of pigment glycoprotein)	Ocular albinism type 1 gene	Expressed intracell in melanosome
Lymphotoxin (LT-α)	Enhanced tumor growth and metastais	Lymphotoxin LT-α	Knock-out lacks lymph nodes and Payer patches
Mouse *ApC* gene [targeted mutation at codon 1638 (ApC 1638 T)]	Colorectal tumors	Adenomatous polyposis coli (APC)	An early mutation in GI cancer; see Chapter 10
Nkx2.1 locus Targeted disruption	Tracheoesophageal fistula	NKX 2.1 (transcription factor)	Expressed in thyroid, lung, brain
Col 6 *al* gene (inactivation)	Bethlem myopathy	Type VI collagen	Connective tissue protein expressed in muscle

the model analysis should be compared to relevant observations of the disease in humans, to assess how closely the animal and human conditions resemble one another at the biochemical, anatomical, and physiological levels. Additional studies should proceed only after phenotypic analysis has been appropriately performed.

MOUSE MODELS FOR GENE THERAPY: WHAT MAKES A GOOD MODEL OF HUMAN DISEASE?

The ideal mouse model of a human genetic disease should recapitulate exactly the genotypic and phenotypic characteristics of the human disease. This, in reality, rarely occurs. All animal models of human disease have limitations. The limitations, however, should not preclude the use of the model. It is important to recognize the strengths and weaknesses of any model, as well as to use the model to address testable hypotheses and answer questions. For example, spontaneous, ENU-induced or targeted mutants have gene deficiencies possibly correctable by somatic gene replacement. Animal models of monogenic disorders created by these techniques, such as cystic fibrosis, Duchenne's muscular dystrophy, and hemophilia, can be used in gene therapy experiments. Specifically, studies should investigate the addition of a normally functioning gene(s) into somatic cells relevant to the inherited gene deficiency. Less straightforward is the modeling and evaluation of gene therapy for polygenic diseases (see Chapter 1) such as cancer, diabetes mellitus, and cardiovascular disease. These diseases, by definition, do not have a single genetic cause. However, hypotheses regarding pathogenesis and treatment can be addressed using mice that have been genetically manipulated such that they exhibit altered susceptibility to that disease.

For many monogenic disorders, the inherited mutant allele has decreased or absent function (hypo- or nullimorph) compared to the normal allele. Such diseases hold the greatest promise for use of somatic gene therapy. Here, a normal copy of the gene could be introduced into affected cells to correct the genetic basis of disease. For monogenetic diseases, either targeted mutagenesis or high-efficiency germline mutagenesis are generally the most efficient methods for creating appropriate models. Other genetic diseases result from increased or novel function of a mutant allele (hyper- or neomorph). For these diseases, a transgenic approach may produce a phenotypic model of the disease. However, both normal alleles need to be present.

Transgenic animal models also are useful for introducing highly expressed candidate target genes into a mouse for testing of gene therapy strategies. These mice may not model a particular disease but produce a particular protein that can serve as a gene therapy target. For example, transgenic mice overexpressing transforming growth factor alpha (TGF-α) in mammary epithelium provide a uniform population of experimental animals with a defined genetic lesion that efficiently causes cancer. These animals can be used to evaluate the effectiveness of gene therapy strategies that interfere, directly or indirectly, with increased TGF-α growth signaling and thereby potentially inhibit mammary carcinogenesis.

In many instances, mutation of a locus in the mouse does not produce the same phenotype observed in patients with mutation of the corresponding human locus. The reasons are varied. The precise character of the murine and human mutations may differ. That is, different sites in the gene may have been mutated in the mouse and human. Thus, the level of residual mutant protein function varies. There may be different patterns of expression of the target gene or modifier genes between species. Finally, there may exist biochemical or physiological differences between species that affect the resulting phenotype. Although in principle, one desires models that closely mimic the disease in humans, models that fall short of this ideal are still useful. For example, humans carrying germline mutations in the tumor suppressor gene retinoblastoma (*rb*) develop the ocular tumor retinoblastoma. However, mice deficient for the same gene develop tumors of the intermediate lobe of the pituitary. In spite of this difference, *rb*-null mice can provide general information about mechanisms of *rb*-mediated tumor genesis. Such studies allow an evaluation of gene therapy protocols designed to restore *rb* function to deficient cells regardless of specific tumor.

MODELS OF MONOGENIC DISORDERS

Modeling monogenic disorders is conceptually straightforward. Nonetheless, development of a model to evaluate gene-based treatment may be difficult. Two genetic diseases that are promising candidates for molecular therapeutics are Duchenne's muscular dystrophy and cystic fibrosis. The following sections discuss the genetic bases of these diseases and evaluate the strengths and shortcomings of current models being used to study both disease pathogenesis and treatment.

Duchenne's Muscular Dystrophy

Duchenne's muscular dystrophy (DMD) is an X-linked recessive disorder with a worldwide incidence of 1 in 3500 male births. DMD is characterized clinically by severe, progressive weakness and fibrosis of muscle tissue that eventually leads to respiratory or cardiac failure. DMD is caused by mutations within a 2.3-Mb gene comprised of 79 exons located on the short arm of the X chromosome. Because it is present on the X chromosome, all carrier males (with only one X chromosome) are affected. The transcribed mRNA encodes a large subsarcolemma cytoskeleton protein, dystrophin. Dystrophin is tightly associated with a large oligomeric complex of membrane glycoproteins, the dystrophin–glycoprotein complex (DGC). The DGC spans the sarcolemma of skeletal and cardiac muscle, linking the actin cytoskeleton and the extracellular matrix. Structurally, the dystrophin protein is composed of four polypeptide domains. They are (1) an α-actinin-like actin binding domain at the amino terminus, (2) the rod domain, composed of a series of 24 spectrinlike α-helical repeats, (3) a cysteine-rich region, and (4) a variable C-terminal domain that is subject to alternative transcript splicing. At the molecular level, dystophin deficiency results in loss of the DGC and weakening of the muscle cell membrane. The exact function of the protein is poorly understood, but it is presumed to serve a structural function in force transmission or stabilization of the sarcolemma. The characteristic lesions of DMD patients include muscle cell necrosis and regeneration and elevated serum levels of muscle creatinine kinase (an indicator of muscle damage). As the disease progresses, muscle fibers are replaced by fat and connective tissue.

Mutations resulting in DMD and the clinically milder Becker's muscular dystrophy can cause complete or partial loss of dystrophin or production of a truncated, nonfunctional dystrophin protein. Resulting phenotypes are variable and depend on the precise mutation involved. The *mdx* mouse, a spontaneous mutant, displays many of the biochemical and pathological features of DMD. The *mdx* mice have a stop codon mutation in the mRNA transcript of the dystrophin gene. The biochemical and histopathological defects observed in *mdx* mice are similar to those present in DMD patients. Histologically, these mice display muscle necrosis, fibrosis and phagocytic infiltration within muscle tissue, variation in myofiber size, an increased proportion of myofibers with centrally located nuclei (an indicator of regeneration), and elevated serum levels of muscle creatinine kinase. However, mice do not display severe progressive myopathy. In mice the only muscle to undergo progressive myopathy is the diaphragm. Clinically, these animals do not exhibit visible signs of muscle weakness or impaired movement.

DMD is a condition that appears well suited for treatment by gene therapy. It is a monogenic disorder, and the distinctive properties of skeletal muscle favor delivery of gene targeting vectors. Myofibers are formed as a syncytium of embryonic myoblasts. The nuclei migrate to the periphery of the plasma membrane, and each contributes mRNA transcripts to the entire myofiber. Furthermore, muscles also contain *satellite cells*, which are myoblast precursors. They lie at intervals along the outside surface of the myofibers. If these cells could be genetically manipulated, they could serve as a future and potentially unlimited source of targeting vector expression in regenerating muscle. For this disease, gene therapy has been attempted using virtually every gene transfer technique developed. These include retroviral and adenoviral vector infection, direct gene transfer, receptor-mediated gene transfer, and surgical transfer of genetically manipulated muscle cells.

The general feasibility of gene therapy for DMD was demonstrated using the transgenic mouse approach. Full-length human or murine dystrophin cDNAs were expressed in *mdx* mice under the control of skeletal muscle-specific gene promoters. Expression of as little as 5% of the normal level of dystrophin was able to partially reverse the histopathological lesions. Expression of approximately 20 to 30% of the normal level prevented essentially all dystrophic histopathology and restored diaphragmatic muscle function. Unfortunately, the full-length 14-kb cDNA exceeds the cloning capacity of current viral delivery vectors. Therefore, a truncated gene with an in-frame deletion (exons 17 to 48) in the rod domain (which produces a very mild phenotype in humans with the corresponding mutation) was expressed as a transgene in *mdx* mice. Truncated human and murine dystrophin cDNAs were capable of restoring most of the normal muscle phenotype and function, although a higher level of expression may be required compared to the full-length cDNA. When an adenovirus capable of expressing a recombinant truncated dystrophin was injected into muscles of newborn *mdx* mice, reduction in the histological evidence of muscle degeneration was noted. Also, protection from stretch-induced mechanical damage in these mice as adults were seen. More recently, it was found that truncated utrophin, a structurally similar protein present in skeletal muscle, could substitute for dystrophin as a therapeutic molecule when expressed in transgenic *mdx* mice. This finding is noteworthly because DMD patients have a functional utrophin gene. Thus, it may be possible to reverse or prevent muscle damage by up-regulating utrophin expression.

Cystic Fibrosis

Cystic fibrosis is a common recessive disorder in the Caucasian population that affects about 1 in 2500 live births in populations of northern European ancestry. Clinical manifestations of this devastating disease include chronic pulmonary obstruction, bacterial colonization of the airways, pancreatic enzyme insufficiency, meconium ileus, elevated sweat electrolytes, and reduced fertility in males. The gene causing cystic fibrosis is the cystic fibrosis transmembrane conductance regulator (*cftr*) gene, a transmembrane protein that functions as a cyclic adenosine 5′-monophosphate (cAMP)-regulated chloride channel in the apical membrane of respiratory and intestinal epithelial cells. Elevation of cAMP within normal cells results in opening of the chloride channel and subsequent chloride secretion onto the mucosal surface. Water follows by osmosis. This flushing process is thought to be important in maintaining proper mucociliary clearance in the airways. Mutations in the *cftr* gene result in reduced or absent cAMP-mediated chloride secretion because the protein is either mislocalized or functions with reduced efficiency. Cystic fibrosis (CF) mutations have other primary effects in addition to chloride conductance dysfunction. The CF transmembrane regulator (CFTR) also may be involved in regulation of an outwardly rectified chloride channel, sodium reabsorption, and sulfation. Little is known about these other functions or their relevance to CF. The CF gene is large, spanning approximately 230 kb and consisting of 28 exons. The most common human mutation results in a single amino acid deletion, phenylalanine 508 (ΔF508). The ΔF508 accounts for 70% of the disease alleles in the human population. Hundreds of additional mutant alleles have been identified, each occurring at a much lower frequency. The ΔF508 mutant protein is mislocalized in epithelial cells, presumably because of improper folding. Other mutations prevent proper synthesis of full-length normal protein because of either nonsense, frameshift mutations, aberrant mRNA splicing, or they alter protein function thereby affecting chloride channel regulation, conductance, or gating.

The initial animal models of this disease were created using ES cells to target disruption of either exon 10 or exon 3 of the murine *cftr* gene. Disruption of exon 10 gave rise to a truncated protein similar to that seen with several types of human CF mutations. For all four of the initial CF mouse models, affected animals displayed defective cAMP-mediated chloride transport, consistent with CFTR dysfunction. However, despite producing an apparent phenocopy of the biochemical and electrophysiological defect, the histopathological features of the human disease were only partially reproduced in these models. The most striking phenotypic abnormality in mouse homozygotes in three of the four mutant lineages was a high incidence of death between birth and weaning. The causative lesion resembles meconium ileus (an intestinal obstruction caused by failure to pass a thick, viscous meconium), also found in 10 to 15% of CF patients. Additional human pathology was not seen in the mouse model. Approximately 85% of human CF patients have pancreatic insufficiency from birth. However, histologically, none of these CF mutants reported severe pancreatic pathology. It was also unclear whether the relatively mild pancreatic lesions observed in some mutant mice were primarily effects of the *cftr* mutations or secondary to intestinal disease. Also in humans, lung disease accounts for virtually all of the morbidity and mortality in CF patients. But no histological abnormalities are observed at birth. Over time CF patients develop progressive inflammatory lung disease. None of the mouse mutants developed lung lesions when housed under standard conditions. One *cftr* mutant lineage displayed a low level (<10% of expected) of residual normal CFTR mRNA due to aberrant mRNA splicing and exon skipping, resulting in 30% of normal cAMP-mediated chloride transport and >90% of affected animals surviving to adulthood. Interestingly, mice in this mutant lineage, when repeatedly exposed to the common bacterial pathogens *Staphylococcus aureus* and *Pseudomonas aeruginosa*, displayed a significant incidence of lung disease, consistent with findings in human CF patients.

In view of the major differences outlined above between the human and mouse diseases, how good are the animal models of CF? As is often true for a disease with a complex of clinical features and associated lesions, some characteristics are faithfully reproduced, others less so, some not at all. Furthermore, other new and apparently unique phenotypes may appear in the mouse. All four lineages of mutant mice display the expected alterations in electrophysiology, specifically cAMP-mediated chloride conductance, as predicted based on the proposed role of the CFTR in chloride conductance. Thus, at the level of electrophysiology pathogenesis is reproduced. With respect to reproducing the clinical and histopathological manifestations of the human disease, however, the models are less satisfying. The differences in reproduction of clinical features of the disease may be related to species differences in *cftr* gene expression, protein function, or organ-specific differences in physiology. There may also be differences in the specific mutation introduced. However, the models are providing useful information. An important finding relevant to gene therapy is that a small amount of CFTR can have profound phenotypic consequences. When nullimorph *cftr* mutants are crossed with the lineage displaying slight residual *cftr* expression, the double heterozygotes express <5% of wild-type levels of CFTR mRNA. This low level of mRNA expression partially restores CFTR-mediated chloride transport and, more importantly, prevents most of the perinatal lethality caused by intestinal disease. This relatively low-level expression of *cftr* may be effective in reversing a disease phenotype. In the context of gene therapy this is an encouraging observation. Can these models be improved? Of course! Two possibilities can be explored. The introduction of the AF508 mutation, the most common CF mutation into the mouse gene, would more closely mimic the human disease at the genetic level. Second, potential genetic modifiers of lung disease progression in mutant mice should be identified and explored. This can be done through studying the phenotypic effects of the mutation on different inbred mouse backgrounds.

MODELS OF POLYGENIC AND MULTIFACTORIAL DISORDERS

As shown above, the strategy for creating models of monogenetic disorders is conceptually straightforward. But is this true for diseases that have a polygenic basis? Will gene therapy addressing one component of a multigenic disease be of benefit? Can gene therapy approaches directed at one gene influence other genes related to disease pathology? Can gene therapeutic approaches be devised to target multiple genes involved in pathogenesis? What types of models are useful for studying such diseases? There are many questions and the answers are evolving. Cancer and diabetes mellitus, are two multigenic diseases that also are influenced by epigenetic factors. Below is a discussion of animal models in the study of cancer and diabetes. Chapter 10 presents the use of gene therapy in the treatment of cancer.

Cancer

Inherited and/or spontaneous somatic mutations characterize the pathogenesis of all cancers. However, the specific mutations involved vary greatly from tumor to tumor. Similarly, as noted in Chapter 1, epigenetic influences, such as the patient's hormonal environment, diet, and exposure to environmental carcinogens can vary tremendously. A hallmark of malignant tumors is genomic instability. If a mutation occurs in a gene involved in the maintenance of DNA integrity, the cell may become susceptible to further genetic changes. These events make gene therapy especially problematic for the treatment of cancer because the molecular characteristics of both normal, precancerous cells and cancer cell population are perpetually changing. Nonetheless, gene-based treatments may be of clinical benefit for some types of cancer. Mouse modeling of a variety of genetic alterations associated with cancer has provided insight into cancer molecular pathogenesis. Furthermore, by creating a mouse model with a defined genetic lesion, the effects of gene therapy directed at that specific lesion can be evaluated. Gene therapy for the treatment of cancer has been directed at: (1) replacing mutated tumor suppressor genes, (2) inactivating overexpressed oncogenes, (3) delivering the genetic component of targeted prodrug therapies, and (4) modifying the antitumor immune response (see Chapter 10).

Tumor suppressor genes are a genetically distinct class of genes involved in suppressing abnormal growth. Loss of function of tumor suppressor proteins results in loss of growth suppression. Thus, tumor suppressor genes behave as recessive oncogenes. Study of "cancer families" predisposed to distinct cancer syndromes has led to the identification of mutated tumor suppressor genes transmitted through the germline. Individuals from these families are more susceptible to cancer because they carry only one normal allele of the gene, so that loss of function requires only one mutagenic event instead of the usual two (Knudson's "two-hit hypothesis"). In recent years, mouse models of most of the known tumor suppressor genes have been created by targeted mutagenesis. In most cases tumor suppressor gene-deficient mice (heterozygotes and/or homozygotes) have an increased incidence of spontaneous tumors. In some cases, homozygotes are not viable, presumably because suppressor gene function has critical involvement in normal development. This is not unexpected, given that many tumor suppressor genes appear to be involved in cell cycle regulation (see Chapter 10) or the maintenance of DNA integrity.

Mice carrying mutations in tumor suppressor gene alleles vary with respect to how closely their phenotype resembles the corresponding phenotype in humans. For example, in humans, mutation in the retinoblastoma gene (*rb*) is the underlying genetic defect causing the ocular tumor retinoblastoma. *Rb* mutations also have been found in other human tumor types, including sarcomas and prostate, breast, and lung cancers. Generation of heterozygous *rb*-deficient mice by targeted mutagenesis in ES cells results in tumor development with a 100% penetrance by 18 months of age. However, most tumors that develop are of the pituitary intermediate lobe, apparently the result of species differences in susceptibility of differentiated cell types to *rb* loss. Homozygous *rb*-null mice are not viable. Despite of these differences in phenotype, the fundamental genetic lesion is the same between humans and mice. Furthermore, as in humans, loss of the remaining normal *rb* allele is the crucial mechanistic step in progression to cancer.

Heterozygous *rb*-deficient mice establish a model for therapeutic growth suppression by an exogenous *rb* gene. A recombinant adenovirus carrying a human *rb* cDNA under control of its own promoter was delivered by transauricular injection into spontaneously occurring pituitary intermediate melanotroph tumors. Intratumoral *rb* gene transfer decreased tumor cell proliferation, allowed reestablishment of innervation by growth regulatory dopaminergic neurons. The reestablished innervation further inhibited the growth of tumors and prolonged the life span of treated animals relative to heterozygous null littermates that were untreated or that received vector alone. This study modeled a realistic therapeutic scenario: Adenoviral vectors delivered in vivo targeting of spontaneously arising tumors in humans. Importantly, this study also indicates the therapeutic benefit of replacing *rb* in tumor cells. This is despite the likely presence of additional mutations in treated tumor cells, supporting a role for gene therapy in treatment of polygenic or multifactorial diseases.

A second example concerns human patients with familial adenomatous polyposis, an inherited disease predisposing to colorectal neoplasms and caused by a germline mutation in the tumor suppressor gene adenomatous polyposis coli (APC). Mutations in the APC also play a major role in the early development of spontaneous colorectal neoplasms (see Chapter 10) and possibly other tumors. The *min* (multiple intestinal neoplasia) mouse model for this condition was identified as part of an ENU mutagenesis program. *Min* is a fully penetrant dominant mutation leading to the development of multiple intestinal adenomas throughout the small intestine and colon. Linkage analysis shows the murine homolog of the human *apc* gene is tightly linked to the *min* locus. Furthermore, a nonsense muatation in the murine gene was found to co-segregate with the Min phenotype. Thus, the *min* mouse has proven to be an excellent model for studying the role of APC in intestinal tumorigenesis. Here, the molecular lesion and resulting phenotype are similar to those of inherited and sporadic forms of human colorectal tumorigenesis. As a result, this model is being used extensively to study pathogenesis, chemoprevention, and treatment of intestinal/colorectal neoplasia. In particular, the feasibility of introducing a normal human *apc* cDNA into *min* mice as a means to either prevent or reverse adenoma formation has been demonstrated. In this study, the presence of plasmid DNA and expression of human APC could be detected in *min* mice treated with enemas containing lipofectant and a normal human *apc* cDNA-encoding plasmid. The therapeutic efficacy of such an approach remains to be determined.

A limitation of the *min* mouse model is that most tumors develop in the small intestine rather than the colon as in humans. Colonic tumors do occur in the *min* mouse, but the mortality associated with the large number of small intestinal tumors renders the study of colon tumors difficult. This limitation recently has been addressed in the following manner. By using a conditional gene targeting system, based on the Cre-*loxP* recombination system (see Chapter 5), *apc* inactivation and subsequent adenoma formation can be directed specifically to the colorectal epithelium. Based on the Cre-*loxP* recombination system, a pair of 34 nucleotide viral *loxP* sites were introduced into introns 13 and 14 of the *apc* gene by targeted mutagenesis of ES cells. This phenotypically silent allele can undergo recombination in the presence of Cre recombinase, deleting *apc* exon 14, thereby introducing a frameshift mutation at codon 580. Gene-targeted mice were produced using these cells, and heterozygous and homozygous gene-targeted offspring displayed no observable phenotype. However, when an adenoviral vector expressing the Cre recombinase gene was injected via the anus into the colon of homozygotes, only colorectal tumors were observed within 4 weeks of infection. Thus, by conditionally targeting the colorectal epithelium, the mouse model of APC-induced neoplasia has been refined so that (1) the phenotype is confined to the colorectal region, as predominantly seen in humans, and (2) the genetic lesion is inducible, facilitating the temporal analysis of the role of *apc* mutations in a multistep, polygenic, multifactorial disease.

Gene replacement is the goal of gene therapy for most monogenetic disorders and perhaps in cancer for the treatment of tumor suppressor gene deficiencies. However, as suggested in Chapter 1, for polygenic or multifactorial disorders, gene therapy may be directed at genetic or nongenetic factors that influence the pathogenesis of the disease. For example, several gene therapy trials have attempted to enhance the immune system's response against cancer using transgenic mice and cytokines that mediate antitumor activity in vitro and in vivo. The cytokine inter-leukin-2 (IL-2) can stimulate cell-mediated killing activity by cytotoxic T lympho-cytes, induce lymphokine-activated killer cells, and activate tumor-infiltrating lymphocytes. In a representative study, an adenoviral vector containing the human IL-2 gene was injected intratumorally into mammary tumors from transgenic mice in an attempt to achieve high intratumoral concentrations of IL-2. The treated mice expressed the viral polyoma middle T antigen under control of the mouse mammary tumor virus long terminal repeat, thereby targeting expression to mammary epithe-lium. By 8 to 10 weeks of age, untreated mice develop mammary carcinomas that phenotypically resemble beast cancers of women. Adeno-IL-2 treatment resulted in regression or elimination of 87% of the treated tumors, demonstrating that modifi-cation of the immune response via gene therapy is an experimental approach to treatment of a complex disease such as cancer (see Chapter 10).

A final approach to gene therapy against cancer is tumor-directed delivery of a gene that activates a nontoxic prodrug to a cytotoxic product. This approach should maximize toxicity at the site of vector delivery while minimizing toxicity to other, more distant cells. The potential effectiveness of this approach was demonstrated using transgenic mice that express HSVtk under control of its own gene regulatory elements (see also Chapter 10). These mice were crossed with transgenic mice over-expressing the activated rat *neu* oncogene, the rat homolog of the *her-2/erbB* gene, an oncogene strongly associated with breast cancer in women. Mice carrying both *neu* and HSVtk transgenes developed mammary tumors with the same latency as mice expressing *neu* alone. Tumor bearing double transgenic mice treated in-tratumorally with ganciclovir showed inhibition of tumor growth relative to saline-treated double transgenic mice, or *neu*-only transgenic mice administered ganciclovir. Extending these findings to a clinically relevant model, a retroviral vector bearing the HSVtk gene was injected into spontaneous mammary tumors present in the *neu* transgenc mice. Regression, though not eradication, of the tumors was achieved following subsequent ganciclovir treatment. No HSVtk expression was detected in the residual tumor, suggesting that incomplete antitumor response was related to low transduction efficiency and/or inefficient bystander killing of non-transduced neoplastic cells. Other suicide gene/prodrug systems are under study, including the *Escherichia coli* nitroreductase (NTR) gene/ CB1954 (5-aziridin-1-yl-2-4-dinitrobenzamide) system.

Insulin-Dependent Diabetes Mellitus (IDDM)

The two major forms of diabetes are classified as insulin-dependent diabetes melli-tus (IDDM) and non-insulin-dependent diabetes mellitus (NIDDM). The patho-genesis of these disorders differ. However, each is characterized by an inability to produce and release insulin in an appropriately regulated manner to control glucose homeostasis. IDDM, or type I diabetes, is an autoimmune disorder characterized by immune cell infiltration into the pancreatic islets of Langerhans (insulitis) and destruction of insulin-producing β cells. NIDDM, or type 2 diabetes, is character-ized by β-cell dysfunction and systemic insulin resistance. Both diseases have complex pathophysiology, with significant genetic and environmental components. Mutant mice have been used to study the pathogenesis of IDDM as well as develop gene-based therapies for treatment.

The first mouse model of IDDM to be studied in detail was the spontaneous mutant nonobese diabetic (NOD) mouse. IDDM in humans and in NOD mice is a T-cell-mediated disease under polygenic control. This includes, in humans, an association of IDDM with certain human lymphocyte antigen haplotypes. The NOD mouse carries a diabetes-sensitive allele at the *Idd1* locus located in the mouse major histocompatability complex. In both humans and NOD mice, infiltration of pancreatic islets of Langerhans by T and B lymphocytes, dendritic cells, and macrophages (insulitis) precedes autoimmune destruction of β cells and diabetes. Thus, disease pathogenesis in both humans and NOD mice is similar.

Transgenic mouse models also are being used to study mechanisms of IDDM pathogenesis. The studies relate specifically to the roles of (1) cytokines and inflammation, (2) T-lymphocyte subsets, (3) autoantibodies, and (4) antigen presentation. Proinflammatory cytokines are thought to be critically involved in autoimmunity. Because cytokine networks are complicated and tightly regulated, it is difficult to identify the roles of individual cytokines in the pathogenesis of spontaneous disease. Transgene-driven overexpression of individual cytokines can address this issue. For example, IL-2, IL-4, IL-6, IL-10, TNF-α, TNF-β, and IFN-γ each have been targeted to islets, and the pathophysiological responses have been examined. Overexpression of IL-2 in islet cells results in insulitis, β-cell destruction, and diabetes. Overexpression of IL-2 in NOD mice accelerates the development of diabetes. Thus, local overproduction of IL-2 amplifies existing autoimmune mechanisms. This suggests an anti-IL-2 molecular therapeutics may be effective in controlling the early stages of IDDM. A related experimental approach was based on the observation that different T-cell subsets (Th1 and Th2) secrete different cytokines. Transgene-driven overexpression in islet cells of IL-4, a Th2 cytokine, protects NOD mice from spontaneous diabetes. In contrast, overexpression in islets of IFN-γ, a Th1 cytokine, leads to inflammatory cell infiltration and diabetes. Finally, to test the roles of tolerance and autoantibody production, transgenic mice were created that expressed certain antigenic molecules in islet cells at different stages of development. When expression of antigen occurred during early embryonic development, tolerance was established. However, as expected, when expression occurred during adult life, tolerance was not induced and diabetes resulted. Since autoantigenicity appears to be a component of IDDM, identifying the relevant autoantigen becomes important for understanding disease pathogenesis. β-cell-specific proteins, such as insulin and its precursor proinsulin, are likely candidates. NOD mice were created expressing the mouse proinsulin gene under control of the major histocompatability class II promoter. This promoter is transcribed during development and therefore should cause deletion of proinsulin-reactive T cells. Insulinitis and diabetes were prevented in these mice, suggesting that autoimmunity to proinsulin is an important aspect of IDDM pathogenesis in the NOD mouse. Thus, the transgenic approaches provide the ability to manipulate, one at a time, specific components or steps in the pathogenesis of a complex, multifactorial disease like IDDM.

Gene-based therapies hold great promise for treatment of IDDM. Various approaches are being investigated. These include in vitro genetic manipulation and transplantation of β cell as well as neuroendocrine cell lines, introduction of genes into (non-β) cell types for transformation into β-like cells, and in vivo delivery of gene targeting vectors. Transplantation of genetically engineered cells or modification of endogenous cells can provide cells appropriately producing and secreting insulin in response to physiological concentration of glucose. Tumor-derived β cells or neuroendocrine cell lines generally do not display appropriately regulated glucose-stimulated insulin production. To correct this defect, molecules that regulate insulin secretion such as the GLUT2 glucose transporter can be transfected into cell lines. Engineering of correct secretory responses makes these cells an attractive source of transplantable cells.

The modification of the hepatocyte genome for treatment of diabetes is being explored using transgenic models. Transgenic mice have been created expressing the human proinsulin gene under control of the phosphoenolpyruvate carboxykinase (PEPCK) gene regulatory elements. PEPCK is a gluconeogenic enzyme and its gene is transcriptionally activated in liver by fasting and cAMP. When transgenic and nontransgenic mice were treated with β-cell toxin streptozotocin to induce diabetes, blood glucose levels were significantly lower (i.e., better regulated) in transgenic mice. Unfortunately, there are difficulties associated with transforming hepatocytes into insulin-producing cells. Hepatocytes lack enzymes necessary for processing proinsulin into insulin. To address this problem, mutated proinsulin genes have been constructed with novel cleavage sites that can be processed by hepatocytes. By following expression of this modified gene in transgenic mouse hepatocytes, human C-peptide (the expected proinsulin cleavage fragment) was detected in serum. An additional issue related to gene therapy in the liver and diabetes is that most hepatocyte-specific enhancer/promoters, including PEPCK, do not display optimal transcriptional activation under hyperglycemic conditions or inhibition under hypoglycemic conditions. Here, the engineering of genes with multiple regulatory elements combined from different genes has been proposed. A related approach involves ex vivo gene therapy. The goal is to introduce new genes into autologous cells in culture and return the modified cells to the patient. Gene therapy ex vivo with autologous hepatocytes is well suited for study in mouse systems. The techniques for stimulating hepatocyte proliferation and repopulation by donor cells (autologous or allogeneic) are well established, and the approach, in principle, is reasonable from a clinical standpoint.

MOUSE MODELS OF MOLECULAR THERAPEUTICS: DEVELOPING AND TESTING GENE THERAPY METHODOLOGY

In addition to generating models of human disease, genomic modification technology can be used in other ways that support research into molecular medicine methodology. For these approaches the goal is not to recreate a human disease but rather to create genetic alterations that permit (1) the identification of potentially important targets for gene therapy, (2) the optimization of gene targeting expression vectors, (3) the optimization of gene therapy protocols, and (4) recreation of the in vivo context for human tissues using immunodeficient mice.

Identification of Gene Therapy Targets

Appropriate molecular targets for gene therapy should have significant causal role(s) in disease pathogenesis as well as be amenable to manipulation. For monogenic disorders, gene replacement may be, in principle, curative. However, for polygenic disorders such as cancer, appropriate targets may not be obvious and the transgenic approach holds great promise to solve this difficulty. For example, transforming growth factor alpha (TGF-α) is overproduced by cells of several human malignancies, including those of breast, liver, and pancreas. TGF-α is a ligand for epidermal growth factor receptor (EGFR). By itself, overexpression of TGF-α does not prove involvement in causation, nor does it identify the strength of any causative role a molecule may possess. In transgenic mice, where expression of TGF-α can be targeted to either mammary, liver, or pancreatic epithelial cells, the consequences were found to differ. TGF-α was potently oncogenic in the mammary gland, moderately oncogenic in liver, and only weakly oncogenic in pancreas. Thus, overexpression of TGF-α produced variable pathogenicity among tissues. However, when bitransgenic mice were generated targeting both the oncogene c-*myc* and TGF-α to each tissue, there was strong synergy between transgenes and a dramatic acceleration in onset of c-*myc*-induced neoplasia in all tissues including the pancreas. Although certain effects of TGF-α overexpression may be tissue specific, synergistic interaction with epithelia TGF-α strongly enhanced tumor cell growth. This finding, together with evidence for overproduction of TGF-α in human cancer, identifies TGF-α and signaling through the EGFR, as important potential targets for molecular therapeutics. Furthermore, these same transgenic lineages are models to develop and test efficacy of anti-EGFR therapy. The posttherapeutic slowing of tumor growth and increase in life span of treated c-*myc*/TGF-α bitransgenic mice indicate a potential candidate therapy for use in the treatment of human cancers.

Modeling Therapeutic DNA Constructs

Expression of DNA constructs in trangenic mice can be used to evaluate therapeutic potential. This technique may be especially useful in the modeling of gene therapy for monogenic disorders. Mice can express a transgene encoding a potential therapeutic molecule and mated to a mutant mouse strain displaying the relevant disease. Correction of the disease phenotype in transgene-bearing mutant mice provides strong evidence that the construct has therapeutic potential. Examples of this approach include the use of full-length and truncated dystrophin minigenes in *mdx* mice to treat DMD and the expression of human *cftr* in *cftr*-deficient mice. A second application of transgenic mice in modeling constructs involves promoter analysis. Although viral and mammalian gene regulatory elements with a broad tissue specificity have been used extensively in gene targeting approaches, additional enhancer/promoters are needed. Desperately needed are regulatory elements that provide a pattern of tissue-restricted gene expression that is continuous and at a high level (see Chapter 5). Tissue specificity may be advantageous from a safety perspective through restricting expression of potentially toxic therapeutic gene to the target cell populations. For example, the epidermis is an attractive target for gene therapy. The epidermis can be targeted for treatment of skin diseases as well as an easily accessible and manipulative site for the production and secretion of therapeutic gene products exerting systemic effects. Cytokeratins are a family of epithelial-specific intermediate filament proteins expressed differentially within the epidermis as keratinocytes differentiate. Cytokeratin promoters are available and target transgene expression to specific cell layers of the epidermis. The feasibility of using cytokeratin gene regulatory elements to target expression of therapeutic genes to the skin was illustrated by the creation of transgenic mice expressing human growth hormone (hGH) under the regulatory control of the cytokeratin 14 promoter. In those mice, production of recombinant hGH was confined to specific layers of the epidermis, yet the protein could be detected at a physiologically significant concentration in the serum. In addition, the mice grew larger than nontransgenic littermates. Experiments of this type can be useful as an aid to designing and testing efficacy of therapeutic gene targeting strategies.

GENERATION OF CHIMERIC TISSUES

Transgenic animals display the phenotypic consequences of transgene expression when 100% of the target cells carry the transgene. Unfortunately, current gene delivery systems fall short of this rate of transduction. Relative to transgenic approaches, clinically relevant questions may be: What are the consequences of gene transfer and expression in 1, 5, or 10% of the target cell population? Will these levels of transduction restore function to a genetically deficient tissue or organ? Can expression of the therapeutic gene in one cell benefit a neighboring nontransduced cell, that is, are there juxtacrine, paracrine, or endocrine effects of foreign gene expression or are transgene effects strictly cell autonomous? These questions can be addressed by creating chimeric tissues, which are composed of two genetically distinct cellular populations in variable proportion to one another. Chimeric tissues can be created by injection of ES cells into blastocysts, as described above (see Fig. 3.5), or by embryo aggregation. Embryo aggregation is performed by physical aggregation of two distinct preimplantation embryos at the 4- to 8-cell stage, followed by transfer of the chimeric embryo to the oviduct of a pseudopregnant recipient mouse. In either case, the two populations of cells can associate with one another and develop into a chimeric mouse, which possess in each tissue a variable proportion of the two donor genotypes. By manipulating (or selecting for) the level of chimerism in each animal, it is possible to identify the phenotypic effect of a minority population of cells of one genotype upon the majority of cells of a second genotype. For example, the therapeutic consequences to the *cftr*-null mouse chimeric with 5% of cells with normal *cftr* genes could be addressed using this approach. Analysis is facilitated by marking one or both genotypes with reporter genes so that each genotype can be precisely localized in microscopic tissue sections. A related approach involves reconstitution of a tissue by cell transplantation using a mixed population of donor cells of two genotypes. Both mammary gland and liver can be reconstituted as chimeric organs using transplantation of mammary epithelial cells into the caudal mammary fat pads or of hepatocytes into the portal vein. Chimera analysis is being used more frequently to ask fundamental biological questions regarding cellular interactions. It also can be a powerful technique for evaluating the clinical effects of incomplete transduction of a target cell population in a patient.

HUMAN CELL XENOGRAFT MODELS IN IMMUNODEFICIENT MICE

The best mouse models of human disease have an inherent limitation. The tissues studied are of murine, not human, origin, and these do not always reproduce a model of human disease. This is true even though there are substantial similarities in biochemical and physiological functions in mice and humans. A unique model to study human pathology in animals as well as murine/human biochemistry and physiology is the chimeric animal. Chimeric animals possess either cells, tissues, or organs derived from human stem cells, but limitations in these animals result from interactions with systemic autologous growth factors and other biological molecules on cells. Chimeric animals can be generated through xenotransplantation, the transfer of tissue from one species into another species. Xenotransplantation broadens the range of experimental manipulations and tissue samplings that can be performed relative to using human subjects. The principal factor limiting xenotransplantation is immune rejection, the destruction of donor tissue by the host immune system. Xenotransplant recipients have been rendered immunodeficient by irradiation, drug therapy, or surgical thymectomy in an attempt to inhibit the rejection process. Alternatively, genetically immunodeficient hosts have been used. The more commonly used immunodeficient mouse strains include the *nude*, *scid*, and *beige* genotypes. *Nude* mice are athymic animals and thus T-lymphocyte-deficient. *Scid* (severe combined immunodeficiency) mice are B- and T-lymphocyte-deficient. *Beige* mice have reduced natural killer cell activity. Mice displaying combined immunodeficiencies (e.g., *scid-beige*) also have been generated. More recently, targeted mutations in genes involved in B- and T-cell development have produced new models of immunodeficiency that resemble *scid* mice. Because *scid* mice display a major immune defect, they provide a unique biological setting that can be used to address major questions in the fields of gene therapy and xenotransplantation.

Scid mice are deficient in both mature T and B lymphocyte. This phenotype is the result of expression of a recessive gene mutation maping to mouse chromosome 16. The *scid* mutation results in defective rearrangement of immunoglobulin and T-cell receptor genes during differentiation of the respective cell lineages, thereby blocking the differentiation of B- and T-lymphocytic lineage committed progenitors. Older *scid* mice express leakiness and produce a small amount of murine immunoglobulin. *Scid* mice retain functional macrophages and natural killer cells. The immune phenotype also can be influenced dramatically by genetic background, age, and microbial flora, complicating comparisons of experimental outcomes among different laboratories. A fade-out use of immunodeficient mice has been as a repository for human tissue, particularly human tumors. Both *nude* and *scid* mice can support transplantation and growth of a variety of human tumors. However, *nude* mice will not support the growth of all tumors grown in *scid* mice, possibly due to the presence of competent B cells in *nude* mice. The adopted transfer of human cells is followed by a period of growth and expansion with experimental manipulation in a manner not possible with human patients. Specific gene therapy protocols, employing varying target genes and delivery vehicles, can be systematically evaluated for efficacy directly on human tissue in an in vivo setting. More sophisticated manipulations using immunodeficient mice also have been performed. The engraftment of a functional human immune system into *scid* mice has provided a powerful tool for studying the role of the human immune system in cancer, autoimmunity, and infectious disease. Several protocols involving engrafting thymus, liver, bone marrow, cord blood, and/or peripheral blood lymphocytes have produced xenotransplant models where engrafted human hematopoietic cells reconstitute a human immune system in the mouse. These models are particularly useful for developing gene therapy strategies targeted at correction of human disorders of the hematopoietic system. The successful ex vivo transduction of hematopoietic (see Chapter 6) progenitor cells and subsequent engraftment into *scid* mice has resulted in novel animal models for use in gene therapy research.

MOUSE MODELS: THE NEXT GENERATION

In the future, emerging and new technologies will permit increasingly sophisticated manipulation of gene expression in the living animal. Currently, for certain applications, the usefulness of transgenic and gene-targeted mice has been limited based on the occassionally deleterious effects of engineered changes on gene expression and subsequent mouse development. Some mice with targeted mutations die in utero, suggesting that the affected gene plays a critical role in fetal development. Similarly, overexpression of certain transgenes can cause embryonic death. This obviously is problematic in attempting to model a disease that occurs postnatally in humans. A solution is to generate models in which transgene expression or gene deletion can be targeted to specific tissues in adult animals. Tissue-specific transgene expression can be achieved by use of tissue-specific gene regulatory elements.

Developmental expression of stage-specific gene expression can be produced in animals. However, temporal pattern of transgene expression may be dictated by the multiregulatory elements. At present, this is a concern not easily manipulated. In some cases, transgene expression can be induced by virtue of regions within the gene regulatory elements that bind to molecules and enhance transcription. For example, the metallothionein (MT) promoter can be up-regulated by administration of heavy metals (Zn^{2+} or Cd^{2+}), although the basal level of expression remains high. Recently, several additional inducible systems have been examined where there is minimal trangene expression in the uninduced state and high-level trangene expression following induction. The best established of these new systems employs tetracycline (Tc) as the inducing agent. The administration of Tc (or withdrawl of Tc, depending on the specific DNA elements selected) results in transgene expression in a tissue where the Tc binding protein has been targeted. Thus, a transgene whose expression would otherwise result in embryonic death would remain "silent" in utero until tetracycline was administered via injection or drinking water. The transgene becomes silent again when tetracycline is removed. Similar systems employing the *lac*-operon inducer Isopropyl-beta-D-thiogalacto pyranoside (IPTG) or the insect hormone ecdysone are also being developed.

In an additional approach, the viral cre/*lox* system recently has been employed to knock out specific genes in selected cell types of the adult animal (see also Chapter 5). In brief, this technology is based on the ability of the bacteriophage P1 virion cre recombinase to bind 34 nucleotide DNA sequences called *loxP* sites. When cre encounters two *loxP* sites, the enzyme splices out the intervening DNA, leaving one *loxP* site. Using this maipulation, gene deletion can be limited to a particular cell type in the mouse, rather than affecting all cells throughout development. A further refinement of this technique would involve placing *cre* gene expression under control of an inducible gene regulatory element. In this manner, the targeted gene would function normally in all tissues during development. But, cre expression and targeted gene deletion could be induced in specific adult tissues at a precisely selected time.

A final approach to model development that will certainly gain future prominence is large-scale modification of the mouse genome. This will involve changing the pattern of expression of multiple genes in a single animal. Currently, breeding between different transgenic and/or gene-targeted lineages has been used to produce animals with two or three gene changes. This approach, although in principle is unlimited, is inefficient and time consuming. Instead, it is now possible to introduce large changes into the genome in one step. Large pieces of DNA, carried on yeast artificial chromosomes (YACs) potentially carrying multiple independent trangene units, can be introduced into mouse eggs. Similarly, gene targeting approaches can be used to delete or replace chromosome-sized pieces of DNA. At some point, it will be possible to introduce complete chromosomes into mouse cells. An advantage of large-scale genetic engineering is that multigenic disorders can be more effectively modeled in animals.

Finally, there are many other animal species that have been used to create models of human diseases. Each has its own set of anatomical, biochemical, or physiological characteristics that make them well suited to examine specific human conditions. In view of the recent advances in animal cloning using somatic cells (see Chapter 2), it is certain that genetic manipulation of these species will become easier and each species will find an increasingly important place in studies involving molecular medicine.

KEY CONCEPTS

- The existence of inbred strains of mice with a unique but uniform genetic background, the increasingly dense map of the murine genome, and well-defined experimental methods for manipulating the mouse genome make the development of new models of human disease relatively straightforward in the mouse.
- In mice, genetic mutations may occur spontaneously or they can be induced by experimental manipulation of the mouse genome via high-efficiency germline mutagenesis, via transgenesis, or via targeted gene replacement in ES cells.
- If mouse models of human disease are to assist in the establishment or testing of somatic gene therapies, then the mutated gene must be identified. This usually requires genetic mapping and positional cloning of the mutated gene. The ideal model for the study of somatic gene therapy should exhibit the same genetic deficiency as the disease being modeled. In general, the greater the similarity between the mouse mutation and the mutation as it occurs in humans, the greater the likelihood that the mouse will produce a reliable model of the human disease.
- A strength of ENU mutagenesis, producing DNA lesions that are typically single nucleotide changes, is that models can be generated for diseases caused by mutations at unidentified loci. These new mutations then can be mapped in the mouse genome, and perhaps the human gene location inferred through synteny homologies.
- Transgenic animals carry a precisely designed genetic locus of known sequence in the genome. Foreign DNA, or transgenes, can be introduced into the mam-

malian genome by several different methods, including retroviral infection or microinjection of ES cells and microinjection of fertilized mouse eggs. Most transgenes contain three basic components: the gene regulatory elements (enhancer/promoter), mRNA encoding sequence, and polyadenylation signal. Transgenes generally permit assessment of the phenotypic consequences of dominant acting genes because the mouse retains normal copies of all endogenous genes.

· Gene targeting in ES cells involves inserting a mutant copy of a desired gene into a targeting vector, then introducing this vector into the ES cell. With a low frequency, the vector will undergo homologous recombination with the endogenous gene. Using this approach, we can identify the phenotypic consequences of deleting or modifying endogenous mouse DNA versus adding new DNA as in the transgenic approach.

· The phenotype of a genetically altered mouse will be determined not only by the specific molecular consequences of the mutation (e.g., loss of gene expression, increased gene expression, production of a mutant protein, etc.), but also by how that mutation influences (and is influenced by) cellular biochemistry, tissue- and organ-specific physiology, and all the organism-wide homeostatic mechanisms that regulate the adaptation of an individual to its surroundings.

· *Mdx* mice have a stop codon mutation in the mRNA transcript of the dystrophin gene. The biochemical and histopathological defects observed in *mdx* mice are similar to those present in DMD patients. For this disease, gene therapy has been attempted using virtually every gene transfer technique developed, including retroviral and adenoviral vector infection, direct gene transfer, receptor-mediated gene transfer, and surgical transfer of genetically manipulated muscle cells.

· The affected gene causing cystic fibrosis is the cystic fibrosis transmembrane conductance regulator (*cftr*) gene, a transmembrane protein that functions as a cAMP-regulated chloride channel in the apical membrane of respiratory and intestinal epithelial cells. Mutations in the *cftr* gene result in reduced or absent cAMP-mediated chloride secretion because the protein is either mislocalized or functions with reduced efficiency. In all four of the initial CF mouse models, affected animals displayed defective cAMP-mediated chloride transport, consistent with CFTR dysfunction. However, despite producing an apparent phenocopy of the biochemical and electrophysiological defect, the histopathological features of the human disease were only partially reproduced in these models.

· Diabetes mellitus is characterized by an inability to produce and release insulin in an appropriately regulated manner to control glucose homeostasis. IDDM, or type I diabetes, is an autoimmune disorder characterized by immune cell infiltration into the pancreatic islets of Langerhans (insulitis) and destruction of insulin-producing β cells. The first mouse model of IDDM to be studied in detail was the NOD mouse. NOD mice exhibit a T-cell-mediated disease under polygenic control and carries a diabetes-sensitive allele at the *Idd1* locus located in the mouse major histocompatability complex. As in humans, infiltration of pancreatic islets of Langerhans (insulitis) by T and B lymphocytes, dendritic cells, and macrophages precedes autoimmune destruction of β cells and diabetes in NOD mice. Thus, disease pathogenesis in both humans and NOD mice is very similar. These mice have been used to identify the effects of immunological modulation upon disease progression.

· In addition to the creation of models of human disease, genomic modification technology can be used in other ways that support research into molecular medicine methodology. For these approaches, the goal is not to recreate a human disease but rather to create genetic alterations that permit (1) identification of potentially important targets for gene therapy, (2) optimization of gene targeting expression vectors, (3) optimization of gene therapy protocols, and (4) recreation of the in vivo context for human tissues using immunodeficient mice as recipients of human cell transplants.

SUGGESTED READINGS

Gene Therapy in Animal Models

Addison CL, Braciak T, Ralston R, Muller WJ, Gauldie J, Graham FL. Intratumoral injection of an adenovirus expressing interleukin 2 induces regression and immunity in a murine breast cancer model. Proc Natl Acad Sci USA 92:8522–8526, 1995.

Akkina RK, Rosenblatt JD, Campbell AG, Chen ISY, Zack JA. Modeling human lymphoid precursor cell gene therapy in the SCID-hu mouse. Blood 84:1393–1398, 1994.

Deconinck N, Ragot T, Marechal G, Perricaudet M, Gillis JM. Functional protection of dystrophic mouse (mdx) muscles after adenovirus-mediated transfer of a dystrophin minigene. Proc Natl Acad Sci USA 93:3570–3574, 1996.

Docherty K. Gene therapy for diabetes mellitus. Clin Sci 92:321–330, 1997.

Mitanchez D, Doiron B, Chen R, Kahn A. Glucose-stimulated genes and prospects of gene therapy for type I diabetes. Endocr Rev 18:520–540, 1997.

Pagel CN, Morgan JE. Myoblast transfer and gene therapy in muscular dystrophies. Micro Res Tech 30:469–479, 1995.

Riley DJ, Nikitin AY, Lee W-H. Adenovirus-mediated retinoblastoma gene therapy suppresses spontaneous pituitary melanotroph tumors in *Rb+/–* mice. Nat Med 2:1316–1321, 1996.

Mutagenesis

Nagy A, Rossant J. Targeted mutagenesis: Analysis of phenotype without germ line transmission. J Clin Invest 97:1360–1365, 1996.

Russell WL, Kelly EM, Hunsicker PR, Bangham JW, Maddux SC, Phipps EL. Specific-locus test shows ethylnitrosourea to be the most potent mutagen in the mouse. Proc Natl Acad Sci USA 76:5818–5819, 1979.

Transgenic Mice

Grewal I, Flavell RA. New insights into insulin dependent diabetes mellitus from studies with transgenic mouse models. Lab Invest 76:3–10, 1997.

Jacenko O. Strategies in generating transgenic mamals. Meth Molec Biol 62:399–424, 1997.

Phelps SF, Hauser MA, Cole NM, Rafael JA, Hinkle RT, Faulkner JA, Chamberlin JS. Expression of full-length and truncated dystrophin mini-genes in transgenic mdx mice. Hum Mol Genet 4:1251–1258, 1995.

Sacco MG, Benedetti S, Duflot-Dancer A, Mesnil M, Bagnasco L, Strina D, Fasolo V, Villa A, Macchi P, Faranda S, Vezzoni P, Finocchiaro G. Partial regression, yet incomplete eradication of mammary tumors in transgenic mice by retrovirally mediated HSVtk transfer "in vivo". Gene Therapy 3:1151–1156, 1996.

Sacco MG, Mangiarini L, Villa A, Macchi P, Barbieri O, Sacchi MC, Monteggia, Fasolo V, Vezzoni P, Clerici L. Local regression of breast tumors following intramammary ganciclovir administration in double transgenic mice expressing neu oncogene and herpes simplex virus thymidine kinase. Gene Therapy 2:493–497, 1995.

Disease Pathogenesis

Bilger A, Shoemaker AR, Gould KA, Dove WF. Manipulation of the mouse germline in the study of Min-induced neoplasia. Sem Cancer Biol 7:249–260, 1996.

Dorin JR. Development of mouse models for cystic fibrosis. J Inher Metab Dis 18:495–500, 1995.

Grubb Br, Boucher RC. Pathophysiology of gene-targeted mouse models for cystic fibrosis. Physiol Rev 79(1 Suppl):S193–S214, 1999.

Macleod KF, Jacks T. Insights into cancer from transgenic mouse models. J Pathol 187:43–60, 1999.

Sandhu JS, Boynton E, Gorczynski R, Hozumi N. The use of SCID mice in biotechnology and as a model for human disease. Crit Rev Biotech 16:95–118, 1996.

Shibata H, Toyama K, Shioya H, Ito M, Hirota M, Hasegawa S, Matsumoto H, Takano H, Akiyama T, Toyoshima K, Kanamura R, Kanegae Y, Saito I, Nakamura Y, Shiba K, Noda T. Rapid colorectal adenoma formation initiated by conditional targeting of the *Apc* gene. Science 278:120–123, 1997.

Tinsley JM, Potter AC, Phelps SR, Fisher R, Trickett JI, Davies KE. Amelioration of the dystrophic phenotype of *mdx* mice using a truncated utrophin transgene. Nature 384:349–353, 1996.

Wells DJ, Wells KE, Asnate EA, Turner G, Sunada Y, Campbell KP, Walsh FS, Dickson G. Expression of full-length and minidystrophin in transgenic *mdx* mice: Implications for gene therapy of Duchenne muscular dystrophy. Hum Mol Genet 4:1245–1250, 1995.

Williams SS, Alosco TR, Croy BA, Bankert RB. The study of human neoplastic disease in severe combined immunodeficient mice. Lab Anim Invest 43:139–146, 1993.

Zielenski J, Tsui L-C. Cystic fibrosis: Genotypic and phenotypic variations. Annu Rev Genet 29:777–807, 1995.

CHAPTER 4

Vectors of Gene Therapy

KATHERINE PARKER PONDER, M.D.

INTRODUCTION

Currently, gene therapy refers to the transfer of a gene that encodes a functional protein into a cell or the transfer of an entity that will alter the expression of an endogenous gene in a cell. The efficient transfer of the genetic material into a cell is necessary to achieve the desired therapeutic effect. For gene transfer, either a messenger ribonucleic acid (mRNA) or genetic material that codes for mRNA needs to be transferred into the appropriate cell and expressed at sufficient levels. In most cases, a relatively large piece of genetic material (>1 kb) is required that includes the promoter sequences that activate expression of the gene, the coding sequences that direct production of a protein, and signaling sequences that direct RNA processing such as polyadenylation. A second class of gene therapy involves altering the expression of an endogenous gene in a cell. This can be achieved by transferring a relatively short piece of genetic material (20 to 50 bp) that is complementary to the mRNA. This transfer would affect gene expression by any of a variety of mechanisms through blocking translational initiation, mRNA processing, or leading to destruction of the mRNA. Alternatively, a gene that encodes antisense RNA that is complementary to a cellular RNA can function in a similar fashion.

Facilitating the transfer of genetic information into a cell are vehicles called vectors. Vectors can be divided into viral and nonviral delivery systems. The most commonly used viral vectors are derived from retrovirus, adenovirus, and adeno-associated virus (AAV). Other viral vectors that have been less extensively used are derived from herpes simplex virus 1 (HSV-1), vaccinia virus, or baculovirus. Nonviral vectors can be either plasmid deoxyribonucleic acid (DNA), which is a circle of double-stranded DNA that replicates in bacteria or chemically synthesized compounds that are or resemble oligodeoxynucleotides. Major considerations in determining the optimal vector and delivery system are (1) the target cells and its characteristics, that is, the ability to be virally transduced ex vivo and reinfused to the patient, (2) the longevity of expression required, and (3) the size of the genetic material to be transferred.

An Introduction to Molecular Medicine and Gene Therapy, Edited by Thomas F. Kresina
ISBN 0-471-39188-3 © 2001 Wiley-Liss

VIRAL VECTORS USED FOR GENE THERAPY

Based on the virus life cycle, infectious virions are very efficient at transferring genetic information. Most gene therapy experiments have used viral vectors comprising elements of a virus that result in a replication-incompetent virus. In initial studies, immediate or immediate early genes were deleted. These vectors could potentially undergo recombination to produce a wild-type virus capable of multiple rounds of replication. These viral vectors replaced one or more viral genes with a promoter and coding sequence of interest. Competent replicating viral vectors were produced using packaging cells that provided deleted viral genes in trans. For these viruses, protein(s) normally present on the surface of the wild-type virus were also present in the viral vector particle. Thus, the species and the cell types infected by these viral vectors remained the same as the wild-type virus from which they were derived. In specific cases, the tropism of the virus was modified by the surface expression of a protein from another virus, thus allowing it to bind and infect other cell types. The use of a protein from another virus to alter the tropism for a viral vector is referred to as pseudotyping.

A number of viruses have been used to generate viral vectors for use in gene therapy. The characteristics of these viruses and their virulence are shown in Table 4.1. Characteristics of viral vectors that have been generated from these viruses are shown in Table 4.2. Important features that distinguish the different viral vectors include the size of the gene insert accepted, the duration of expression, target cell infectivity, and integration of the vector into the genome.

RETROVIRAL VECTORS

Retroviruses are comprised of two copies of a positive single-stranded RNA genome of 7 to 10 kb. Their RNA genome is copied into double-stranded DNA, which integrates into the host cell chromosome and is stably maintained. A property that allowed for the initial isolation was the rapid induction of tumors in susceptible animals by the transfer of cellular oncogenes into cells. However, retroviruses can also cause delayed malignancy due to insertional activation of a downstream oncogene or inactivation of a tumor suppressor gene. Specific retroviruses, such as the human immunodeficiency virus (HIV), can cause the immune deficiency associated with the acquired immunodeficiency syndrome (AIDS) see Chapter 12. Retroviruses are classified into seven distinct genera based on features such as envelope nucleotide structure, nucleocapsid morphology, virion assembly mode, and nucleotide sequence.

Retroviruses are ~100 nm in diameter and contain a membrane envelope. The envelope contains a virus-encoded glycoprotein that specifies the host range or types of cells that can be infected by binding to a cellular receptor. The envelope protein promotes fusion with a cellular membrane on either the cell surface or in an endosomal compartment. The ecotropic Moloney murine leukemia virus (MLV) receptor is a basic amino acid transporter that is present on murine cells but not cells from other species. The amphotropic MLV receptor is a phosphate transporter that is present on most cell types from a variety of species including human cells. There are co-HIV receptors, CD4, and a chemokine receptor. After binding to the

TABLE 4.1 Characteristics of Viruses That Have Been Used to Generate Viral Vectors

Virus	Size and Type of genome	Viral Proteins	Physical Properties	Disease in Animals
Retrovirus	7–10 kb of single-stranded RNA	Gag, Pro, Pol, Env	100 nm diameter; enveloped	Rapid or slow induction of tumors; acquired immunodeficiency syndrome (AIDS)
Adenovirus	36-kb double-stranded linear DNA	Over 25 proteins	70–100 nm in diameter; nonenveloped	Cold; conjunctivitis; gastroenteritis
Adenovirus-associated virus	4.7-kb single-stranded linear DNA	Rep and Cap	18–26 nm in diameter; nonenveloped	No known disease
Herpes simplex virus 1 (HSV-1)	152 kb of double-stranded linear DNA	Over 81 proteins	110 nm in diameter	Mouth ulcers and genital warts; encephalitis
Vaccinia virus	190 kb of double-stranded linear DNA	Over 198 open reading frames	350 by 270 nm rectangles; enveloped	Attenuated virus that was used to vaccinate against smallpox
Baculovirus	130 kb of double-stranded circular DNA	Over 60 proteins	270 by 45 nm rectangles; enveloped	None in mammals; insect pathogen

cellular receptor, the viral RNA enters the cytoplasm and is copied into double-stranded DNA via reverse transcriptase (RT) contained within the virion. The double-stranded DNA is transferred to the nucleus, where it integrates into the host cell genome by a mechanism involving the virus-encoded enzyme integrase. This activity is specific for each retrovirus. For MLV, infection is only productive in dividing cells, as transfer of the DNA to the nucleus only occurs during breakdown of the nuclear membrane during mitosis. For HIV, infection can occur in nondividing cells, as the matrix protein and the *vpr*-encoded protein have nuclear localization signals that allow transfer of the DNA into the nucleus to occur.

Moloney Murine Leukemia Virus: MLV Proteins

Retroviral proteins are important in the manipulation of the system to develop a vector. MLV is a relatively simple virus with four viral genes: *gag*, *pro*, *pol*, and *env* (Fig. 4.1). The *gag* gene encodes the group specific antigens that make up the viral core. The Gag precursor is cleaved into four polypeptides (10, 12, 15, and 30 kD) by the retroviral protease (PR). The 15-kD matrix protein associates closely with the membrane and is essential for budding of the viral particle from the membrane. The 12-kD phosphoprotein (pp12) is of unresolved function. The 30-kD capsid protein

TABLE 4.2 Summary of Relative Advantages and Disadvantages of Vectors Used for Gene Therapy

Vector	Infects Nondividing Cells?	Maximum Size of Insert	Stability of Expression	Titer
Retroviral vectors	No (yes for lentiviral vectors)	≤8 kb	Stable (random DNA insertion)	1×10^6 cfu/ml unconcentrated; 1×10^8 cfu/ml concentrated
Adenovirus	Yes	8 kb for E1/E3 deleted vectors; 35 kb for "gutless" vectors	Expression lost in 3–4 weeks in normal animals; expression can last weeks to months with immunosuppression. No integration	1×10^{12} pfu/ml
Adenoassociated virus (AAV)	Yes	<4.5 kb	Stable; it is unclear if DNA integrates in vivo	1×10^6 infectious particles/ml unconcentrated; 1×10^{10} infectious particles/ml concentrated
Herpes simplex virus (HSV)-1	Yes	>25 kb	Stable; maintained as episome	1×10^{10} pfu/ml
Vaccinia	Yes	>25 kb	Expression transient due to an immune response; replicates in cytoplasm	1×10^8 pfu/ml
Baculovirus	Yes	>20 kb	Unstable	1×10^{10} pfu/ml

forms the virion core while the 10-kD nucleocapsid protein binds to the RNA genome in a viral particle. The PR and polymerase (Pol) proteins are produced from a Gag/Pro/Pol precursor. This precursor is only 5% as abundant as the Gag precursor and is produced by translational read-through of the *gag* termination codon. The number of infectious particles produced by a cell decreases dramatically if PR and Pol are as abundant as the Gag-derived proteins. PR cleaves a Gag/Pro/Pol precursor into the active polypeptides, although it is unclear how the first PR gets released from the precursor. The *pol* gene product is cleaved into 2 proteins, the amino terminal 80-kD reverse transcriptase (RT) and the carboxy terminal 46-kD integrase (IN). The RT has both reverse transcriptase activity (which functions in RNA- or DNA-directed DNA polymerization) and RNase H activity (which degrades the RNA component of an RNA:DNA hybrid). The IN protein binds to double-stranded DNA at the viral *att* sites located at the ends of each long terminal repeat and mediates integration into the host cell chromosome.

The *env* gene is translated from a subgenomic RNA that is generated by splicing between the 5′ splice site in the 5′ untranslated region and the 3′ splice site present just upstream of the *env* coding sequence. The *env* precursor is processed

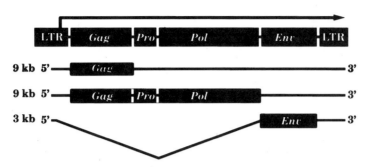

FIGURE 4.1   Diagram of a Moloney murine leukemia retrovirus (MLV). The proviral form with two complete long terminal repeats (LTRs) and the genomic RNA that is expressed from the provirus are shown at the top. The genomic RNA can be translated to produce the *Gag* gene products, or produce a Gag/Pro/Pol precursor by reading through the translational stop codon at the 3' end of the *Gag* gene. The genomic RNA can also be spliced to generate a smaller subgenomic RNA, which is translated into the Env protein. The regions that are translated are shown as black boxes, while the untranslated regions of the RNA appear as a black line.

into 3 proteins: SU, transmembrane (TM; or p15E), and p2. The 70-kD SU protein binds to a cell surface receptor. Neutralizing antibodies directed against SU can block infection. The 15-kD TM plays a role in fusion of the virus and cellular membrane. In many retroviruses, the association between the SU and TM proteins is rather tenuous and SU is rapidly lost from virions. This contributes to poor infectivity of viral preparations and instability to manipulations such as concentration by ultracentrifugation. Envelope proteins from different retroviruses, or even from viruses of other families, can be used to produce infectious particles with altered tropism and/or greater stability.

Sequences Required in *cis* for Replication and Packaging

The term provirus refers to the form of the virus that is integrated as double-stranded DNA into the host cell chromosome. Genetic sequences are needed in *cis* to develop a provirus that can transfer genetic information into a target cell. Four important sequences are required in *cis* for replication and infection in the context of gene therapy. They are (1) the long terminal repeats (LTRs), (2) the primer binding site (PBS), (3) the polypurine (PP) tract, and (4) the packaging sequence. These sequences and their function are shown in Figure 4.2. LTRs are approximately 600 nucleotide sequences present at both the 5' and the 3' end of the provirus. They initiate transcription at the 5' end, perform polyadenylation at the 3' end, and integrate a precise viral genome into a random site of the host cell chromosome by virtue of the *att* sites at either end. The LTR-initiated transcripts serve as an mRNA for the production of viral proteins and as the RNA genome for producing additional virus. The PBS is located just downstream of the 5' LTR. It binds to a cellular transfer RNA (tRNA), which serves as a primer for the polymerization of the first DNA strand. The PP tract contains at least nine purine nucleotides and is located upstream of the U3 region in the 3' LTR. The RNA within this sequence is resistant to degradation by RNase H when hybridized with the first DNA strand.

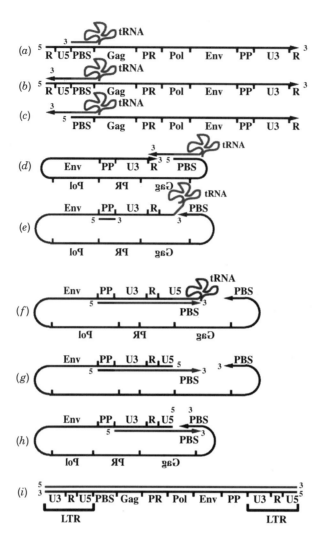

FIGURE 4.2 Mechanism of reverse transcription and integration of the genomic RNA into the host cell chromosome. (*a*) Genomic RNA with a tRNA primer. The genomic RNA has a 60-nt R region (for redundant) at both the 5′ and the 3′ end. The 5′ end has the 75-nt U5 region (for unique to 5′ end) and the 3′ end has the 500-nt U3 region (for unique to 3′ end). The PBS of the genomic RNA (shown in black) hybridizes to the terminal 18 nt at the 3′ end of a tRNA. (*b*) Reverse transcription of the 5′ end of the genomic RNA. The tRNA primer enables the RT to copy the 5′ end of the genomic RNA, to generate a portion of the first DNA strand. (*c*) Degradation of the RNA portion of an RNA : DNA hybrid by RNase H. RNase H degrades the RNA portion that was used as a template for synthesis of the first DNA strand. Although shown as a separate step here, this occurs ~18 nt downstream of where polymerization is occurring. (*d*) First strand transfer. The portion of the first strand that represents the R region hybridizes with the R region in the 3′ end of the genomic RNA. (*e*) Reverse transcription of the remainder of the genomic RNA. The RT copies the genomic RNA up to the PBS. As elongation occurs, RNase H continues to degrade the RNA portion of the RNA : DNA hybrid. The RNA in the PP tract (shown in black) is resistant to cleavage by RNase H and remains associated with the first DNA strand. (*f*) Initiation of second strand synthesis. The primer at the PP tract initiates polymerization of the second strand. Polymerization up to the 3′ end of the PBS continues. Additional sequences in the tRNA are not copied, as the 19th nucleotide is blocked by a methyl group in the base pairing region of the tRNA. (*g*) RNase H digestion of the tRNA. The RNase H degrades the tRNA, which is present in an RNA : DNA hybrid. (*h*) Second strand transfer. The second DNA strand hybridizes to the first DNA strand in the PBS region. (*i*) Completion of the first and second strands. RT copies the remainder of the first and the second DNA strands, to generate a double-stranded linear DNA with intact LTRs at both the 5′ and the 3′ end. The integrase binds to the *att* sequence at the 5′ end of the 5′ LTR and at the 3′ end of the 3′ LTR (not shown) and mediates integration into the host cell chromosome. Upon integration, the viral DNA is usually shortened by two bases at each end, while 4 to 6 nt of cellular DNA is duplicated. Although integration is a highly specific process for viral sequences, integration into the host chromosome appears to be random.

The PP tract therefore serves as the primer for synthesis of the second DNA strand. The packaging signal binds to the nucleocapsid protein of a retroviral particle allowing the genomic RNA to be selectively packaged. Although the encapsidation sequence was initially mapped to the region of the virus between the 5′ LTR and the *gag* gene, vectors that only contained this sequence were packaged inefficiently, resulting in low titers of viral vector produced. Subsequent studies demonstrated that inclusion of some *gag* sequences (the extended packaging signal) greatly increased the titer of the vector produced. Most vectors that are currently in use utilize the extended packaging signal.

Use of Retroviral Sequences for Gene Transfer

All of the genomic sequences that are necessary in cis for transcription and packaging of RNA, for reverse transcription of the RNA into DNA and for integration of the DNA into the host cell chromosome need to be present in the retroviral vector. It is, however, possible to remove the coding sequences from the retroviral genome and replace them with a therapeutic gene to create a retroviral vector. The deletion of viral coding sequences from the retroviral vector makes it necessary to express these genes in *trans* in a packaging cell line. Packaging cell lines that stabily express the *gag*, *pro*, *pol*, and *env* genes have been generated. The transfer of a plasmid encoding the retroviral vector sequence into packaging cell results in a retroviral particle capable of transferring genetic information into a cell (assuming appropriate tropism). However, upon transfer of the retroviral vector into a cell, infectious particles are not produced because the packaging genes necessary for synthesizing the viral proteins are not present. These vectors are therefore referred to as replication incompetent. Figure 4.3 diagrams how retroviral vectors and packaging cells are generated.

Commonly used retroviral vectors and their salient features are summarized in Table 4.3. Plasmid constructs that resemble the provirus and contain a bacterial origin of replication (see Chapter 1) outside of the LTRs can be propagated in bacteria. The therapeutic gene is cloned into a vector using standard molecular biology techniques. Upon transfection into mammalian cells, the 5′ LTR of the vector DNA initiates transcription of an RNA that can be packed into a viral particle. Although a packaging cell line can be directly transfected with plasmid DNA, the integrated concatemers are unstable and are often deleted during large-scale preparation of vector. To circumvent this problem, most cell lines used in animals are infected with the vector rather than transfected. This involves transfection into one packaging cell line, which produces a vector that can infect a packaging cell line with a different envelope gene. The infected packaging cell line generally contains a few copies of the retroviral vector integrated into different sites as a provirus.

Most vectors have genomic RNAs that are less than 10 kb, to allow for efficient packaging. N2 was the first vector using an extended packaging signal that, as noted earlier, greatly increased the titer of vector produced. In LNL6, the AUG at the translational initiation site was mutated to UAG, which does not support translational initiation. This mutation prevents potentially immunogenic *gag* peptides from being expressed on the surface of a transduced cell. In addition, it decreases the possibility that a recombination event would result in replication-competent virus since the recombinant mutant would not translate the *gag* gene into a protein. The LN

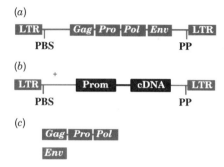

FIGURE 4.3   Retroviral vectors. (*a*) Wild-type retrovirus. The proviral form of a retrovirus is shown. Long-terminal repeats (LTRs) are present at both ends and are necessary for reverse transcription of the RNA into a double-stranded DNA copy and for integration of the DNA into the chromosome. The packaging signal (Ψ) is necessary for the RNA to bind to the inside of a viral particle, although sequences in the *Gag* region increase the efficiency of packaging. The primer binding site (PBS) and the polypurine tract (PP) are necessary for priming of synthesis of the first and second strands of DNA, respectively. The retroviral packaging genes *gag*, *pro*, *pol*, and *env* code for proteins that are necessary for producing a viral particle. (*b*) Retroviral vector. Retroviral vectors have deleted the retroviral coding sequences and replaced them with a promoter and therapeutic gene. The vector still contains the LTR, a packaging signal designated as Ψ^+, which contains a portion of the *Gag* gene, the PBS, and the PP tract, which are necessary for the vector to transmit its genetic information into a target cell. (*c*) Packaging cells. The retroviral vector alone cannot produce a retroviral particle because the retroviral coding sequences are not present. These packaging genes, need to be present in a packaging cell line along with the vector in order to produce a retroviral particle that can transfer genetic information into a new cell.

series is similar but has deleted the sequences 3′ to the *env* gene, thereby limiting recombination events to generate wild-type virus. Double copy vectors place the promoter and coding sequence within the 3′ LTR. As shown in Figure 4.2, the 3′ U3 region is copied into both the 5′ and the 3′ LTRs when the genomic RNA is copied into double-stranded DNA. This results in two complete copies of the transgene in the target cell. The self-inactivating (SIN) vectors were created to address concerns regarding insertional mutagenesis. A deletion in the 3′ U3 region is incorporated into both the 5′ and the 3′ LTR of the provirus. However, insertion into the 3′ U3 region often results in deceased titers. The MFG vector uses the retroviral splice site and the translational initiation signal of the *env* gene resulting in a spliced mRNA that is presumably translated with high efficiency.

Packaging Cells Lines

Commonly used packaging cell lines are summarized in Table 4.4. Initially, packaging cell lines simply deleted the packaging sequence from a single packaging gene plasmid that contained all four genes and both LTRs. These lines occasionally generated replication-competent virus due to homologous recombination between the vector and the packaging constructs. Development of replication-competent virus is a serious concern since it leads to ongoing infection in vivo and ultimately may cause malignant transformation via insertional mutagenesis. Several approaches

TABLE 4.3 Summary of Retroviral Vectors Used for Gene Therapy in Animals or Humans

Name	Salient Features
N2	Contains an intact 5′ and 3′ LTR, an extended packaging signal with 418 nt of coding sequence of the *gag* gene, and an intact translational start codon (AUG) of the *gag* gene. Can recombine to generate wild-type virus.
LNL6	Contains intact 5′ and 3′ LTRs, an extended packaging signal with 418 nt of coding sequence of the *gag* gene, a mutation in the translational start codon (AUG) of the *gag* gene to the inactive UAG, and the 3′ portion of the *env* gene.
LN series	Similar to LNL6 except all *env* sequences are deleted to decrease the chance of recombination with the packaging genes. This series includes LNSX, LNCX, and LXSN, where L stands for LTR promoter, N for neomycin resistance gene, S for SV40 promoter, C for CMV promoter, and X for polylinker sequences for insertion of a therapeutic gene.
Double copy	Places the promoter and the therapeutic gene in the U3 region of the 3′ LTR. This results in two copies of the therapeutic gene within the 5′ and 3′ LTRs after transduction.
Self-inactivating (SIN)	Deletes the enhancer and part of the promoter from the U3 region of the 3′ LTR. This deletion is present in both the 5′ and the 3′ LTRs after transduction. This decreases the chance of transcriptional activation of a downstream oncogene after transduction of a cell.
MFG	Contains an intact 5′ and 3′ LTR, an extended packaging signal with an intact 5′ splice site, a 380-nt sequence with the 3′ end of the *pol* gene and the 3′ splice site, and 100 nt of the 3′ end of the *env* gene. The therapeutic gene is translated from a spliced RNA and uses the *env* gene translational start site.

have been taken to reduce the generation of replication-competent virus. One strategy is to separate the packaging genes into two plasmids integrated into different chromosomal locations. Examples of this approach include the GP + E86, GP + *env*AM12, Ψ-CRIP, and Ψ-CRE packaging cell lines. For these cell lines, the *gag/pro/pol* genes are expressed from one piece of DNA while the *env* gene is expressed from a second piece of DNA. Then each DNA piece is introduced into the cell independently. Another strategy is to minimize homology between the vector and packaging sequences. Some packaging systems use transient transfection to produce high titers of retroviral vector for a relatively short period of time for use in animal experimentation.

Recently developed packaging cell lines are of human origin and are advantageous. The presence of human antibodies in human serum results in rapid lysis of retroviral vectors packaged in murine cell lines. The antibodies are directed against the α-galactosyl carbohydrate moiety present on the glycoproteins of murine but not human cells. This murine carbohydrate moiety is absent from retroviral vectors that are produced by human cells, which lack the enzyme α_1-3-galactosyl transferase. Human or primate-derived packaging cell lines will likely be necessary to produce retroviral vectors for in vivo administration to humans. To this point, the produc-

TABLE 4.4 Summary of Retroviral Packaging Cell Lines Used for Animal and Human Studies

Line	Plasmids That Contain Packaging Genes	Envelope Protein	Detection of Wild-Type Virus?
Ψ-2, Ψ-Am, and PA12	All contain a 5′ LTR, a deletion in the packaging signal, the *gag, pro, pol*, and *env* genes, and the 3′ LTR.	Variable	Yes
PA317 PE501	The 5′ LTR has a deletion 5′ to the enhancers, the Ψ sequence is deleted, *gag, pro, pol*, and *env* genes are present on one plasmid with intact splice signals, the PBS is deleted, and the 3′ LTR is replaced with the SV40 poly A site.	PA317: amphotropic; PE501: ecotropic	Some detected with N2; none with LN-based vectors
Ψ-CRE Ψ-CRIP	One plasmid contains a 5′ LTR, has a deletion of Ψ, expression of *gag-pro-pol* from a construct that also contains an inactive *env* gene, and has an SV40 polyadenylation site. The second plasmid has a 5′ LTR, deletion of Ψ, expression of *env* from a construct that also contains inactive *gag, pro*, and *pol* genes, and an SV40 polyadenylation site.	Ψ-CRE: ecotropic; Ψ-CRIP: amphotropic	Not reported
GP + E-86 GP + *env*AM 12	One plasmid has an intact 5′ LTR, the 5′ splice site, a deletion in the packaging signal Ψ, the *gag-pro-pol* gene with a small amount of the *env* gene, and the SV40 polyadenylation site. A second plasmid has an intact 5′ LTR, the 5′ splice site, the 3′ splice site, and the *env* gene.	GP + E-86: ecotropic; GP + *env*AM12: amphotropic	Reported but not verified

tion of retroviral vectors for clinical use is simple but not without challenges. A suitable stable packaging cell line containing both the packaging genes and the vector sequences is prepared and tested for the presence of infectious agents and replication-competent virus. This packaging cell line can then be amplified and used to produce large amounts of vector in tissue culture. Most retroviral vectors will produce ~1×10^5 to 1×10^6 colony forming units (cfu)/ml, although unconcentrated titers as high as 1×10^7 cfu/ml have been reported. The original vector preparation can be concentrated by a variety of techniques including centrifugation and ultrafiltration. Vectors with retroviral envelope proteins are less stable to these concentration procedures than are pseudotyped vectors with envelope proteins from other viruses. The preparations can be frozen until use with some loss of titer on thawing.

Use of Retroviral Vectors for Gene Therapy

Retroviral vectors have been extensively used in animals and substantially used in humans to determine the efficacy of gene therapy. They are the major vector that has been used for ex vivo gene therapy. Cells that have been modified ex vivo with a retroviral vector include hematopoietic stem cells, lymphocytes, hepatocytes, fibroblasts, keratinocytes, myoblasts, endothelial cells, and smooth muscle cells. Retroviral vectors have also been used for in vivo delivery. For many organs, the requirement of cellular replication for transduction poses a problem since terminally differentiated cells in organs are not proliferative. Thus, retroviral organ-based gene therapy approaches necessitate the induction of cell replication for in vivo transfer into cell types such as hepatocytes, endothelial cells, or smooth muscle cells. Alternatively, the use of viral vectors that do not require cellular replication could be used to transfer genes into nondividing cells in vivo. Studies using HIV have been initiated since that virus does not require replicating cells for transduction. Retroviral vectors have been directly injected into malignant cells in various locations, as malignant cells are highly proliferative. Efficient in vivo delivery will likely require human or primate-derived packaging cell lines or pseudotyping to prevent complement-mediated lysis in all clinical applications of retroviral gene therapy.

After transfer into a replicating cell, the expression of the retroviral vector is critical to achieve a therapeutic effect. In the application of retroviral vectors for gene therapy, the relatively low levels of gene expression achieved in animals are problematic. For currently selected genes used for gene therapy, the level of expression of the gene product does not need to be tightly regulated for clinical effectiveness. However, for diseases such as diabetes mellitus or thalassemia, the level of expression of insulin or β-globin, respectively, requires precise control. Thus, a specific clinical condition may not only require a threshold level for therapeutic effectiveness but may also require a narrow window of concentration for physiological effect. There is a paucity of quantitative data in animals regarding the levels of expression per copy from different vectors, particularly in the context of organ-specific gene expression. This is a major challenge for the field of gene therapy. The difficulties in this area are many. First, current delivery systems make the experimental determination of surviving transduced cells in situ difficult. Accurate determation of the copy number present in vivo is necessary since overall protein expression is a function of both the number of transduced cells and the gene expression per cell. Second, direct comparison of expression levels of different proteins cannot be determined for current delivery systems because of the marked differences in mRNA half-life, protein translation, and protein half-life for different genes. Third, the genomic integration site can dramatically influence the expression level. For delivery systems that modify a small number of stem cells, such as in bone marrow stem-cell-directed gene therapy (see Chapter 7), considerable variation in expression occurs based on animal species. This variation makes it essential to quantitate expression in a large number of animals and report the average results. Thus, an improved understanding of the regulatory controls of gene expression from retroviral vectors remains essential for the clinical application of gene therapy in humans. Unfortunately, expression of vectors in differentiated cell types in vitro does not accurately predict expression levels that can be achieved in vivo. In vitro screening for expression levels provides only limited information on different retroviral vector systems in the context of human application.

An important genetic sequence or element in the gene expression from a retroviral vector is the LTR. The in vivo transcriptional activity of the LTR in bone-marrow-derived cells, liver, and muscle often attenuates over the first few weeks after transfer. However, long-term expression in some cases has been achieved. The attenuation of the LTR reflects the absence of transcription factors that are essential for expression of the LTR promoter in nondividing cells, the presence of inhibitory proteins that shut off the LTR, methylation of the LTR, or deacetylation of the associated histones. Retroviral sequences from the U3 region and the PBS can inhibit expression of the LTR in embryonic carcinoma cells by binding to proteins that inhibit transcription. These inhibitory sequences may contribute to the poor expression observed from the LTR in vivo. Retroviral vectors that alter these inhibitory sequences are expressed in vitro in embryonic carcinoma cells and may also be expressed in vivo. Methylation of the LTR is associated with loss of promoter activity. It is unclear, however, whether methylation per se is responsible for inactivation of the promoter or if methylation is a by-product of binding to the promoter.

Retroviral vectors can include an internal promoter located immediately upstream of the therapeutic gene. These "internal promoters" can be viral promoters, housekeeping promoters, or organ-specific promoters. Viral promoters were components of many first-generation vectors because they are active in most cell types in vitro. However, many of the viral promoters, such as the cytomegalovirus (CMV) promoter, are attenuated or completely shut-off in vivo in organs such as the liver. This loss of function could reflect the absence of transcription factors that are essential for expression of the promoter or the presence of inhibitory proteins that terminate viral promoter activity in nonreplicating cells. Internal promoters may also comprise the ubiquitously expressed housekeeping promoters that direct the expression of proteins required by all cells. However, housekeeping genes are often expressed at relatively low levels, and their promoters have been shown to be relatively weak in vitro and in vivo in retroviral vectors constructs. Alternatively, organ-specific promoters have two major advantages: (1) allowing limited expression to specific cell types or tissues and (2) directing high levels of gene expression. Muscle- or liver-specific enhancers and/or promoters, in comparison to housekeeping or viral promoters, direct higher levels of expression in vivo. Gene expression, in these studies, has been stable for over one year. In other studies, however, organ-specific promoters have been inactivated in vivo in transgenic mice or in a retroviral vector by the presence of adjacent retroviral sequences. These inhibitory sequences play a role in attenuation of the LTR promoter. It is also possible that these inhibitory sequences can decrease expression from adjacent internal promoters.

The control of gene expression in vivo may be an appropriate mechanism to decrease variability in expression as well as decrease the chance that the therapeutic gene is overexpressed. In clinical situations, variability or overexpression would have adverse therapeutic effects. Inducible expression systems have been developed to tightly regulate expression from a retroviral vector through responsivness to an orally administered drug. A tetracycline-responsive system can modify expression >200-fold from a retroviral vector in muscle cells in the presence of a drug when compared to the absence of a drug in vivo. However, this system requires the all-important introduction of a drug-responsive transcription factor. This is an additional burden to the individual cell, which needs to receive and express two separate genes.

Other factors, in addition to the choice of the promoter, can influence gene expression from a retroviral vector. For some genes and through an unknown mechanism, the presence of a splice site dramatically increases the level of expression of the protein. Inclusion of genomic splice sites from the therapeutic gene is technically difficult. An intron would be efficiently removed from the RNA genome if the gene were inserted in the forward orientation. However, the gene can sometimes be packaged in the backwards orientation. In this case the mRNA for the therapeutic gene is transcribed from the opposite strand and these constructs are often unstable. Some retroviral vectors such as the MFG vector have used the retroviral splice signals that direct partial splicing of the genomic retroviral RNA.

Co-expression of two genes has many potential advantages. Through the use of a selectable marker gene and a therapeutic gene, it is possible to eliminate cells not expressing the therapeutic gene by either in vitro or in vivo selection methods. Many first-generation vector constructs express one gene from the LTR promoter and a second gene from an internal promoter. Using these vectors, however, cells selected by virtue of expression of one gene product have a lower level of expression of the second gene product. This observation was due to the phenomenon of promoter interference. An improved approach that obtains co-expression of two genes utilizes a bicistronic mRNA with an internal ribosome entry site (IRES). This enables the downstream gene to be translated in a Cap-independent fashion.

Risks of Retroviral Vectors

There are two major concerns in the use of retroviral vectors for gene therapy in humans: (1) insertional mutagenesis and (2) generation of wild-type virus. Insertional mutagenesis occurs when a retroviral vector inserts within or adjacent to a cellular gene. This insertion could result in the development of malignancy through the inactivation of a tumor suppressor gene or by activation of a proto-oncogene. The risk of developing a malignancy through the process of receiving a single copy of a retroviral vector appears to be minimal. The induction of malignancy has not been observed in animals receiving replication-incompetent retroviral vectors. This observed low incidence of mutagenesis indicates that the retroviral vector is unlikely to integrate into a genomic site that will modify cellular growth properties such as cyclins- or cyclin-dependent kinases (see Chapter 10). However, if the vector inserts into a growth-sensitive site, this would represent only the first step in a multistep process. Thus, procedures that introduce multiple retroviral vector integrations into a single cell will only increase the risk of the development of malignancy. A second safety concern regarding retroviral vectors in human use is viral recombination. Viral recombination may result in the development of replication-competent virus. This event can clearly result in the slow onset of malignancy in animals. Technical refinements in vector development have lowered the risk of generating a replication-competent virus. These include elimination of homology between the packaging genes and the vector as well as separation of the packaging genes into two or more separate pieces of DNA. However, if recombination occurs, the extensive testing performed prior to administration of vectors to humans is an added safety measure that identifies recombinant(s). Thus, it is unlikely that replication-competent virus will be administered to humans when the appropriate safety controls are observed. It remains possible, however, that a replication-incompetent retroviral vector could recombine with endogenous viruses in vivo. Endogenous viruses are present in vivo and recombination in the human genome can generate additional pathogenic replication-competent virus(es). The occurrence can only be determined by monitoring individual gene therapy recipients for the appearance of replication-competent virus.

Summary: Retroviral Vectors

Replication-incompetent retroviral vectors can be easily generated by deleting retroviral genes and adding gene(s) of interest. Vectors can be produced in packaging cell lines that express packaging genes. The major advantage of retroviral vectors is the precise integration into a random site in the host cell chromosome. This can result in long-term survival of the gene in the transduced cell. The major disadvantage is the need to transduce dividing cells. This characteristic poses difficulties for the in vivo delivery to quiescent cells. Gene expression at therapeutic levels has been achieved from a retroviral vector in vivo in some studies for over one year, but expression has been problematic in other studies.

Lentiviral Vectors

The lentiviruses are a family of retroviruses comprising seven subgenera with specific biological properties. One such property is an advantage for its use in gene therapy, that is, the ability to transduce nondividing cells. The matrix protein and the *vpr* gene product of the lentivirus contain nuclear localization signals that allow the DNA to be transported to the nucleus without breakdown of the nuclear membrane. These gene products facilitate the infection of nondividing cells. Lentiviruses contain a number of proteins exclusive of the MLV genome (see also Chapter 11). The *tat* gene encodes a protein that stimulates expression via the *tat* response element (TAR) located in the HIV LTR. The *rev* gene encodes a protein that binds to the *rev* response element (RRE) and facilitates the transfer of unspliced RNAs to the cytoplasm. The *nef* gene encodes a protein that is localized to the inner surface of the cell membrane and can decrease the amount of the HIV cell surface receptors, such as CD4. The *nef* gene protein is important for virulence in vivo through as yet undefined mechanisms. The function of the *vif* gene is unclear. The product of the *vpu* gene appears to play a role in processing of the *env* gene product and in the efficient budding and release of virions. The *vpr* gene product contains a nuclear localization signal and may play a role in transporting HIV to the nucleus of nondividing cells. The role of the *vpx* gene product is unclear.

Several replication-defective HIV-based vectors and packaging system has been used to deliver genes to nondividing neurons, muscle, lung, endothelial cells, hematopoietic stem cells, and liver cells in vivo. One HIV packaging system contains a vector with the HIV LTRs at either end (including the TAR), an extended packaging signal, the RRE, and a reporter gene whose expression was directed by the CMV promoter. The packaging construct deleted the packaging signal and mutated the *env* gene. The VSV-G envelope was expressed from a third construct. The supernatant of cells that were transfected simultaneously with all three plasmids contained retroviral particles that infected nondividing cells in vitro and in vivo. More recently, all of the accessory genes except for *tat* and *rev* have been mutated in the packaging construct, and the particles still transduced nondividing cells at the site of injection allowing for multiple exposures. Also, a new series of lentiviral vectors based on HIV-1 have been developed as a self-inactivating vector. Here, the U3 region of the 5′ LTR was replaced by the CMV promotor, resulting in *tat*-independent transcription. The self-inactivating vector was constructed by deleting 133 bp in the U3 region of the 3′ LTR including the TATA box and the binding sites for specific transcription factors. This deletion is transferred to the 5′ LTR after reverse transcription and integration into the genome of infected cells resulting in transcriptional inactivation of the LTR of the provirus. Such a self-inactivating virus transfected brain cells at a comparable level to wild-type virus.

Transduction of nondividing cells is a major advance for retroviral vector technology. Furthermore, lentivirus vectors pseudotyped with vesticular stomatitis virus G glycoprotein can transduce a wide range of nondividing cells. In addition, no inflammation is observed at the site of injection allowing for multiple exposures. It is possible that the multiple added properities of nonvirulent HIV-based vectors as described above will revolutionize human gene therapy procedures for non-replicating cells in vivo. Three major concerns regarding these vectors remain, however. The first is the absolute assurence that recombination to generate wild-type HIV that causes immunodeficiency syndrome in a patient will not occur. Many of the HIV accessory genes can be mutated to prevent production of a functional protein. But, the complicated nature of the HIV genome and the high mutagenic rate currently made it impossible to completely assure that these accessory genes will remain nonpathogenic. Stringent tests regarding the generation of wild-type virus will be necessary prior to human use. A second concern regards the possibility of promiscuous transduction of all cell types in vivo. This may cause the unnecessary transduction of cell types where expression of the vector does not have a therapeutic effect. As noted above pseudotyping of the viral vector may limit or broaden the spectrum of cells infected. The third concern is the production of sufficient quantities of these vectors for in vivo delivery. The packaging cells currently using a transient expression system need to be enhanced.

ADENOVIRAL VECTORS

The adenovirus is a 36-kb double-stranded linear DNA virus that replicates extra-chromosomally in the nucleus. The virus was first isolated from the adenoids of patients with acute respiratory infections, although it can also cause epidemic conjunctivitis and infantile gastroenteritis in humans. In patients with an intact immune system, infections are mild and self-limited. In immunosuppressed patients, however, infections can result in dissemination to the lung, liver, bladder, and kidney and can be life-threatening. Although human adenovirus type 12 can induce malignant transformation after inoculation into newborn hamsters, adenoviral DNA has not been associated with human tumors.

Adenoviral particles are 70 to 100 nm in diameter and do not contain membrane.

Over 100 different adenoviruses have been identified that infect a wide range of mammalian and avian hosts. Initial attachment of adenoviruses to cells is mediated by the fiber protein that binds to a cellular receptor. The cellular receptor has yet to be identified and may be different for different serotypes. Type-specific viral neutralization results from antibody binding to epitopes on the fiber protein and the virion hexon protein. Subsequent to initial binding, the penton base protein binds to members of a family of heterodimeric cell surface receptors known as integrins. The adenovirus:receptor complex then enters the cell via coated pits and is released into the cytoplasm from an endosomal compartment. The viral particles are transported to the nucleus via nuclear localization signals embedded in the capsid proteins. There the DNA is released in part by proteolytic degradation of the particle. The viral DNA persists during an active infection and for long periods of time in lymphocytes as a nonintegrated episome, although integration can occur during the process of transformation. Adenoviruses can transfer genetic information to a variety of cell types from many species, although they only replicate in human cells. For wild-type adenovirus, DNA replication begins ~5 h after infection and is completed at 20 to 24 h in HeLa cells, a human cervical carcinoma-derive cell line. Each cell produces 10,000 progeny virus and is lysed by their release. The production of large numbers of adenoviral particles facilitates the preparation of very high titers of adenoviral vectors.

Adenoviral Genes and Sequences Required in cis for Replication

Adenoviral genes can be transcribed from either strand of DNA and have a complex splicing pattern. There are five early transcription units, E1A, E1B, E2, E3, and E4, all of which are transcribed shortly after infection and encode several different polypeptides. Two delayed early units and the major late unit generate five families of late mRNAs. Adenoviruses also contain one or two VA genes that are transcribed by RNA polymerase III and serve to block host cell translation.

The E1A region codes for two E1A polypeptides. E1A polypeptides can activate transcription by binding to a variety of different cellular transcription factors and regulatory proteins, including the retinoblastoma gene product *Rb*. E1A induces the cell to enter the cell cycle, which is necessary for replication of adenoviral DNA. The E1B 55-kD protein binds to p53 and prevents p53 from blocking progression through the cell cycle or inducing apoptosis. The E1B 19-kD protein blocks apoptosis by an as yet unknown mechanism. The E2 region encodes three different proteins, all of which function directly in DNA replication. The E2-encoded terminal protein is an 80-kD polypeptide that is active in initiation of DNA replication. It is found covalently attached to the 5′ ends of the viral DNA. The other E2-encoded proteins include a 140-kD DNA polymerase and a 72-kD single-stranded DNA binding protein. The E3 region encodes proteins that modify the response of the host to the adenovirus. The E3-gp 19-kD protein binds to the peptide-binding domain of MHC class I antigens and causes retention of class I antigen in the endoplasmic reticulum. The E3 14.7-kD protein, or the complex of E3 14.5-kD/E3 10.4-kD proteins prevent cytolysis by tumor necrosis factor. The E4 unit encodes proteins that regulate transcription, mRNA transport, and DNA replication. Of the 11 virion proteins, 7 are located in the outer shell and 4 are present in the core of the virion. These are primarily encoded by the late genes.

There are two sequences that need to be supplied in *cis* for viral replication: (1) the 100- to 140-bp inverted terminal repeats at either end of the linear genome and (2) the packaging signal, which is adjacent to one of the inverted terminal repeats. The 5′ ends of the viral DNA have a terminal protein of 80 kD covalently attached via a phosphodiester bond to the 5′ hydroxyl group of the terminal deoxycytosine residue. The terminal protein serves as a primer for DNA replication and mediates attachment of the viral genome to the nuclear matrix in cells. Inverted repeats enable single strands of viral DNA to circularize by base pairing of their terminal sequences. The resulting base-paired panhandles are thought to be important for replication of the viral DNA. The packaging sequence, located at nucleotide 194 to 358 at the left end of the chromosome, directs the interaction of the viral DNA with the encapsidating proteins.

Use of Adenoviral Sequences for Gene Transfer

The observation that E1A- and E1B-deficient adenoviruses are propagated in 293 cells paved the way for the development of adenoviral vectors. The 293 cells are a human embryonic kidney cell line that contains and expresses the Ad5 E1A and E1B genes. Early first-generation adenoviral vectors replaced a 3-kb sequence from the E1 region with a promoter and a gene of interest, as shown in Figure 4.4. In addition to providing space for the therapeutic gene, deletion of the E1 region removed oncogenes that might contribute to malignancy. Although the early

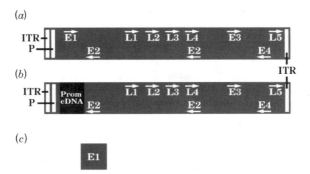

FIGURE 4.4   Adenoviral vectors. (*a*) Wild-type adenovirus. Adenoviruses contain a double-stranded linear DNA genome of ~36 kb. The inverted terminal redundancies (ITRs) of ~100 base pairs at either end are necessary for replicating the DNA. The packaging signal (P) is necessary for the viral DNA to get packaged into a viral particle. Multiple early (E) and late (L) genes code for proteins that are necessary for replicating the DNA and producing an infectious adenoviral particle. (*b*) Adenoviral vector. Most adenoviral vectors have deleted the E1 gene and replaced it with a promoter and therapeutic gene. This results in a vector that still contains most of the adenoviral genes. Other adenoviral vectors that are not shown here have deleted additional adenoviral genes from the E2, E3, or E4 region. (*c*) Packaging cells. The adenoviral vector alone cannot produce adenoviral particles because it does not contain the E1 gene. Packaging cells that express E1 and contain the adenoviral vector sequences are necessary for producing adenoviral particles that can transmit information to a new cell. E2 or E4 also need to be expressed in packaging cells that are used to produce E2- or E4-deleted adenoviral vectors.

adenoviral vectors resulted in high levels of expression in a variety of organs at early time points in animals, expression was transient. The transient expression was primarily a result of an immune response targeted to cells that express the residual adenoviral vector proteins. This observation led to further manipulations of the adenoviral vector genome in an attempt to stabilize the vector in vivo and reduce the inflammatory response.

Later generations of adenoviral vectors have deleted E2, E3, or E4 in addition to E1 in an attempt to decrease the expression of late genes and the subsequent immune response. An added advantage of the manipulation is the additional space for the therapeutic gene. E2- or E4-deleted adenoviral vectors require cell lines that express E2 or E4 in addition to E1. The E3-deleted adenoviral vectors can still be produced in 293 cells, since the E3 region does not encode any genes that are essential for replication in vitro. The products of the E2 gene include a 72-kD single-stranded DNA binding protein, which plays a role in both DNA replication and viral gene expression. An adenoviral vector that contained a mutation in the E2A gene has resulted in the generation of a temperature-sensitive single-stranded DNA binding protein. Use of this vector construct results in prolonged expression of the therapeutic gene, decreased expression of the late adenoviral vector genes, and a delayed inflammatory response. However, even in the latter case expression still did not extend beyond 100 days. Deletion of the E4 region has led to increased stability of the adenoviral DNA in vivo, with a loss of expression from the CMV promoter in the liver. Deletion of the E3 region has decreased the stability of the adenoviral vector in vivo. This E3 region helps the virus to avoid the immune system of the host by blocking class I MHC presentation of viral antigens, and thus deletion of this region promotes antigen presentation and host immunity.

The removal of all adenoviral proteins creates a so-called gutless adenoviral vector. The purpose of this line of investigation is to eliminate the expression of the adenoviral proteins in vivo in order to prevent a host immune response. Gutless adenoviral vectors have been generated in which the inverted terminal repeats and the packaging signal remains, but all adenoviral coding sequences have been removed and replaced with the therapeutic gene. Unfortunately, these vectors have not resulted in prolonged expression in vivo. It is possible that the adenovirus contains other sequences that are necessary for long-term extrachromosomal maintenance of the DNA in cells.

Preparation of recombinant adenoviral vectors for clinical use is somewhat more complicated than is the production of retroviral vectors. The 293 cells are a human embryonal kidney cell line that expresses the E1 genes and are commonly used to propagate E1-deficient adenoviral vectors. The large size of the adenovirus (~36 kb) makes cloning by standard methods difficult due to the paucity of unique restriction sites. Most genes are inserted into the adenoviral vector by homologous recombination between a transfer vector and the helper vector in cells that express any necessary proteins in trans. The transfer vector contains the therapeutic gene flanked by adenoviral sequences on a plasmid that contains a bacterial origin of replication, and this can be propagated in bacteria. The helper virus contains all of adenoviral genes except those that are supplied in trans by the packaging cells. In some cases, the helper virus can be propagated in 293 cells and therefore must be restricted prior to co-transfection with the transfer vector to decrease the number of nonrecombinants that are obtained. For other helper vectors, such as pJM17, the helper vector is present on a plasmid with a bacterial origin of replication inserted in the E1 region. This can be propagated in bacteria but is too large to be packaged into an adenoviral particle. After co-transfection of the transfer vector and the helper vector into 293 cells, homologous recombination results in the insertion of the therapeutic gene and deletion of the bacterial origin of replication. The resulting vector can be packaged. Recombinants that replicate in 293 cells result in cell death that appears as a plaque on a lawn of viable cells. Plaques are screened for the presence of the therapeutic gene and the absence of the helper vector. A therapeutic gene of up to 8 kb can be inserted into an adenoviral vector.

To produce large amounts of the adenoviral vector, packaging cells are infected with the plaque-purified adenoviral vector. When a cytopathic effect is observed, the cells are broken up and the adenoviral vector is purified from the cellular debris using a variety of techniques including $CsCl_2$ gradients and column chromatography. Titers of up to 10^{12} plaque forming units (pfu)/ml can be obtained and are stable to freezing. Preparations must be tested for the presence of wild-type adenovirus or other pathogens prior to use in humans.

Use of Adenoviral Vectors for Gene Therapy

Adenoviral vectors have been used to transfer genes in vivo into the lung, liver, muscle, blood vessel, synovium, eye, peritoneum, brain, and tumors in animals. The titers that can be achieved enable a high percentage of the cells to be transduced as well as express elevated levels of the transgene. A major limitation of adenoviral vectors is the transgene expression for less than one month primarily due to an immune response to the remaining viral proteins. This targeted specific immune response rapidly eliminates the transduced cells. This immune response can also result in severe inflammation at the site of delivery and organ dysfunction. Furthermore, the vigorous host immune response to the surface proteins of the adenovirus diminishes the efficacy of repeat administration.

A strategy to prolong gene express is to inhibit the immune response to the adenoviral vector. Studies in immunodeficient mice have demonstrated that in the absence of antigen-specific immunity, gene expression is prolonged and secondary gene transfer is possible. MHC class I-restricted $CD8^+$ cytotoxic T lymphocytes are the primary effector cells for the destruction of adenoviral infected cells in the mouse. The use of immunosuppressive therapy could provide persistent gene expression following adenovirus-mediated gene transfer and allow secondary gene transfer. A variety of approaches to suppress the immune response have been taken. These include immunosuppression with drugs such as cytoxan or cyclosporine, or inhibition of the CD28:B7 costimulatory response using a soluble form of murine CTLA4Ig. Injection of adenoviral vector into neonates or into the thymus, resulting in tolerization, allows subsequent injection of an adenoviral vector into adults without immune rejection.

Evaluation of gene expression from adenoviral vectors has been complicated by its instability. Many studies have not differentiated between loss of DNA and loss of gene expression. Some studies have demonstrated relatively long-term expression from the CMV promoter of an adenoviral vector in the liver in vivo. These studies contradict the results obtained using a retroviral vector, in which the CMV promoter was rapidly shut-off. However, it was subsequently demonstrated that deletion of the E4 region of the adenovirus led to loss of expression from a CMV promoter in an adenoviral vector in the liver in vivo. It is therefore likely that the deletion of other early genes might modulate expression of an adenoviral vector in vivo. Studies have demonstrated that the housekeeping promoter elongation factor 1 was more active than the CMV promoter. The CMV-enhancer–β-actin-promoter combination was more active than the SRα promoter. Additional experiments in which transgene expression is followed over time and normalized to the adenoviral vector copy number in various organs will be necessary to optimize expression levels in vivo.

Risks of Adenoviral Vectors

There are three potential risks of adenoviral vectors: (1) the development of organ inflammation and dysfunction due to the immune response to adenoviral vector-transduced cells, (2) the development of tolerance to an adenoviral vector that could result in fulminant disease upon infection with wild-type virus, and (3) the development of wild-type virus. Early generation adenoviral vectors were toxic when administered at high doses. For example, one patient with cystic fibrosis who received an adenoviral vector to the lung had a severe inflammatory response. It is likely that decreasing the immunogenicity of adenoviral vector-transduced cells or modulating the immune response will decrease this inflammation. Whether limited organ-based inflammation will be acceptable is an open question. The risks of modulating the immune response to an adenoviral vector have not been adequately studied. It is possible that immunomodulation will predispose to fulminant disease upon infection with wild-type adenovirus of the same serotype. These risks cannot be assessed in animal models where the adenovirus does not replicate. The third risk of using adenoviral vectors is the generation of wild-type virus in vivo. This also could lead to fulminant infection if immunomodulation has led to tolerance. It is less likely that development of wild-type adenovirus would contribute to malignancy since the virus does not integrate.

Summary: Adenoviral Vectors

In summary, adenoviral vectors result in high-level expression in the majority of cells of many organs for 1 to 2 weeks after transfer. Gene transfer occurs in nondividing cells, a major advantage over most retroviral vectors. However, expression is transient in most studies. This is due primarily to an immune response. The instability of expression is a serious impediment to the use of adenoviral vectors in the treatment of monogenic deficiencies. It is less of a problem for gene therapy approaches for cancer that require short-term expression. The immune response to adenoviral-transduced cells can lead to organ damage and has resulted in death in some animals. Any preexisting or induced antiadenovirus neutralizing antibodies could prevent an initial or subsequent response to adenoviral treatment. Modification of the adenoviral vector to decrease its immunogenicity or suppression of the recipient's immune response may prolong expression and/or allow repeated delivery to patients.

ADENOVIRUS-ASSOCIATED VIRUS

Adenovirus-associated virus (AAV) is a 4.7-kb single-stranded DNA virus that replicates in the nucleus in the presence of adenovirus and integrates into the chromosome to establish a latent state. It was first discovered as a satellite contaminant in human and simian cell cultures infected with adenovirus. AAV has not been associated with disease in humans, although up to 90% of all humans have evidence of prior infection with some serotypes of AAV. Humans are frequently seropositive for AAV2 and AAV3, while evidence of prior AAV5 infection is infrequent. AAV particles are 18 to 26 nm in diameter and do not contain membrane. They enter the cell by receptor-mediated endocytosis and are transported to the nucleus. Although the receptor has not yet been cloned, entry occurs in a wide range of mammalian species. Wild-type AAV integrates as double-stranded DNA into a specific region of chromosome 19. AAV can also be maintained in an extrachromosomal form for an undefined period of time.

AAV Genes

The AAV genome has two major open reading frames, as shown in Figure 4.5. The left open reading frame extends from map position 5 to 40 and encodes the Rep proteins. The right open reading frame extends from map position 50 to 90 and encodes the AAV coat proteins. The *rep* gene was so named because its products

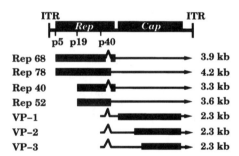

FIGURE 4.5 Map of the AAV genome. The AAV2 genome of 4.8 kb has 100 map units. AAV has inverted terminal repeats (ITRs) of 145 nt at either end, which contain sequences necessary for DNA replication and packaging into virions. There are 3 promoters at map position 5, 19, and 40, which are designated p5, p19, and p40, respectively. These is an intron at map position 42 to 46, which may or may not be utilized, resulting in 2 transcripts that derive from each promoter. There is a polyadenylation site at map position 96, which is used by all transcripts. The p5-initiated proteins Rep 68 and Rep 78 are necessary for replication and for transcriptional regulation of AAV and heterologous viral and cellular promoters. The p19-derived proteins Rep 40 and Rep 52 are required for accumulation of single-stranded DNA. For the p5- and p19-derived transcripts, the unspliced species is the major mRNA. The AAV *cap* gene encodes the structural AAV capsid proteins, which are transcribed from the p40 promoter. VP-1 is derived from an alternatively spliced mRNA that uses an AUG for translational initiation. VP-2 is derived from the more common splice product and utilizes the nonconsensus ACG as the translational initiation site. VP-3 is derived from the most common splice product and uses the consensus AUG for translational initiation. The size of each RNA is shown on the right.

are required in trans for DNA replication to occur. Rep 68/78 is an ATPase, helicase, site-specific endonuclease and transcription factor. Rep 68/78 plays a critical regulatory role in several phases of the AAV life cycle. It is necessary for site-specific integration into the host cell chromosome and to establish a latent infection. Rep 68/78 binds to a dodecamer sequence $(GCTC)_3$ in the stem of the ITR and causes a nick in the DNA. The latter is essential for replication of the DNA. A region of chromosome 19 also contains the AAV Rep protein binding sequence $(GCTC)_3$ responsible for region-specific integration. Integration can occur within several hundred nucleotides of this recognition site. In the presence of helper virus, Rep 68/78 is a transactivator at all three AAV promoters, p5, p19, and p40. In the absence of co-infection with a helper virus, Rep68/78 negatively regulates AAV gene expression. Although the functions of the smaller 52- and 40-kD Rep proteins are not totally clear, each are necessary for the accumulation of single-stranded genomic DNA. The *cap* gene codes for the capsized proteins, VP-1 of 87 kD, VP-2 of 73 kD, and VP-3 of 62 kD. VP-2 and VP-3 are initiated from different transnational start codons of the same mRNA, while VP-1 is translated from an alternatively spliced mRNA. Although VP-3 is the most abundant protein, VP1, 2, and 3 are required for infectivity.

Sequences Required in *cis* for Replication

AAV has an inverted terminal repeat of 145 nt at both ends that is required in cis for DNA replication, encapsidation, and integration. The first 125 bases contains a palindromic sequence that forms a T-shaped structure, as shown in Figure 4.6. Replication begins in the ITR where a stable hairpin is formed, leading to self-priming from the 3' end and replication using a cellular DNA polymerase. Rep 68/78 nicks the parental strand in the ITR as shown in Figure 4.6c, which allows filling in of the bottom strand. When capsid proteins are expressed, capsid assembly leads to displacement and sequestration of single-stranded AAV genomes. Single stands of either polarity can be packaged into AAV particles.

Helper Functions of Other Viruses

AAV are unique in that they usually require co-infection with another virus for productive infection. The helper (co-infection) virus is usually adenovirus or herpes simplex virus. Cytomegalovirus and pseudoradies virus can also function as a helper virus. Treatment of cells with genotoxic agents such as ultraviolet irradiation, cycloheximide, hydroxyurea, and chemical carcinogens can also induce production of AAV, albeit at low levels. The helper functions of adenovirus requires the early but not late genes. E1A is required for AAV transcripts to be detected and presumably activates transcription of the AAV genes. The E4 35-kD protein forms a complex with the E1B 55-kD protein and may regulate transcript transport. The E2A 72-kD single-stranded DNA binding protein stimulates transcription of AAV promoters and increases AAV DNA replication, but it is not absolutely required for AAV replication. The adenovirus VAI RNA facilitates the initiation of AAV protein synthesis. The helper functions provided by HSV-1 have been less clearly defined. Two studies indicate that the ICP-8 single-stranded DNA protein is required.

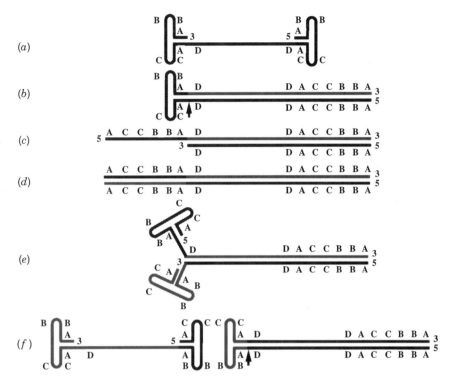

FIGURE 4.6 Mechanism of replication of AAV DNA. AAV has a single-stranded DNA genome (shown in black) with inverted terminal repeats (ITRs) at either end. (*a*) Structure of the single-stranded genomic DNA. The ITRs are palindromic and form a T-shaped structure at either end. The 3′ end is double stranded and thus can serve as a primer for the initiation of DNA synthesis. (*b*) Elongation of the 3′ end. A cellular DNA polymerase initiates DNA synthesis at the 3′ end and copies the DNA up until the 5′ end of the genomic DNA. The arrow designates the site at which Rep will cleave the DNA. (*c*) Endonucleolytic cleavage of the genomic DNA. The viral protein Rep performs an endonucleolytic cleavage of the DNA. The T-shaped structure can be unfolded to result in the structure shown. (*d*) Elongation of the DNA to generate a double-stranded unit length intermediate. DNA polymerase initiates polymerization at the free 3′ end, resulting in the synthesis of a full-length double-stranded intermediate. Note that the B and C sequences have become inverted relative to their initial orientation. This is designated as the "flop" orientation, while the initial structure shown in (*a*) in which the B sequence was closer to the terminus is designated as the "flip" orientation. Either orientation can be packaged into a viral particle. (*e*) Isomerization. The left end of the double-stranded intermediate can isomerize to form the structure shown. Alternatively, the right end of the double-stranded intermediate could isomerize to form a similar structure (not shown here). (*f*) Continued DNA synthesis to release a single-stranded genomic DNA and a covalently linked double-stranded intermediate. The free 3′ end primes synthesis of new DNA. This results in the release of a single-stranded genomic DNA that can be packaged into a viral particle. The double-stranded DNA intermediate shown here is homologous to the intermediate shown in (*b*) and can be cleaved by Rep to generate a free 3′ end and undergo the subsequent steps shown in (*c*) through (*f*). These steps would return the DNA to the original "flip" orientation.

There are discrepancies as to the function of the ICP4 transactivator, the DNA polymerase, and various submits of the helicase–primase complex.

Use of AAV Sequences for Gene Transfer

AAV vectors, like retroviral vectors, can be deleted of all coding sequences and replaced with a promoter and coding sequence of interest, as shown in Figure 4.7. This process eliminates the immune response to residual viral proteins. The most common method for packaging AAV vectors involves co-transfection of an ITR-flanked vector-containing plasmid and a *rep-cap* expression plasmid into adenoviral-infected 293 cells. A cloned duplex forms containing ITRs and results in the production of the single-stranded DNA genome. *Rep* and *cap* genes are expressed from a packaging plasmid not containing ITRs and thus cannot replicate or be packaged into a viral particle.

Wild-type AAV integrates within a specific region of several hundred nucleotides on chromosome 19. AAV vectors do not integrate specifically because they do not express the Rep protein. Upon integration, the viral termini are extremely heterogeneous, and significant deletions are common. AAV vectors can also integrate as a tandem head-to-tail array. Episomal forms of AAV have been found after up to 10 passages.

The production of large quantities of AAV vector for clinical use has been problematic. Large-scale preparation of the ITR-containing plasmids in bacteria is difficult since the palindromic sequences are subject to deletion. The toxicity of the

(a)
    ITR   ITR

(b)
    ITR   ITR

(c)

FIGURE 4.7   Adenovirus-associated virus (AAV) vectors. (*a*) Wild-type AAV. AAV contain a single-stranded DNA genome of 4.7 kb. The inverted terminal repeats (ITRs) are necessary for conversion of the single-stranded genome to double-stranded DNA, packaging, and for integration into the chromosome. The protein products of the *rep* and *cap* genes are necessary for replicating the AAV genome and for producing an AAV particle. (*b*) AAV Vector. AAV vectors have deleted the AAV coding sequences and replaced them with a promoter and therapeutic gene. They still contain the ITRs which are necessary for the vector to transmit its genetic information into a target cell. (*c*) Packaging Cells. The AAV vector alone cannot produce an AAV particle because the *rep* and *cap* genes are not present. These AAV genes need to be present in a packaging cell line along with the AAV vector in order to produce an AAV particle that can transfer genetic information into a cell. In addition, another virus such as an adenovirus needs to be present for the production of infectious particles.

Rep proteins limits the generation of stable mammalian packaging lines that can be used to propagate the vector. To produce AAV vectors, most investigators have used transient transfections with two plasmids in combination with infection with an adenoviral vector. However, the number of recombinant AAV vector particles produced by packaging cells is lower than the amount of wild-type AAV that can be produced. The lack of production may reflect the fact that Rep and Cap proteins are limiting since their plasmid does not contain ITRs and is not amplified. After recombinant AAV particles are produced, they must be separated from adenovirus and cellular components for the isolation of a nontoxic vector. Methods for separation of AAV vector from adenovirus include heat inactivation of adenovirus, $CsCl_2$ banding, and ion-exchange chromatography. AAV vector preparations are stable to freezing and must be tested for wild-type AAV, adenovirus, and other pathogens prior to use.

Use of AAV Vectors for Gene Therapy

A major advantageous characteristic of AAV vectors is their ability to transduce nondividing cells. AAV vectors have been used to transfer genes into a variety of cell types including hematopoietic stem cells in vitro and hepatocytes, brain, retina, lung, skeletal, and cardiac muscle in vivo. Stable expression has been observed for up to one year in several organs. It is not yet clear if the AAV vectors integrate into the host cell chromosome or are maintained episomally. Studies in a variety of animal models indicate that AAV-transduced cells do not elicit an inflammatory reaction or a cytotoxic immune response.

Some studies have suggested that AAV transduction efficiency increases when cells are replicating, or treated with cytotoxic agents, or co-infected with an adenoviral vector. However, such procedures did not increase the copy number of the AAV vector in experimental studies. The data indicate the techniques increase the number of cells that expressed the reporter gene through activation of the viral promoter of the AAV vector rather than increasing the transfer of genetic material into the cells.

Little information is available regarding the level of expression per copy from an AAV vector in various cell types in vivo. ITRs have transcriptional activity and have been utilized to direct expression of the cystic fibrosis transmembrane receptor. Most AAV vectors utilize an internal promoter to direct expression of the therapeutic gene. The CMV promoter functions at levels sufficient to produce detectable protein product in muscle and brain. But it is poorly functional in the liver in vivo. Use of the LTR promoter from the MFG retroviral vector resulted in a high-level expression in the liver. However, an LTR promoter in another context was much less active. It is possible that the presence of a splice site in the MFG-derived vector accounts for this discrepancy. These studies indicate that it will be necessary to empirically test different constructs in vivo for their relative efficacy.

It is possible that residual AAV sequences will not have the inhibitory effect that occurs for some internal promoters of a retroviral vector. However, expression from an internal promoter of an AAV vector can attenuate in vitro by a process that involves histone deacetylation. In addition, the ITRs have transcriptional activity and may be subject to inhibitory factors. Recently, a protein has been identified as the single-stranded D-sequence-binding protein whose phosphorylation and ITR-binding activity is modulated by the cell cycle. Binding of the phosphorylated protein to the ITR inhibited replication of the DNA and might influence transcription. It is therefore possible that AAV vector sequences will attenuate expression from some internal promoters.

A noteworthy feature of AAV vectors is the slow increase of gene expression over several weeks after delivery to the liver or the muscle. Such an increase may represent conversion of the single-stranded DNA genome to double-stranded DNA. The DNA is maintained as a concatemer in an episomal or integrated state. Expression has been stable for up to one year in liver and muscle, implying that the DNA is not lost or inactive. Longer-term evaluation and determination of the status of the DNA is required in future studies.

Risks of AAV Vectors

There are three potential risks of AAV vectors: (1) insertional mutagenesis, (2) generation of wild-type AAV, and (3) administration of contaminating adenovirus. It is theoretically possible that AAV vectors could activate a proto-oncogene or inactivate a tumor suppressor gene by integration into the chromosome in vivo. However, AAV vectors have not been reported to result in malignancy. Wild-type AAV could be produced when recombination between the vector and the packaging plasmid occurred. However, since AAV has not been shown to be pathogenic and is not capable of efficient replication in the absence of a helper virus, the generation of wild-type AAV may not be a serious concern in human gene therapy. A final potential problem is a helper virus contaminating preparations of AAV vector and causing adverse effects. Careful testing of AAV vectors for the presence of the helper virus would reduce this risk. It therefore appears that AAV vectors can be considered relatively safe, although further long-term studies in animals are necessary.

Summary: AAV Vectors

AAV vectors can be generated by removing viral genes and replacing them with a promoter and therapeutic gene. They can be produced in cells expressing the AAV *rep* and *cap* genes and that have been co-infected with a helper virus such as adenovirus. Production of large amounts of AAV vector is difficult. The major advantage of AAV in gene therapy is the ability to transfer genetic information into nondividing cells in vivo. In addition, expression has been maintain for up to one year. Further experiments to determine if AAV vectors integrate or are maintained in an episomal state are necessary. A current major limitation of AAV vectors is that they cannot accommodate more than 4.5 kb of exogenous genetic material.

HERPES SIMPLEX VIRUS 1

Herpes simplex virus 1 (HSV-1) has a 152-kb double-stranded linear DNA genome that can be maintained episomally in the nucleus of cells. It can cause mucocutaneous lesions of the mouth face, and eyes and can spread to the nervous system and cause meningitis or encephalitis. The related HSV-2 can cause lesions in the genitalia. HSV can establish a lifelong latent infection in neurons without integrating into the host cell chromosome.

The HSV-1 virion is enveloped and ~110 nm in diameter. Viral infection is initiated in epithelial cells of the skin or mucosal membranes by binding of the viral envelope glycoproteins to heparin sulfate moieties on the plasma membrane. Specific attachment can be mediated by a novel member of the tumor necrosis factor (TNF)/nerve growth factor (NGF) receptor family, which triggers fusion of the virus envelope with the plasma membrane. After the initial rounds of replication, the virus is taken up into the axon terminals of neurons innervating the site of primary infection. The viral capsid is transported to the nucleus via a process that probably involves the cytoskeleton. For neurons, this process results in the retrograde transport of viral particles long distances within the axon. Upon entering a cell, the virus can enter a lytic cycle, resulting in cell death within 10 h, or can enter a latent phase in the nucleus.

HSV Genes

The viral genome consists of a long and short unique region, designated U_L and U_S, respectively, each flanked by inverted repeats. Transcription of early genes is initiated by VP16, a potent transcription factor present in the virion. These early gene products lead to replication of the viral DNA, followed by expression of the late genes. HSV-1 has at least 81 gene products, 43 of which are not essential for replication in vitro but contribute to the virus life cycle in vivo. During the latent state, however, no HSV proteins are detected. Instead, a family of RNAs, the latency-associated transcripts (LAT), are present in the nucleus. The roles of these transcripts are unknown. The virus can establish latency without the LATs.

Sequences Required in cis for Replication

HSV-1 contains three origins of replication (see Chapter 2). One is located in the middle of the U_L region (OriL) and two are within the inverted repeats flanking the U_S region (OriS). Only one replication origin needs to be present on a circular piece of double-stranded DNA to support replication. The viral DNA is packaged via the packaging signal, a sequence which is located in the genomic termini. Both origin of replication and the packaging signal are sufficient to allow a circular piece of DNA to be replicated and to be packaged by cells that express the remainder of the essential HSV-1 proteins in trans.

Use of HSV Sequences for Gene Transfer

Most vectors based upon HSV-1 have deleted one or more genes necessary for replication. Genes coding for proteins necessary for replication such as infected cell polypeptide (ICP)4 can be deleted. HSV-1 particles are produced in cells that express these proteins in trans. HSV-1 vectors can accommodate up to 25 kb of foreign DNA sequences and can establish latency. However, these viral vectors are toxic for some cells in vitro and can cause encephalitis when administered to the brain at high doses.

An alternative type of HSV-1-based vector is an amplicon. Amplicons contain bacterial and HSV origins of replication, as well as the packaging sequence. If an amplicon is present in cells that also contain wild-type HSV, the amplicon will be replicated along with the wild-type virus and then packaged into viral capsids. It is difficult, however, to separate the amplicon from the wild-type virus. One approach to circumvent this problem is to co-transfect cells with an amplicon and a series of cosmids that contained all the HSV-1 coding sequences, except for the packaging signals. Amplicons have been used to express genes for up to 1 month in the brain (see Chapter 9).

The insertion of a therapeutic gene into HSV-1 vectors requires homologous recombination, using procedures that are similar to those described for adenoviral vectors. HSV-1 vectors that have deleted HSV genes are produced in cell lines that express the deficient protein in trans. HSV-1 amplicons genes are expressed in cells that are co-infected with a replication-competent HSV-1 genome or have the HSV-1 genes introduced on multiple cosmids.

Use of HSV Vectors for Gene Therapy

HSV vectors have been used to transfer genes into the brain, spinal cord, and muscle but have not been used in humans for gene therapy. Delivery into the central nervous system has utilized stereotactic injection. Transduced cells have been limited to a relatively small region because the virus does not readily diffuse. Delivery of HSV-1-based vectors to the muscle has resulted in only short-term expression due to the cytopathic effects and/or the immune response to the residual HSV-1 proteins. These results with HSV-1 in muscle are similar to what has been observed with the adenoviral vectors in many organs.

Expression from an HSV Vector in vivo

A number of promoters are active in vivo when lytic infection occurs. However, stable expression during latency from an HSV-1-based vector has only been detected in the brain using the LAT promoter. A variety of others including viral, RNA polymerase III-activated, housekeeping, and neuronal promoters are shut down in vivo. The LTR, LAP, or a neuronal-specific promoter have resulted in stable expression in dorsal root ganglion neurons of the spinal cord.

Risks of HSV Vectors

There are two major risks of HSV-1-based vectors: (1) toxicity due to the cytopathic effect of relatively unattenuated virus and (2) the development of wild-type virus. Administration of high doses of HSV-1 vectors with only a single gene deleted had a considerable cytopathic effect. HSV-1 vectors with deletion of four genes had less toxicity. The development of wild-type HSV-1, which can cause serious infections such as encephalitis, is a concern.

Summary: HSV Vectors

HSV-1 vectors can be generated by deleting genes that are essential for replication, inserting a therapeutic gene into a nonessential region, and transferring the DNA into cells that supply the essential HSV-1 protein(s) in trans. HSV-1 amplicons can be generated by placing the therapeutic gene on a plasmid with the HSV-1 origin of replication and packaging signal and transferring the DNA into a cell along with the essential wild-type HSV-1 genes. HSV-1 vectors have resulted in stable expression in the brain with the LAT promoter. Toxicity due to the HSV-1 vector and the generation of wild-type virus are a concern.

OTHER VIRAL VECTORS

There are other viruses that have been used as vehicles for gene transfer. The baculovirus is an 80- to 230-kb double-stranded circular DNA virus that replicates in insect cells in vitro or in vivo. The vaccinia virus is a 191-kb double-stranded DNA that was used in the past to vaccinate humans against the related smallpox virus. It has over 198 open reading frames and ~50 kb of the genome is not essential for replication in vitro. Genes can be inserted into a vaccinia genome by homologous recombination. Recombinant vaccinia has been used for immunization against proteins that play an important role in the pathogenesis of encephalitis, rabies, and other infectious diseases. It has also been used to express cytokine genes in animals in an attempt to boost the immune response to a cancer. An advantage of vaccinia-derived vectors is their ability to accommodate a large amount of exogenous genetic material. Disadvantages include the fact that an immune response to the vector will preclude gene transfer in some patients and will limit the duration of gene expression.

Baculoviruses can transfer genetic information into hepatocytes but do not express the baculoviral genes, which require transcription factors that are only present in insect cells. A mammalian promoter and gene of interest can be expressed, however. Baculoviral vectors have been used to express genes in hepatocytes in vitro and have been delivered to intact livers using an isolated perfusion system. Advantages of baculoviral vectors include the ability to accept large amounts of genetic material and the absence of expression of baculoviral proteins in mammalian cells. Disadvantages include the transient gene expression and the sensitivity of the vector to complement.

NONVIRAL VECTORS

Nonviral vectors include any method of gene transfer that does not involve production of a viral particle. They can be divided into two classes: (1) RNA or DNA that can be amplified in bacteria or eukaryotic cells, and whose transfer into a cell does not involve a viral particle, or (2) oligodeoxynucleotides or related molecules synthesized chemically. Nonviral vectors amplified in cells generally encode a gene that expresses the therapeutic protein, although they can encode antisense RNA that acts by blocking expression of an endogenous gene. Oligonucleotides act by altering expression of endogenous genes in cells by a variety of mechanisms (see Figure 4.8).

There are three important factors regarding nonviral vectors that can be amplified in prokaryotic or eukaryotic cells for gene therapy: (1) the size of the insert accepted, (2) how to get the genetic material into cells efficiently, and (3) how to maintain the genetic material inside the cell in order to achieve long-term expression.

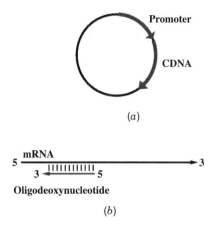

(a)

mRNA
5 ——————————————————→ 3
3 ←|||||||||| 5
Oligodeoxynucleotide

(b)

FIGURE 4.8   Nonviral vectors for gene therapy. Nonviral vectors are any type of vector that does not involve a viral particle that can alter gene expression in a cell. (*a*) Plasmid DNA. Plasmids are double-stranded circles of DNA that replicate efficiently in bacteria. They can contain up to 15 kb of exogenous genetic information. They generally contain a promoter and coding sequence that results in expression of a therapeutic protein. Although plasmid DNA does not enter cells efficiently because of its large size, cationic liposomes, or receptor-mediated targeting can be used to facilitate its entry into cells. (*b*) Oligonucleotide vectors. Oligonucleotides, or more stable analogs such as phosphorothioates, contain 10 to 25 bases. An oligonucleotide is shown hybridized with RNA in this panel, which can affect the processing, translation, or stability of the RNA. Oligonucleotides can also form a triple helix with DNA and alter transcription, or serve as a decoy by binding to transcription factors and prevent them from binding to endogenous genes.

Size of Insert

The size of the insert accepted varies considerably among the different nonviral vectors that replicate in cells. Bacteria can amplify plasmids, bacteriophage, cosmids, or bacterial artificial chromosomes. All can be purified from cells as nucleic acid devoid of proteins. Plasmids are double-stranded circular DNA molecules that contain a bacterial origin of replication. They can accommodate up to 15 kb of exogenous genetic information. Bacteriophage is a double-stranded linear DNA virus that can accommodate up to 20 kb of foreign DNA. Cosmids are modified plasmids that carry a copy of the DNA sequences needed for packaging the DNA into a bacteriophage particle. They can accommodate up to 45 kb of genetic information. Bacterial artificial chromosomes (BACs) contain elements from a normal chromosome that allow it to replicate and to segregate appropriately and can accommodate up to 100 kb of exogenous genetic material. Yeast artificial chromosomes (YACs) contain telomeres, replication origins, and sequences that ensure appropriate segregation in yeast cells. They can accommodate up to 1000 kb of exogenous genetic material. They do not replicate in mammalian cells. More recently, the production of a human artificial minichromosome was reported, although its transfer into cells was very inefficient. The most successful use of artificial chromosomes is the recent report of the generation of transgenic mice (Chapter 3) via germline transmission of a mammalian artificial chromosome using nuclear microinjection (Chapter 2). Thus, artificial chromosomes could theoretically be used for gene therapy. To date, most studies have used plasmid DNA for gene transfer using nonviral vectors because they are easily amplified to a high copy number in bacteria, and their smaller size makes them easier to insert into cells.

Transfer of Nonviral Vectors into Cells

A major problem of nonviral vectors is the difficulty to efficiently transfer the highly charged DNA molecule into a cell. Transfer of nonviral vectors into cells can be performed ex vivo or in vivo. For ex vivo transfer, genes are usually transferred into the cell by using calcium phosphate co-precipitation, electroporation, cationic lipids, or liposomes. For most cell types, 5 to 10% of the cells can be modified, and transfected cells can often be selected by virtue of a selectable marker that is also present on the piece of DNA. Larger pieces of DNA are transferred less efficiently than smaller pieces of DNA.

Efficient in vivo transfer is somewhat more difficult to achieve than ex vivo gene transfer. Many investigators have utilized liposomes, cationic lipids, or anionic lipids that promote entry of the DNA into the cell. A variety of such molecules have been synthesized. Another effective method for promoting entry into the cell is to complex the DNA with an inactivated viral particle containing plasma membrane fusions proteins. For example, association of DNA with heat-inactivated Sendai virus [also known as the hemagglutinating virus of Japan (HVJ)] dramatically increases the expression of the DNA in vivo. Similarly, inactivated adenovirus greatly potentiates the entry of DNA into a cell. A third approach is to attach the DNA to a small particle delivered to the inside of the cell using a ballistic device referred to as a DNA gun (see Chapter 5).

Selective delivery (targeting) of a nonviral vector to a specific organ or cell type would be desirable for some applications. For example, DNA has been targeted to the asialoglycoprotein receptor of hepatocytes by complexing the DNA with polylysine-conjugated asialoglycoprotein or targeted for cells that express a transferrin or folate receptor (see Chapter 7).

Stabilization of Nonviral Vectors in Cells

A major problem with nonviral vectors is transient gene expression, since the genetic material transferred into the cell is unstable. Methods for stabilizing the DNA in the cell would prolong the clinical effect in vivo. Some investigators have placed origins of replication derived from viruses into nonviral vectors. Plasmids must be engineered to express any proteins necessary to activate the origin of replication. The human papilloma virus (HPV) E1 protein supports replication of the HPV origin of replication, while the Ebstein–Barr virus nuclear antigen 1 (EBNA-1) supports replication of an EBV origin of replication. Plasmids containing these replication origins and relevent appropriate proteins activating origins are maintained longer in cells in vitro and in vivo than plasmids that do not contain these sequences. Artificial chromosomes have elements that stabilize genetic material in a cell and should not have problems of instability. If difficulties in amplifying and transferring artificial chromosomes into cells can be overcome, such vectors should be maintained stably in a cell.

Use on Nonviral Vectors for Gene Therapy

Plasmid DNA has been delivered into muscle in vivo as naked DNA, into a variety of organs complexed with cationic lipids, with HJV liposomes, or by using a DNA gun. Expression has been detected in several organs, although it is usually both transient and at a relatively low level because the DNA is not stable in cells. There is little quantitative data regarding the efficacy of expression from different promoters in vivo. Gene therapy with plasmid vectors has been used to attempt to treat cystic fibrosis (see Chapter 3) and cancer in humans (see Chapter 10).

Risks of Nonviral Vectors for Gene Therapy

There are two major risks of using nonviral vectors for gene therapy: (1) insertional mutagenesis could activate oncogenes or inhibit tumor suppressor genes if the plasmid integrates and (2) the compounds that are used to facilitate the entry of DNA into a cell might have some toxicity. A major advantage of using nonviral vectors is the lack of risk of generating a wild-type virus via recombination. In addition, episomal plasmids do not pose the risk of insertional mutagenesis since they do not integrate into the chromosome. However, some plasmids can integrate into the genome particularly when a procedure is used to select clones exhibiting long-term expression. This is often done with ex vivo gene therapy procedures. Indeed, transplantation of myoblasts transfected with a plasmid DNA and selected in vitro has led to the development of tumors in the muscle. It, therefore, appears that selection of cells with an integrated plasmid vector poses some risks in animals, although maintenance of episomal DNA should be relatively safe. A second potential risk for nonviral vectors is that certain compounds can facilitate entry into a cell and exert a toxic effect in vivo. For example, many cationic lipids have considerable toxicity when administered at high doses to cells in vitro. These could be toxic at high doses in vivo as well. It will be necessary to assess the toxicity of such compounds carefully in vivo.

Summary: Nonviral Vectors

Nonviral vectors can be amplified to high copy numbers in bacterial cells as well as readily engineered to express a therapeutic gene from a mammalian promoter. These plasmids can be efficiently introduced into cells ex vivo and introduced somewhat less efficiently into cells in vivo. Their major advantages are the ease of production and that they cannot recombine to generate replication-competent virus. They can, however, integrate at a low frequency into the chromosome and, therefore, do pose some risk of insertional mutagenesis. A major disadvantage is the transient nature of gene expression that is observed.

OLIGONUCLEOTIDES

The second major class of nonviral vectors are oligodeoxynucleotides and related polymers of nucleotides that have different backbones. Oligodeoxynucleotides are 15 to 25 nt long pieces of DNA that can modulate gene expression in cells in a variety of ways including: (1) formation of triplex DNA, (2) acting as an antisense molecule to block processing or expression of mRNA or to promote its degradation, and (3) forming a transcription factor binding site that serves as a decoy.

Triplex DNA is the colinear association of three deoxynucleotides strands and usually involves binding of an oligodeoxynucleotide in the major groove of a DNA double helix. This binding can block access of transcription factors, thus inhibiting transcription of a gene. The triplex-forming oligodeoxynucleotide binds to the purine-rich strand of the double helix via Hoogsteen hydrogen bonds. Potential target sites for triplex formation are limited to regions that contain homopurine on one strand. The relatively weak binding affinity and the instability of oligodeoxynucleotides in cells results in a transient effect.

A second mechanism by which oligodeoxynucleotides alter gene expression involves binding to an mRNA via standard Watson–Crick base pairing. This can block splicing by binding to a pre-mRNA splice signal or block translational initiation by binding to the 5′ Cap region or the translational initiation codon region. They can also result in degradation of the mRNA by RNase H, an enzyme that degrades the RNA portion of an RNA:DNA hybrid. A third mechanism by which oligodeoxynucleotides can alter gene expression is to bind transcription factors, which prevents them from associating with endogenous genes.

Natural antisense oligodeoxynucleotides consist of phosphodiester oligomers, are sensitive to nucleases, and have a half-life in serum of 15 to 60 min. Modifications to the backbone have increased the stability of oligonucleotides to allow a prolonged biological effect on targeted cells in vivo. Substitution of a nonbridge oxygen in the phosphodiester backbone with a sulfur molecule results in phosphorothioate nucleotides, which are resistant to nucleases. Substitution of a nonbridge oxygen with a methyl group results in methylphosphonate nucleotides. These are also resistant to nucleases, although they do not allow RNase H to act upon hybridized RNA. Peptide nucleic acids have an achiral amide-linked backbone homologous to the phosphodiester backbone that can form standard Watson–Crick base pairs with RNA. Modified oligonucleotides are stable in culture and serum and have resulted in prolonged biological effects.

For oligonucleotides to exert a biological effect, they must enter the cell. Oligonucleotides appear to enter the cell via receptor-mediated endocytosis. Permeabilization of the cell membrane can potentiate entry. In vivo delivery of oligonucleotides can be increased by HVJ liposome complexes. Improved delivery to cells should result in a biological effect at lower doses.

Use and Safety of Oligonucleotides for Gene Therapy

Oligonucleotides have been administered in vivo for gene therapy. They have successfully inhibited intimal hyperplasia of arteries. Oligonucleotides that served as a decoy for a transcription factor have been used to inhibit proliferation of smooth muscle cells in blood vessels in vivo. Antisense oligonucleotides have blocked expression of oncogenes, slowed replication in cells in vitro, and had a modest but transient effect upon growth of tumor cells in vivo.

The major toxicity of oligonucleotides relates to the administration of large doses to achieve a clinical effect. Administration of high doses of phosphorothioate oligonucleotides resulted in cardiovascular toxicity and death in some primates.

Mechanisms to promote the entry of oligonucleotides into cells should decrease their toxicity. Oligonucleotides are unlikely to have any long-term adverse effects since they do not integrate into the chromosome.

Summary: Oligonucleotides

In summary, oligodeoxynucleotides can be used to alter expression of an endogenous gene by blocking transcription, blocking mRNA processing or translation, potentiating mRNA degradation, or through serving as a decoy for a transcription factor. Modified oligonucleotides can function in a similar fashion and are more stable. Oligonucleotides can alter gene expression in vitro and to a lesser extent in vivo. Their effects are short-lived due to their instability in cells and in blood. Their use for gene therapy will probably be limited to diseases where transient expression is sufficient.

KEY CONCEPTS

- Viral vectors can be produced by removing some or all of the genes that encode viral proteins, and replacing them with a therapeutic gene. These vectors are produced by cells that also express any proteins that are necessary for producing a viral particle. A risk of all viral vectors is that they might recombine to generate replication-competent virus that could cause disease in humans.

- Nonviral vectors are plasmids that can be propagated in bacteria or oligonucleotides that can be synthesized chemically. Plasmids can transfer a therapeutic gene into a cell, while oligonucleotides inhibit the expression of endogenous genes. Transfer of nonviral vectors into cells is inefficient and the effect is generally transient. These vectors do not carry the risk of recombining to generate wild-type virus.

- Retroviral vectors are devoid of any retroviral genes and result in long-term expression due to their ability to integrate into the chromosome. Their major disadvantage is the fact that they only transduce dividing cells. Recently developed lentiviral vectors do transduce nondividing cells, but there are concerns regarding the safety of these vectors.

- Adenoviral vectors generally contain many adenoviral genes, although "gutless" vectors in which all coding sequences have been deleted have been developed recently. Adenoviral vectors transduce nonreplicating cells very efficiently, although expression is short-lived. This transient expression is primarily due to the immune response to residual adenoviral genes or the transgene in early generation vectors and may be due to the deletion of sequences that stabilize the DNA in cells for the gutless vectors.

- AAV vectors are devoid of any AAV genes and can transduce nondividing cells. They have resulted in long-term expression, although it is unclear if they remain episomal or integrate into the chromosome in nondividing cells. Production of large amounts of AAV vector is problematic.

SUGGESTED READINGS

Adenovirus

Armentano D, Zabner J, Sacks C, Sookdeo CC, Smith MP, St. George JA, Wadsworth SC, Smith AE, Gregory RJ. Effect of the E4 region on the persistence of transgene expression from adenovirus vectors. J Virol 71:2408–2416, 1997.

Christ M, Lusky M, Stoeckel F, Dreyer D, Dieterle A, Michou AI, Pavirani A, Mehtali M. Gene therapy with recombinant adenovirus vectors: Evaluation of the immune response. Immunol Lett 57:19–25, 1997.

Ilan Y, Droguett G, Chowdhury NR, Li Y, Sengupta K, Thummala NR, Davidson A, Chowdhury JR, Horwitz MS. Insertion of the adenoviral E3 region into a recombinant viral vector prevents antiviral humoral and cellular immune responses and permits long-term gene expression. Proc Natl Acad Sci USA 94:2587–2592, 1997.

Kiwaki K, Kanegae Y, Saito I, Komaki S, Nakamura K, Miyazaki JI, Endo F, Matsuda I. Correction of ornithine transcarbamylase deficiency in adult spf(ash) mice and in OTC-deficient human hepatocytes with recombinant adenoviruses bearing the CAG promoter. Hum Gene Therapy 7(7):821–830, 1996.

Adeno-Associated Virus

Qing KY, Wang XS, Kube DM, Ponnazhagen S, Bajpai A, Srivastava A. Role of tyrosine phosphorylation of a cellular protein in adeno-associated virus 2-mediated transgene expression. Proc Natl Acad Sci USA 94:10879–10884, 1997.

Snyder RO, Miao C, Patijn GA, Spratt SK, Danos O, Nagy D, Gown AM, Winther B, Meuse L, Cohen LK, Thompson AR, Kay MA. Persistent and therapeutic concentrations of human factor IX in mice after hepatic gene transfer of recombinant AAV vectors. Nat Genet 16:270–275, 1997.

Epstein-Barr Virus

Kieff E. Epstein-Barr virus and its replication. In Fields BN, Knipe DM, Howley PM (Eds.), Fundamentals of Virology, 3rd ed. Lippincott-Raven, New York, 1996.

Herpes Simplex Virus

Glorioso JC, DeLuca NA, Fink DJ. Development and application of herpes simplex virus vectors for human gene therapy. Annu Rev Microbiol 49:675–710, 1995.

Huard J, Krisky D, Oligini T, Marconi P, Day CS, Watkins SC, Glorioso JC. Gene transfer to muscle using herpes simplex virus-based vectors. Neuromusc Disord 7:299–313, 1997.

Lachmann RH, Efstathiou S. The use of herpes simplex virus-based vectors for gene delivery to the nervous system. Mol Med Today 3:404–411, 1997.

Lentivirus Vectors

Kafri T, Blomer U, Peterson DA, Gage FH, Verma IM. Sustained expression of genes delivered directly into liver and muscle by lentiviral vectors. Nat Genet 17:314–317, 1997.

Naldini L, Blomer U, Gallay P, Ory D, Mulligan R, Gage FH, Verma IM, Trono D. In vivo gene delivery and stable transduction of nondividing cells by a lentiviral vector. Science 272:263–267, 1996.

Zufferey R, Nagy D, Mandel RJ, Naldini L, Trono D. Multiply attenuated lentiviral vector achieves efficient gene delivery in vivo. Nat Biotech 15:871–875, 1997.

Baculovirus Vectors

Sandig V, Hofmann C, Steinert S, Jennings G, Schlag P, Strauss M. Gene transfer into hepatocytes and human liver tissue by baculovirus vectors. Hum Gene Therapy 7:1937–1945, 1996.

Oligonucleotides

Scanlon KJ, Ohtat Y, Ishida H, Kijima H, Ohkawa T, Kaminshi A, Tsai J, Horng G, Kashani-Sabet M. Oligonucleotide-mediated modulation of mammalian gene expression. FASEB J 9:1288–1296, 1995.

Wolff JA. Naked DNA transport and expression in mammalian cells. Neuromusc Disord 7:314–318, 1997.

Gene Therapy and Transfer

Bohl D, Naffakh N, Heard JM. Long-term control of erythropoietin secretion by doxycycline in mice transplanted with engineered primary myoblasts. Nat Med 3:299–305, 1997.

Burns KI. Parvoviridae: The viruses and their replication. In Fields BN, Knipe DM, Howley PM (Eds.), Fundamentals of Virology, 3rd ed. Lippincott-Raven, New York, 1996.

Chen WY, Bailey EC, McCune SL, Dong JY, Townes TM. Reactivation of silenced, virally transduced genes by inhibitors of histone deacetylase. Proc Natl Acad Sci USA 94:5798–5803, 1997.

Kay MA, Liu D, Hoogerbrugge PM. Gene therapy. Proc Nat Acad Sci USA 94:12747–12748, 1997.

Kessler PD, Podsakoff GM, Chen X, McQuiston SA, Colosi PC, Matelis LA, Kurtzman GJ, Byrne BJ. Gene delivery to skeletal muscle results in sustained expression and systemic delivery of a therapeutic protein. Proc Natl Acad Sci USA 93:14082–14087, 1996.

Lee RJ, Huang L. Lipidic vector systems for gene transfer. Crit Rev Therapeut Drug Carrier Sys 14:173–206, 1997.

Limbach KJ, Paoletti E. Non-replicating expression vectors: Applications in vaccine development and gene therapy. Epidemiol Infect 116:241–256, 1996.

Smith AE. Viral vectors in gene therapy. Annu Rev Microbiol 49:807–838, 1995.

Artificial Chromosomes

Co DO, Borowski AH, Leung JD et al. Generation of transgenic mice and germline transmission of mammalian artificial chromosome introduced into embryos by pronuclear microinjection. Chrom Res 8:183–191, 2000.

Harrington JJ, van Bokkelen G, Mays RW, Gustashaw K, Williard H. Formation of de novo centromeres and construction of first-generation human artificial minichromosomes. Nat Genet 15:345–355, 1997.

Kumar-Singh R, Chamberlain JS. Encapsidated adenovirus minichromosomes allow delivery and expression of a 14 kb dystrophin cDNA to muscle cells. Hum Mol Genet 5:913–921, 1996.

CHAPTER 5

Gene Targeting

ERIC KMIEC, PH.D.

BACKGROUND AND CHALLENGES

The availability of cloned genes and deoxyribonucleic acid (DNA) sequences, combined with the ability to transfer and express genes in mammalian cells has revolutionized biology. Already, therapeutic proteins like tissue plasminogen activator (TPA), erythropoietin (EPO), and interferon (IF) have helped thousands of patients realize the benefits of molecular medicine. Recent progress in this field has raised the expectation that genes may be used as therapeutic agents. Such approaches, which rely either on purified proteins or genes, are additive, that is, the defective gene (or gene product) is supplemented by the therapeutic drug while the defective gene and its products are ignored.

The "gene addition" approach, however, is plagued by a variety of problems. The most damaging of these limitations is the inability to control the expression of the newly added gene, due, in part, to the lack of precision in locating the new gene within the genome. The vast expanse of chromosomal space includes many regions that are inhospitable for foreign genes. In these "barren" regions of the genome, the transgene is subject to silencing or extinction. The application of modern gene expression technology employing enhancers, insulators, and locus control regions (LCRs) has helped improve the fate of a randomly inserted gene, but success is still sporadic and expression variable.

An obvious solution to these problems is to attempt to direct or target the transgene toward a specific site in the genome. This simple concept was contemplated several decades ago but was considered unattainable until the early 1980s. Once a recombinogenic transgene localizes to the nucleus, its likely fate is to integrate randomly. Two factors influence this outcome: the recombinogenic termini of the DNA fragment promotes insertion at any available site of entry in the genome (often via breaks in the double strands of the DNA molecule) and the ratio of specific to nonspecific site integration.

Early experiments in human cells suggested that homologous recombination (site-specific integration) was feasible but rare. (In contrast, yeast, specifically

An Introduction to Molecular Medicine and Gene Therapy, Edited by Thomas F. Kresina
ISBN 0-471-39188-3 © 2001 Wiley-Liss

Saccharomyces cerevisiae, is quite proficient in targeted integration.) Attempts at mammalian gene targeting employed a strategy where rare homologous recombination events could be selected from a background of random insertion. In 1985, using the human β-globin locus as a target, Dr. Oliver Smithies and colleagues demonstrated that a targeting event between chromosomal DNA and a transfected construct could be identified at a frequency of 1 in 10^3 to 10^4 selected cells. This technology has now been considerably enhanced and applied to over 300 different genes in murine and human cells. This advance, though heartening for a variety of research applications, has not resulted in a significant improvement of the actual frequency of gene conversion. Low frequency (i.e., where less than 1 cell in 1000 undergoes the targeting event) dictates the need for selection strategies and prevents the direct application of the technology to therapeutic use. However, these studies have helped demonstrate that mammalian cells possess the enzymatic machinery needed to catalyze gene conversion between newly introduced DNA and the genome. Deficiency in one or more rate-limiting steps must be responsible for the inefficiency of targeting. Some obvious barriers to high-efficiency targeting in mammalian cells include the condensed structure of the chromatin, the complexity of genomic DNA sequences, and the relative instability of DNA hybrids mismatched at one or more base pairs.

Since over 2000 human diseases have been mapped at the level of their genetic defects and most of them are caused by mutations in the coding regions of a single gene, the most elegant solution is to repair the gene in situ, that is, correct the defect in a living cell either by repairing a nucleotide mutation or by replacing the entire gene. The reality of that challenge, however, has intimidated workers and hindered progress.

INTRODUCTION OF DNA INTO THE CELL

Before these challenges are even addressed, it is imperative to consider how to introduce foreign DNA into a cell. This process is widely described as "gene transfer," but as with many terms in modern science, it is overused and often abused. For the current purposes, gene transfer simply means the introduction of foreign DNA or ribonucleic acid (RNA) into a targeted cell. Once the DNA has entered the cell it can take many routes, but three are most likely (Fig. 5.1). First, it may be destroyed by cellular enzymes known as nucleases whose normal functions center around DNA recombination and repair. Second, the DNA may be kept in the nucleus or cytoplasm where it survives in an episomal state (extra-chromosomal). Finally, it may integrate into the host cell's chromosome and become a stable, permanent, or in rare cases, an unstable part of the genome.

The first of these possible outcomes often occurs when the DNA is mixed with the cells directly or the molecular form is linear. The termini of each molecule are attractive substrates for nucleases, and their action may lead to complete degradation. Alternatively, the combined action of nucleases and a DNA ligase result in the connection of linear DNA, end-to-end, to form long multimers known as concatamers. Hence the transfer of unprotected DNA in the linear form into cells directly is generally unsuccessful.

To solve some of the problems outlined above, other topological forms of DNA are used, that is, supercoiled or fully relaxed DNA. In this case, the DNA fares better

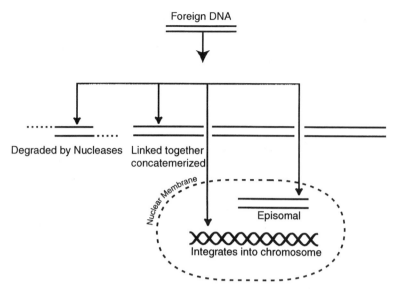

FIGURE 5.1 Fates of foreign DNA entering a mammalian cell. Exogenous DNA may follow several pathways upon entering the cell. First, the molecule may be degraded by nucleases and destroyed. Second, it may be linked together to form long strings of DNA known as comcatemers. Upon entering the nucleus it could remain episomal or become integrated into the chromosome, an event that occurs rarely at the homologous site in the genome.

after mixing with target cells. In fact, many supercoiled plasmids are introduced successfully into cells using a methodology that employs either $CaCl_2/CaPO_4$ or dextran. These two groups of compounds alter the electrophysiological environment of the cell's membrane, reducing the electrostatic repulsion and increasing membrane pore size. Such manipulation permits entry of the DNA into the cells. Although these methods are somewhat labor intensive, they are quite effective and used routinely in many laboratories. More often though, lipid formulations, known as liposomes are used in gene transfer protocols when viral delivery is not an option. The transfer of supercoiled or relaxed DNA into cells by any of these methods results in the DNA becoming episomal more often than integrated. This second outcome of DNA transfer has some advantages in terms of the transient expression of certain foreign genes.

The third possible fate of DNA after entering the cell is to integrate directly into the host chromosome. As mentioned above, DNA packaged in liposomes or mixed with specific compounds can become integrated, but these events require a special "selective pressure," and the frequency of such an event is very low. There is, however, an efficient way to have DNA integrate into the host chromosome that involves the use of viruses as transfer vehicles. Certain viruses insert themselves into a host's chromosomes and become contiguous with the host genome. Retroviruses (RVs) are good examples. The integrative action of retroviral DNA can have significant, yet adverse, effects on the cell since the integration sites are often random. In fact, one of the challenges facing workers in the gene targeting field is to reduce the randomness of retroviral integration while maintaining the explosive infection rate. Random integration events can cause genetic dysfunction by disrupting active genes, and in rare instances random integration may lead to the activation of quiescent genes by positioning a strong viral promoter element adjacent to the coding regions of genes. However, all things considered, the most efficient way to integrate foreign DNA into a chromosome is through the use of a virus.

In summary, most cells are amenable to gene transfer and generally process the DNA (or RNA in rare cases) in three ways. It is often the endpoint the investigator hopes to achieve that dictates which method will be used.

NONVIRAL TRANSFER VEHICLES

Ultimately, the goal of gene targeting centers around the accurate replacement of a mutated gene with a correct version of the gene. The transfer of a normal gene in the perfect situation will, in all likelihood, be carried out by a viral vector where the number of infectious agents and potential of each cell receiving at least one copy of the gene is high. As described above briefly, there is always a limitation on the production levels of biological material and a chance that genetic exchange or recombination events will create a nondesirable or nonusable vector.

An alternative gene transfer strategy employs lipid-based formulations known as liposomes. The development of this strategy has been driven, in large part, by the biotechnology industry. Among the diverse types of liposomes available are those that fuse with the phospholipid bilayer of the cell's membrane and those that can avoid being sequestered in the cytoplasm by pathways that eliminate their effectiveness in gene delivery to the nucleus. Dimethyl sulfoxide (DMSO), dendimers, and polybrene are examples of the types of synthetic reagents that can be used in gene transfer.

With regard to gene targeting, liposomes represent an important option. To transfer foreign genes into a cell using a viral vector, the gene must be inserted into the viral genome, which often requires complicated cloning strategies. By utilizing liposomes, intact plasmid DNA may be transferred into the cell after simply mixing the DNA with the liposome. Hence, many types of DNA molecules that are not amenable to viral vector insertion can be used in gene targeting experiments.

Beyond liposomes, success has been achieved when nucleic acid is introduced using physical force. Two examples of this strategy are particle bombardment and direct DNA injection. The former method usually involves the attachment of plasmid DNA or oligonucleotides onto the surface of 1- to 3-μm gold particles. These particles are accelerated by a gene delivery system (electrical or gas pulse) and sent into the target tissue. The efficiency of transfer, however, is variable and often dependent on the biophysical nature of the membrane. In most cases, however, tissue bombardment does not lead to integrated DNA in the host genome. The latter method centers around the direct injection of material into the tissue by a fine needle or syringe. Again, the introduced DNA does not integrate, remaining episomal. But, the expression of genes on injected plasmids can persist for 60 days, especially in muscle tissue, and cell regeneration activated at the site of injection can improve efficiency of uptake. Although both methods are important experimental systems, where the aim may be an assessment of plasmid construct expression, it is unlikely that a practical use for these approaches in the current gene therapy world will be found. Finally, electroporation of mammalian cells is becoming a standardized and useful technique. Although many cells are killed by the process, careful analyses suggest that electroporation is a better transfer technology than liposomes, at least for some cell types.

GENE TARGETING

The potential now exists in many experimental systems to transfer a cloned, modified gene back into the genome of the host organism. In the ideal situation the cloned gene is returned to its homologous location in the genome and becomes inserted at the target locus. This process is controlled through the action of endogenous recombination functions whose normal activities are to provide a means for repair of DNA damage and to ensure accurate chromosome disjunction during meiosis. The paradigm for thinking about the mechanism of this process has come primarily from two sources: (1) Principles of reaction mechanics have come from detailed biochemical analyses of proteins purified from *Escherichia coli*. (2) Principles of information transfer have been derived from genetic studies carried out in bacteriophage and fungi. A compelling picture of the process of homologous pairing and DNA strand exchange has been influential in directing investigators interested in gene targeting experiments.

Lessons from Bacteria and Yeast

The ability to find and accurately pair DNA molecules enables accurate gene targeting. Biochemically, the overall process can be thought of as a series of steps in a reaction pathway whereby DNA molecules are brought into homologous register, and DNA strands are exchanged. In *E. coli* the pairing reaction is dependent upon a single protein, the product of the *recA* gene. This versatile protein promotes the search for DNA sequence homology, catalyzes the formation of DNA joint molecules, and helps exchange DNA strands. The role of *recA* protein in homologous pairing has been the subject of a great deal of experimentation over the course of the past three decades beginning with the isolation of the *recA* mutant, followed by the cloning of the *recA* gene, the discovery of the DNA pairing activity of the *recA* protein, and the resolution of the *recA* protein crystal structure. Insight into the mechanism of DNA pairing has come from integration of the knowledge provided by experimentation from several laboratories.

Much less is known about the biochemical pathway leading to homologous recombination in most other experimental systems. Nevertheless, in *S. cerevisiae* a great deal of information has accumulated about the molecular events leading to integration of plasmid DNA into homologous sequences within the genome during transformation. Substantial insight into the mechanism of recombination between plasmid DNA and the genome has come from studies using nonreplicating plasmids containing a cloned gene homologous to an endogenous genomic sequence. Transformation of *S. cerevisiae* at high frequency takes place when the plasmid DNA is cut within the cloned DNA sequence. Almost invariably, transformants contain plasmid DNA integrated into the yeast genome at the homologous site. Autonomously replicating plasmids containing gaps of several hundred nucleotide residues within the cloned gene also transform at high efficiency and are repaired by recombination using chromosomal information as a template.

What has emerged from these studies on transformation of *S. cerevisiae* has been a body of observations that has helped shape strategies for gene targeting in higher organisms. Unfortunately, the limited biochemical data available from yeast and the often confusing and sometimes contradictory results from the genetic studies have not provided a thorough foundation for experimentation. It is not completely clear from the transformation studies carried out that information on genetic control of plasmid integration will be generally applicable to higher eukaryotic systems under study by investigators interested in gene targeting.

Transition to Higher Eukaryotes

Recombination between plasmid and chromosome in higher eukaryotes has been exploited in numerous experimental systems where the aim is to inactivate or to replace a gene of interest (Fig. 5.2). In most organisms the usefulness of this process for genetic manipulations is complicated by interference from an alternative illegitimate pathway of recombination that takes place without regard for DNA sequence homology. This process is often viewed as a nuisance by investigators whose priority, generally speaking, is in "knocking out" the gene of interest rather than in understanding the mechanism of the process. Conversely, the virtual absence of this illegitimate pathway of integration in the more genetically amenable systems of yeast and bacteria has precluded investigation into its molecular mechanism. Therefore, strategies for gene targeting have for the most part evolved by the empirical method with only limited guidance from recombination theory or mechanism. It is likely that the failure to achieve high levels of gene targeting in mammalian cells is related directly to the low frequency of homologous recombination. As described above, efforts to overcome this barrier have focused on the development of genetic enrichment methods; but these methods only eliminate nonhomologous events, and they do not improve the frequency of homologous events. Experimental evidence points to the fact that the enzymatic machinery required to catalyze homologous targeting is limiting in mammalian cells. For example, gene

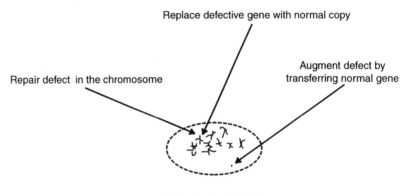

FIGURE 5.2 Strategies of gene targeting. Three prominent options are available in gene targeting. First, one can replace the defective gene. Second, one can add a normal gene into the cell harboring a defective gene. Third, one can repair the defect directly in the chromosome.

conversion events occur with high frequency in avian B cells but not in closely related cells at various stages of B-cell development. Such data lead to the hypothesis that targeting frequencies mammalian cells vary among cell types due to the unpredictable levels of enzymatic components within these cells. It is suspected that gene targeting in mammalian cells is regulated by homologous recombination processes related to DNA repair and that genes known to participate in recombinational repair are likely to be important parts of a specific gene targeting process.

RECOMBINATIONAL AND REPAIR ENZYMES IN GENE TARGETING EFFORTS

A better understanding of DNA repair and recombination mechanisms has been gained recently through the discovery of human homologs of prokaryotic and lower eukaryotic genes known to be involved in these processes. These discoveries provide good examples of how studies in lower organisms impact human biology and contribute to the development of therapeutic strategies. For example, the isolation of the human *MSH2* gene, a gene responsible for major types of human colon carcinoma, arose directly from DNA repair studies conducted in yeast. Homologs of the *recA* protein from yeast to humans have been discovered, although some of these proteins require auxiliary factors for activity and display unique characteristics. This evolution in thinking has arisen from an acquired appreciation for the enzymatic and molecular events surrounding DNA repair and recombination. Clearly, the prototypic organism, *E. coli*, has provided a rich source of enzymes that play critical roles in recombination and in some aspects of DNA repair.

The power of the *recA* protein in promoting homologous recombination in prokaryotes led investigators to outline strategies for gene targeting in other cells based on its activity (Fig. 5.3). By and large, this approach has not proven successful due to the differences between prokaryotic and eukaryotic pathways. Although *recA* protein dominates these events in prokaryotes, it is believed that a complex of proteins, most likely also involved in DNA repair, are required in eukaryotes. There is, however, one approach that does hold promise. The structure of the *recA* protein in absence of DNA appears to contain two disorganized amino acid loops that bind DNA, the essential first step in homologous pairing. If the bound DNA is a synthetic oligonucleotide, a complex is formed that is small enough to transfer into prokaryotic and eukaryotic cells. Further studies using 20 to 30 mer *recA* peptides (4 kD) containing this binding region have some degree of accuracy in positioning the oligonucleotide to its complementary DNA target site in the chromosome. The peptide was found to transport the oligonucleotide to the target site and participate in unstacking the paired bases of the chromosomal DNA.

Most gene targeting experiments use transferred somatic cells such as mouse L cells or Chinese hamster ovary cells. Although useful because of their robustness, the introduction of foreign DNA can often cause unanticipated problems. For example, in cases where nonisogenic DNA is used, existing polymorphisms can lead to DNA mismatches between vector and target and thus stimulate nonhomologous events. Indeed, in cells where homologous recombination events or gene targeting rates increase, a concurrent elevation in nonhomologous (detrimental) events is also seen. One cell line, however, Chicken B cells (DT40) is highly amenable to gene

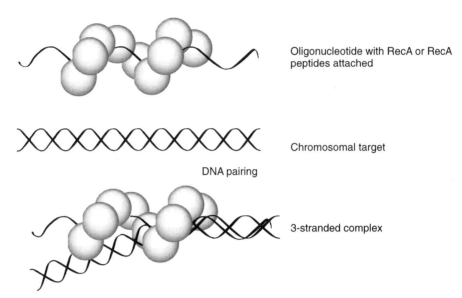

Oligonucleotide with RecA or RecA peptides attached

Chromosomal target

DNA pairing

3-stranded complex

FIGURE 5.3 *RecA* protein-mediated chromosomal targeting. *RecA* protein or a peptide of the *recA* protein bound to an oligonucleotide bearing complementarity to a sequence in the chromosome catalyze DNA paring with the target. The 3-standed complex (triplex) held together by *recA* protein is metastable and eventually the protein dissociates as the third strand anneals to its complement.

targeting reactions. Since the absolute frequency of gene targeting is close to the average (1 to 5×10^{-6}), it is likely that the nonhomologous pathway is suppressed in some fashion. Hence, in one way, actively reducing the rate of nonhomologous recombination may serve to indirectly improve the identification and recovery of correctly targeted cells.

Since the genomic target is part of the targeting equation, it has been suggested that manipulating the allele(s) might improve target frequency. One of the most obvious manipulations is to activate the expression of the gene. Early experiments had shown an effect on absolute frequencies, but subsequent work that took into account the response of the nonhomologous pathway, revealed that no elevation in targeting frequency had actually occurred. Since different genes were targeted, it is plausible that transcription may improve the frequency but may be limited to specific sites in the genome. Other manipulations, such as reagent treatment to loosen chromatin structure, could elevate the number of true events. However, other treatments, such as the addition of sodium butyrate, would change acetylation patterns and thus impact as a generalized effect that may not be beneficial to cell viability and function.

SYNTHETIC OLIGONUCLEOTIDES AS TOOLS FOR TARGETING

The use of synthetic oligonucleotides in recombinase-mediated targeting has been predicated by the natural interaction between proteins like *recA* and single-stranded

DNA, as well as the recombinogenic nature of single strands. As described above, regions of single strandedness within the cell set in motion a cascade of events that include activation of repair genes and recombinational repair events. It is feasible to coat single-stranded DNA fragments with *recA* and introduce them into the cell by electroporation. This dimension, however, has not been examined in detail.

Another application of synthetic oligonucleotides is to chemically modify the molecule so that upon pairing with the target site, the modifier is activated (Fig. 5.4). Such reactivity can lead to an alteration in the target DNA bases and, perhaps, the introduction of a crosslink. Among the most interesting modifications is an alkylation of the target DNA conjugated to chlorambucil, a clinically used nitrogen mustard.

Once paired at the site on the helix, the molecule can alkylate guanine residues nearby, hence inactivating the gene. This is a useful method because it permits accurate quantitation of gene targeting events by ligation-mediated polymerase chain reaction. Once amplified, the targeted gene segment can be electrophoresed on a DNA sequencing gel adjacent to a "G"-ladder, and accurate mapping conducted by single position comparison.

There is, however, a significant limitation to using single-stranded oligonucleotides in targeting: they must often be designed so that they target stretches of homopurimes and/or homopyrimidines. The triplex forming oligonucleotide (TFO) binds the major groove of the duplex segment forming the triple-stranded region. The TFO may bind in a parallel (5′ → 3′) or antiparallel orientation relative to the target strands. Interestingly, purine TFOs form *stable* triplex structures at physiological pH making them useful for gene or promoter ablation strategies. As

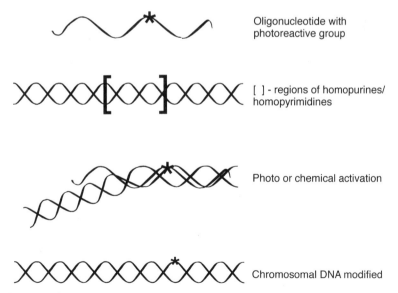

Oligonucleotide with
photoreactive group

[ ] - regions of homopurines/
homopyrimidines

Photo or chemical activation

Chromosomal DNA modified

FIGURE 5.4 Triple helix forming oligonucleotides in chromosomal targeting. Oligonucleotide bearing sequence complementary to the chromosomal target are annealed to the specific site forming a triple helix at regions that are rich in purines or pyrimidines. In some cases, the oligonucleotide may contain a reactive modification that is activated by light. This reaction modifies the target so that block to transcription or replication is blocked.

described above, purine TFOs can be conjugated to DNA damaging agents that are known to stimulate homologous recombination and perhaps gene targeting.

Gene Repair by Novel Oligonucleotides

The sequence constraints placed on TFO effectiveness can be alleviated by the use of oligonucleotides containing a mixture of RNA and DNA residues. Since stretches of RNA can adopt significant secondary structures, these chimeric oligonucleotides are designed in the double-stranded form creating regions of RNA-DNA base-paired hybrid molecules. The ends are capped in a double hairpin conformation to increase stability within the cell and avoid concatemerization reactions that co-join double-stranded (open-ended) DNA fragments after entry into the cell. Hence, the structure is a stable, strong duplex that enters the nucleus efficiently.

These molecules have been shown to catalyze gene targeting by mediating gene conversion events. In mammalian cells, point mutations are converted at a frequency high enough to detect without metabolic selection, and it appears that there is no limitation as to the sites of targeting available to the chimeric oligonucleotides. However, the most important discovery of these molecules comes from their wide-ranging effectiveness in bacterial, plant, and mammalian cells.

The universal application demonstrated by the chimeric oligonucleotide may separate it from other similar approaches, but the mechanism by which it acts is, in all likelihood, similar to TFOs. Due to its intracellular stability, these molecules catalyze gene conversion at a frequency that exceeds most predicted levels. It is not uncommon for bacterial targets to be converted at a rate of 1 to 5%, meaning that 5 cells in 100 receiving the chimera undergo gene conversion. This rate compares favorably with other targeting frequencies, which are often 0.01% or lower, even in bacteria. A simple example using an episomal tetracycline gene as a target serves to illustrate the technique nicely.

A pBR322 plasmid containing a point mutation or single base deletion in the tetracycline (*tet*) gene is transfected into *E. coli* cells containing a wild-type copy of the *recA* gene. A chimeric oligonucleotide designed to mediate the correction is then transferred into the plasmid-containing bacterial cells. After a short recovery in medium containing tetracycline, the cells are grown for 16 h in liquid medium. They are then plated on tetracycline-containing agar plates. The colonies are then "picked," the plasmid DNA isolated, and the targeted nucleotide stretch analyzed by DNA sequencing reactions. This experimental system addresses a series of important questions and concerns of genetic targeting: Is the conversion efficient? Is the conversion stably transmitted to daughter cells and can the genetic change be propagated? Finally, is there a genetic readout and newly functional protein created? The answers to all of these questions is, presumably, "yes," when chimeric oligonucleotides are used in bacterial cells.

INSERTION OF FRAGMENTS OF DNA: GENE DISRUPTION AND REPLACEMENT

Oligonucleotide-based gene targeting may be an effective way to correct single-base mutations or to inactivate genes by inserting or deleting bases. But the true homologous recombination or targeting event wherein a fragment of DNA is integrated into the genome of an organism at the specific, precise site cannot be facilitated by oligonucleotides. Gene targeting in higher eukaryotes using DNA fragments has been tried for many years with varying degrees of success. Early attempts included increasing the length of the homology shared by the fragment and the genomic target. The topology of the targeting vehicle, usually a plasmid construct, was also modified but failed to improve the frequency of targeting specificity.

Genomic Insertion

To keep things in perspective, one must consider naturally occurring events that lead to insertions into the genome. The best example of this molecular process involves the integrative activity of viruses. Among the best examples of these are the retroviruses. As described earlier in this chapter and others, these viruses infect dividing cells at a high frequency but integrate randomly. Such observations have led to frustration among investigators hoping to use retroviruses for gene therapy. In some strategies, for example, precise integration would be helpful to achieve functional results. However, the central issue is that the cell *does not* naturally promote site-specific integration. Whether it is overwhelmed by the biological effort of the virus to integrate frequently or whether the enzymatic machinery driving homologous insertion is naturally suppressed is not clear. In one case, however, site-specific integration by a virus is, in fact, observed. Adenoassociated virus (AAV) appears to integrate with a high frequency into a site on chromosome 19, a DNA site that contains sequence similarity to the viral termini. This integrative event is catalyzed by the virally encoded Rep protein, an enzyme used to replicate the virus in the cell. Thus, a virally encoded protein, not a cellular enzyme, promotes site-specific targeting. An obvious extension of this work is to utilize the AAV-Rep protein to help target other DNA fragments to their homologous chromosomal sites. One significant problem exists, however, in this strategy. Biochemical studies have shown that the Rep protein acts as a dimer, one subunit binding to the viral sequence and the other to the homologous *viral*-like sequence in the chromosome. The requirement for Rep binding sequences in both templates will clearly limit this approach. Hence two examples with naturally integrative elements (retroviruses and adenoassociated virus) have led investigators to conclude that homologous integration in mammalian cells is not a preferred or even a natural reaction.

Gene Targeting: Gene Insertion or Gene Replacement in Mammalian Cells

With this as a background, workers have attempted to translate the genetic observations, and in some cases molecular tricks, found to work in lower eukaryotes or bacteria into the mammalian cell targeting arena. An early observation by yeast geneticists was that a double break in the homologous region of the targeting molecule elevated the frequency of site-specific integration. It had been widely accepted that double-stranded breaks promote homologous recombination even in mammalian cells, but the continual low frequency of specific events has persisted. The fact, however, that some homologous targeting occurs at all established the fact that mammalian cells do have the necessary machinery to catalyze the reaction.

To improve the frequency and develop reliable test systems, several strategies have emerged. Although this group differs in details, the fundamental protocol is to insert specific DNA sequences in the genome and utilize nucleases that make double-stranded breaks only at these sites. This protocol permits the insertion and excision of DNA fragments into the specific site. The prototype for this strategy is the Cre-*lox* system (Fig. 5.5). *LoxP* refers to the DNA sequence at which the bacteriophage (P1) recombinase Cre works. After *loxP* sites are integrated into a mammalian genome, they can be used as integration sites for targeting vectors containing the transgene of choice and a compatible *lox* site, which is required for the specific "docking" effect mediated by Cre. The resulting integrant contains both copies of *lox* as well as the transgene. The importance of this system is really bilateral. On one level the frequency of integration at the "*loxP* site" is high and, on another level, the transgene can be excised since Cre works to promote both integration and excision.

A similar system using a restriction endonuclease from yeast, known as I-SceI, can also be used (Fig. 5.6). The recognition site for I-SceI is 18 base pairs in length, and thus the chances that multiple sites in the genome exist is fairly low. The major difference between I-SceI and Cre-*lox* is that in the I-SceI system, the target sequences are naturally present in the genome, albeit at rare frequency. This fact enables the introduction of the DNA fragments at those rare sites. The yeast I-SceI endonuclease induces double-stranded breaks, and the breaks are repaired by the integrative action of the targeting vectors that provide regions of DNA homologous to the broken I-SceI sites. The entire process makes use of the double-stranded break repair mechanism.

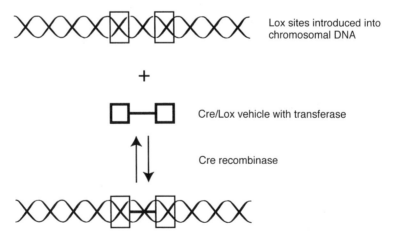

Lox sites introduced into chromosomal DNA

+

Cre/Lox vehicle with transferase

Cre recombinase

FIGURE 5.5 Targeted insertion by the Cre/*lox* system. *Lox* sites are introduced into the chromosomal DNA at a specific site in a particular gene through the process of homologous recombination. The Cre recombinsase (transferase) and the Cre/*lox* vehicle are then added to the cells. In some cases, Cre may be expressed from a co-transfected plasmid containing the gene encoding Cre. By overexpressing Cre recombinase, the vector fragment or sequence can be exchanged in or out.

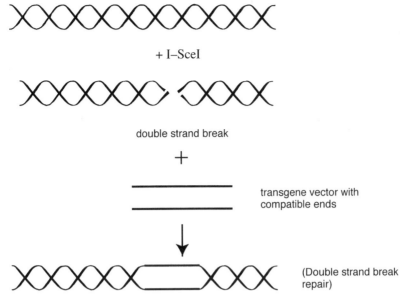

+ I–SceI

double strand break

+

transgene vector with
compatible ends

(Double strand break
repair)

FIGURE 5.6 Targeted insertions via double-strand breaks (I-SCE-I) I-SceI cleaves the chromosomal DNA at its cutting site, and the vector containing the transgene with homologous complemetary ends is integrated stably at the cleaved site through double-strand break repair.

Gene Transfer: A Rate-Limiting Step?

The inefficiency of homologous recombination in mammalian cells may also be directly related to the low quality of gene transfer. Additional problems are frequent nonhomologous events, dependence of length of homologous target, and the lack of correlation between successful events and target copy number. To overcome at least one of these barriers, adenovirus vectors that cannot replicate have been developed. Since this virus infects essentially all of the cells, even a low-frequency event can be amplified if all of the cells receiving the vector undergo at least one homologous recombination reaction.

The use of adenovirus to help in the gene transfer problem amplifies a real problem for all efforts in the use of the homologous recombination; how does one insert the vector into enough cells to make a difference? As outlined previously, electroporation of DNA has serious effects on cell viability while calcium phosphate delivers DNA to only a small fraction of the cells. The solution is to use micro-injection so that a vast majority of the cells receive the molecule. However, this procedure is highly labor intensive and essentially inappropriate to gene therapy strategies. Based on this information, workers have turned to the last alterable component of the gene targeting system: the cell.

The Cells: A Rate-Limiting Step?

A large facet of successful gene targeting for in vitro studies is the culture conditions of the cells. Variations in homologous targeting efficiency are related to the metabolic state of the cell before, during, and after transfection. Clearly, the transfection of mammalian cells with DNA complexed with liposomes can change the condition of the cell, and it is unclear what messages the nucleus gets from the bipolar membrane before, during, and after the liposomal complex encounters the membrane. It is possible that the achievement of high transfer efficiencies may be counterbalanced by the detrimental effects on nuclear metabolism. Simply inserting the vector into the cell is insufficient; delivering it into the nucleus is the ultimate goal. Until liposomes or other delivery vehicles are able to target the vector to the specific site, this problem will persist. Such problems can be accentuated by using tissue culture cell lines that are consistently the same passage and the density at which the cells are plated can also influence the success rate of gene targeting events. Although these issues may seem mundane, they are critical to the development and assessment of the effectiveness of a particular vector prior to the movement of a technology forward with animal models or, ultimately, humans.

GENE TARGETING HAS ALREADY PROVEN USEFUL

Interactions among various disciplines occur with regularity, and the impact of gene targeting on gene therapy has in some ways already been observed. In the mid-1980s, several protocols were established wherein a specific, targeted gene could be rendered dysfunctional through the process of homologous recombination. Gene knock-outs in mice have become almost a routine step in the analysis of newly discovered gene function. It is almost a required step before the scientific community accepts the "definition" of a newly described gene. Although, in principal, generating mouse knock-outs is routine, it is far from straightforward. People who are highly skilled in the art are needed to conduct the technically demanding protocol.

The most significant steps in the entire process may not actually involve homologous targeting; it is the isolation of healthy embryonic stem cells (ES cells) that retain totipotency (the ability to produce a fully developed animal). These cells are isolated from the inner cell mass of early mouse embryos or blastocysts and must be grown under specialized tissue culture conditions. These cells contain all the genetic elements needed to form a new mouse, including coat color. Such an obvious phenotype is helpful in discerning whether the pups, born after the targeting event, are derived from the targeted blastocyst. A traditional strategy follows this pathway:

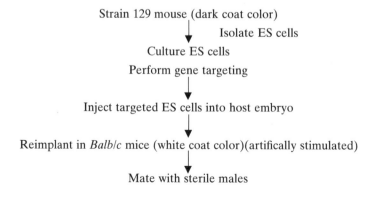

Strain 129 mouse (dark coat color)

Isolate ES cells

Culture ES cells

Perform gene targeting

Inject targeted ES cells into host embryo

Reimplant in *Balb/c* mice (white coat color)(artifically stimulated)

Mate with sterile males

The F1 progeny will have a chimeric coat color often referred to as "agouti." To demonstrate germline transmission, another test breed is undertaken. Genetic inheritance, measured by coat color, will be apparent as a mouse truly derived from the targeted ES cell will be fully dark coated. At this point it is likely that the targeted gene has been transmitted through the germ line. Prior to implantation, ES cells can be checked to see if the targeted gene has truly been disrupted using Southern hybridization or PCR analyses. Hence, the possibility that such a mutation has been carried forward would remain high.

Beyond the biological aspects of ES cell manipulation, there is the strategy for generating the knock-out. The pioneering work of Oliver Smithies and his colleagues formed the basis for most of the protocols used even today. Smithies was able to knock-out one of the β-globin alleles using a gene that renders cells resistant to the antibiotic neomycin (G418). This work demonstrated that it was possible to do targeting in mammalian cells, but the frequency of targeting was somewhere between 10^{-3} and 10^{-4}. Work continues to find ways to enrich for cells containing the specific integration event. This strategy reduces the workload of analyzing many different clones for the correct one.

The most useful strategy of enrichment is known as positive–negative selection. The method uses ES cells with the target being the housekeeping gene, *HPRT*. The selection for mutated, or "knocked-out" *HPRT* is based on the sensitivity of wild-type cells, containing a functional copy (*HPRT* is located on the X chromosome) of the *HPRT* gene, to 6-thioguanine. Hence, when the gene is disrupted and inactivated (these two things are not necessarily linked!), the cells survive and can grow into individual clones. This locus is useful for ascertaining the rate of homologous targeting since selective pressure can be placed on the cells, thereby selecting only those that have dysfunctional *HPRT* genes. This is known as positive selection. Negative selection is provided by the use of the herpes simplex virus thymidine kinase (TK) gene. Expression of this gene within cells renders the cells sensitive to the drug gancyclovir. By coupling that *hTK* gene to the vector containing the *HPRT* targeting sequence, one can estimate the number of random vs. specific integration events.

During the process of integration, the homologous portion of the *HPRT* gene is exchanged with a targeting element rendering the cell *HPRT⁻* and resistant to 6-thioguanine (Fig. 5.7). In these cases the *hTK* gene is lost so the cell is also resistant to gancyclovir. If the vector has been inserted randomly the cell is sensitive to gancyclovir and dies regardless of the activity at the *HPRT* locus. It is important to note that positive–negative selection is an enrichment strategy. Nothing in this procedure is really designed to increase targeting frequency per se. For the most part, workers in the past have accepted the low-frequency or rare event phenomenon for mammalian gene targeting and just wish to enrich for successful targeting events. With the wealth of new techniques, some of which are described above, workers are challenging these paradigms and simply not accepting low-frequency events as the "norm."

How then has homologous targeting in mice influenced gene therapy? The importance of the knock-out strategy centers around the ability of workers to create animal models of human diseases. For example, it is possible to replace the normal mouse gene with a "mutated human gene" assuming that enough homology exists between the two genes. Hence, the mouse now contains a human gene producing a

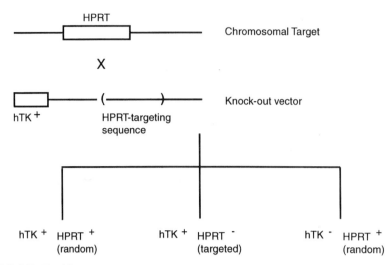

FIGURE 5.7 Positive negative selection. The *tk* and *HPRT* genes can serve as selectable markers for integration events that are either nonhomologous or homologous depending on the phenotype of the surviving colonies.

dysfunctional protein, and studies can ensue to treat this problem. One of the most useful animal models is for cystic fibrosis (CF) where a defect in the cystic fibrosis transmembrane conductance regulator (CFTR) gene/protein causes changes in electrophysiology to predict lung epithelium. This model has been useful in helping to create gene therapy treatments for cystic fibrosis (see Chapter 3).

There are some problems with animal models, and confidence in them as predictors of successful human DNase therapy is waning. For example, none of these CF animal models exhibit the severity of the disease or replicate the symptoms. In fact, it is quite common to augment mouse models with chemicals to recreate more completely the human disease in mouse models. Alternatively, knock-out mice can be valuable by displaying symptoms that are similar to a human condition, whose molecular cause has not been uncovered. Thus, the link between a particular gene and the human disease can be made directly by a cause-and-effect correlation. Such relationships are invaluable for gene therapy strategies as well as defining the function of new genes.

GENE TARGETING: THE FUTURE

Gene therapy is coming of age. Although many significant barriers remain to be overcome, it is apparent that this concept is a part of the future of medicine. The widely held notion that viral vectors and gene addition strategies present more problems than benefits has some basis in fact, but the field is still evolving. There is no consensus even as to the best viral vector, but a consensus opinion may not be an essential requirement for success. There is, however, consensus that the repair or replacement of defective genes in the context of the host chromosome is the ultimate form of gene therapy. But, this pathway is still unclear, and to realize this type of therapy in the next several years, clear clinical benefits must be realized as a direct result of molecular mechanistic studies.

KEY CONCEPTS

- Targeted integration of DNA into the chromosome is a rare event in mammalian cells due in large part to competing pathways of nonhomologous recombination.
- Genetic and biochemical experiments in lower eukaryotes have provided some information regarding the pathways and processes that govern gene targeting in higher organisms that appear to be similar. The most useful similarity may be the conservation of recombination and repair DNA sequences that permit the isolation of the human genes participating in these processes.
- Eukaryotic organisms have the enzymatic machinery to catalyze homologous targeting but utilize the pathway poorly.
- The use of recombinases from prokaryotes, lower eukaryotes, or mammalian cells to improve targeting frequency is a challenging strategy to undertake. Most recombination events occur through the action of protein complexes that require precise stoichiometry, and thus overexpressing a single gene may not simply be sufficient to activate a whole complex.
- Most viruses integrate randomly into the chromosome and are unlikely to be useful for gene targeting events. The single exception, at present, is AAV, which may be useful in future targeting designs.
- Genetic repair of point mutations has been successful using synthetic oligonucleotides, which act in a highly precise fashion. This new strategy enables accurate targeting to single base mutations without concerns about immune responses.
- Triplex-forming oligonucleotides mediate modifications in genomic sequences and may be used to stimulate homologous recombination at adjacent sites.

SELECTED READINGS

Gene Targeting and Transfer

Brenner M. Gene transfer by adenovectors. Blood 94:2965–3967, 1999.

Lanzov VA. Gene targeting for gene therapy: Prospects. Mol Genet Metab 68:276–282, 1999.

Romano G, Pacilio C, Giirdano A. Gene transfer technology in therapy: Current applications and future goals. Stem Cells 17:191–202, 1999.

Templeton NS, Lasic DD. New directions in liposome gene delivery. Mol Biotechnol 11:175–180, 1999.

Yanez RJ, Porter AC. Therapeutic gene targeting. Gene Therapy 5:149–159, 1998.

Homologous Recombination

Chandrasegaran S, Smith J. Chimeric restriction enzymes: What is next? Biol Chem 380:841–848, 1999.

Camerimi-Otero RD, Hsieh P. Homologous recombination proteins in prokaryotes and eukaryotes Ann Rev Genet 29:509–552, 1995.

Essers J, Hendriks RW, Swagemakers SM, Troelstra C, De WJ, Bootsma D, Hoeijmakers JH, Kanaar R. Disruption of mouse RAD54 reduces ionizing radiation resistance and homologous recombination. Cell 89:195–204, 1997.

Faruqi AF, Seidman MM, Segal DJ, Carroll D, Glazer PM. Recombination induced by triple-helix-targeted DNA damage in mammalian cells. Mol Cell Biol 16:6820–6828, 1996.

Lai LW, Lein YH. Homologous recombination based gene therapy Exp Nephrol 7:11–14, 1999.

Templeton NS. Strategies for improving the frequency and assessment of homologous recombination. Methods Mol Biol 133:45–60, 2000.

Umene K. Mechanism and application of genetic recombination in herpesviruses. Rev Med Virol 9:171–182, 1999.

RNA-DNA Oligonucleotides and Hybridization

Cole-Strauss A, Yoon K, Xiang Y, Byrne BC, Rice MC, Gryn J, Holloman WK, Kmiec EB. Correction of the mutation responsible for sickle cell anemia by an RNA-DNA oligonucleotide. Science 273:1386–1389, 1996.

Dicken ES, Epner EM, Fiering S, Fournier REK, Groudine M. Efficient modification of human chromosomal alleles using recombination-proficient chicken/human microcell hybrids. Nat Genet 12:174–182, 1996.

Faruqi AF, Datta HJ, Carroll D, Seidman MM, Glazer PM. Triple-helix formation induces recombination in mammalian cells via a nucleotide excision repair-dependent pathway. Mol Cell Biol 20:990–1000, 2000.

Kren BT, Parashar B, Bandyopadhyay P, Chowdhury NR, Chowdhury JR, Steer CJ. Correction of the UDP-glucuronosyltransferase gene defect in the Gunn rat model of Crigler-Najjar Syndrome Type I with a chimeric oligonucleotide. Proc Natl Acad Sci USA 96:10349–10354, 1999.

Sun JS, Garestier T, Helene C. Oligonucleotide directed triple helix formation. Curr Opin Struct Biol 6:327–333, 1996.

Zhang Z, Eriksson M, Blomback M, Anvret M. A new approach to gene therapy. Blood Coagul Fibrinol 8:S39–S42, 1997.

Gene Targeting

Deng C, Capecchi MR. Reexamination of gene targeting frequency as a function of the extent of homology between the targeting vector and the targeting locus. Mol Cell Biol 12:3365–3371, 1992.

Goncz KK, Gruenart DC. Site-directed alteration of genomic DNA by small-fragment homologous replacement. Methods Mol Biol 133:85–99, 2000.

Ng P, Parks RJ, Cummings DT, Evelegh CM, Sankar U, Graham FL. A high efficiency Cre/loxP-based system for construction of adenovirus vectors. Hum Gene Therapy 10:2667–2672, 1999.

Peng KW, Russell SJ. Viral vector targeting. Curr Opin Biotechnol 10:454–457, 1999.

Reid LH, Shesely EG, Kim H-S, Smithies O. Co-transformation and gene targeting in mouse embryonic stem cells. Molec Cell Biol 11:2769–2777, 1991.

Sauer B, Henderson N. Targeted insertion of exogenous DNA into the eukaryotic genome by Cre recombinase. New Biol 2:441–449, 1990.

Templeton NS, Ronerts DD, Safer B. Efficient gene targeting in mouse embryonic stem cells. Gene Therapy 4:700–709, 1997.

Thomas KR, Cappecchi MR. High frequency targeting of genes to specific sites in the mammalian genome. Cell 44:419–428, 1986.

DNA Repair

Bartlett RJ. Long-lasting gene repair. Nat Biotechnol 16:1312–1313, 1998.

Gura T. Repairing the genome's spelling mistakes. Science 285:316–318, 1999.

Jackson SP, Jeggo PA. DNA double-strand break repair and V(D)J recombination: Involvement of DNA-PK. Trends Biochem Sci 20:412–415, 1995.

Kmiec EB. Targeted gene repair. Gene Therapy 6:1–3, 1999.

Modrich P. Mismatch repair, genetic stability and cancer. Science 266:1959–1960, 1994.

Sargent RG, Brenneman MA, Wilson JH. Repair of site-specific double-strand breaks in a mammalian chromosome by homologous and illegitimate recombination. Mol Cell Biol 17:267–277, 1997.

CHAPTER 6

Gene Therapy for Hematological Disorders

CYNTHIA E. DUNBAR, M.D. and TONG WU, M.D.

INTRODUCTION

Hematopoietic cells are an attractive target for gene therapy for two main reasons. First, it is possible to easily collect and then manipulate hematopoietic cells in vitro. Second, many congenital and acquired diseases are potentially curable by genetic correction of hematopoietic cells, especially hematopoietic stem cells (HSCs, see Fig. 6.1). For hematological disorders, the target cell(s) in which gene expression is required are red blood cells (RBC), lymphocytes, granulocytes, or other mature blood elements. Ideally, the transgene is integrated into the chromatin of pluripotent HSCs, ensuring the continuous production of genetically modified blood cells of the desired lineage for the lifetime of the patient. Other potential cellular targets with potential utility in the treatment of hematologic diseases include dendritic cells, tumor cells, and endothelial cells. Hepatocytes, myocytes, and keratinocytes can be considered as "factories" for soluble factors with clinical utility in hematologic diseases such as hemophilia (see Chapter 7). Relevant targets and applications for gene therapy of hematopoietic or immune system disorders are summarized in Table 6.1.

Many important advances in our understanding of hematopoiesis, stem cell engraftment, and other basic principles have resulted from animal models, in vitro studies, and early clinical trials of gene marking or gene therapy. For example, studies using retrovirally marked murine stem cells show tracking and a quantitative analysis of murine stem cell behavior. Experiments overexpressing oncogenes or cytokines in hematopoietic cells have elucidated the in vivo role of these proteins. Early clinical gene marking trials demonstrated the long-term engrafting capability of peripheral blood stem cells. The observed lack of clinical utility results from several major hurdles, including inefficient gene transfer to desired target cells, especially stem cells, poor in vivo expression of introduced genes, and immune responses against gene products recognized as foreign. Further basic research investigations

An Introduction to Molecular Medicine and Gene Therapy, Edited by Thomas F. Kresina
ISBN 0-471-39188-3 © 2001 Wiley-Liss

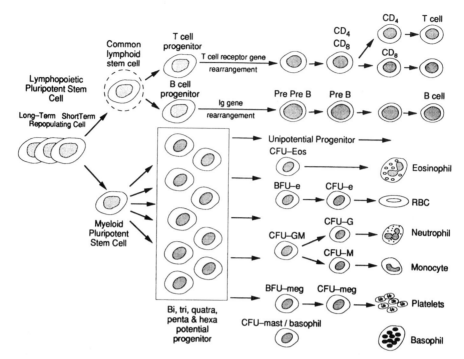

FIGURE 6.1 Hierachal model of lymphohemopoiesis. A primitive lymphohemopoietic cell is capable of producing lymphoid stem cells for lymphopoiesis or myeloid stem cells for hemopoiesis. These stem cells give rise to progressively more differentiated progenitor cells that eventually give rise to lineage-specific terminally differentiated effector cells.

into new or modified vector systems and target cell biology are necessary to move the field forward into real clinical utility.

REQUIREMENTS FOR GENE TRANSFER INTO HEMATOPOIETIC CELLS

Ex Vivo Versus in Vivo Gene Transfer

Specific aspects of gene transfer techniques are advantageous for gene therapy approaches when applied to hematological diseases. Aspects of ex vivo gene transfer as well as certain gene transfer vector systems are particularly useful in the experimental therapy of hematological diseases. Hematopoietic cells such as stem cells or lymphocytes are generally transduced ex vivo because these cells can be easily collected, cultured, and transduced in vitro (see Chapter 1). Subsequently, they can be reinfused intravenously. Ex vivo transduction allows for a controlled exposure of only the desired targets to vector particles. It is less likely to produce an immune response or be impeded by complement-induced vector inactivation. However, limited data indicate that direct in vivo injection of vector into the marrow space can transduce primitive cells. But, there is no evidence that this in vivo method currently has any advantages over the more fully characterized ex vivo transduction approaches. In vivo gene transfer is most appropriate for target cells that cannot

TABLE 6.1 Relevant Targets and Applications for Gene Therapy of Hematopoietic or Immune System Disorders

Target Cell or Lineage	Example of Clinical Application
Hematopoietic stem cells	Fanconi anemia
Red blood cells	Thalassemia, sickle cell anemia
Granulocytes	Chronic granulomatous disease
Lymphocytes	Immunodeficiency diseases Cancer (TIL) AIDS (intracellular immunization)
Macrophage	Gaucher disease
Dendritic cells	Immune therapy
Tumor cells	Tumor suppressing genes Antisense to oncogenes Tumor vaccines Suicide genes
Endothelial cells	Inhibitors of thrombosis Growth factors
Hepatocytes, myocytes Keratinocytes	Hemophilia

be easily harvested or manipulated ex vivo, such as airway epithelium, vascular endothelium, and differentiated muscle cells.

Vector Systems and Nonviral Vectors

The choice of an appropriate vector system depends on the biology of the desired target cell and the need for transient versus prolonged gene expresssion (see Chapter 4). Both viral and nonviral vectors have been utilized to transduce hematopoietic target cells. If prolonged correction or modification of hematopoietic cells is required, then vectors such as retroviruses that efficiently integrate into target cell chromosomes are necessary, otherwise new genetic material will be lost as HSCs or other targets such as lymphocytes proliferate. On the other hand, if transient expression is required, for instance, in the production of leukemic cell tumor vaccines, then nonintegrating but efficiently expressing vectors such as adenoviruses may be preferred. The vast majority of preclinical and clinical investigations of hematopoietic cell gene transfer utilize viral vectors, taking advantage of the characteristics of the virus that have evolved over time to efficiently infect target cells. The viral genes and replication machinery are replaced with nonviral transgene sequences of interest.

For murine retroviruses, the Moloney murine leukemia virus (MuLV) vectors are the vectors of choice since they have not been supplanted by any other vector system for most hematologic applications. Thus, MuLV vectors have been employed in almost every clinical study to date. The main advantages of MuLV vectors are their ability to integrate a stable proviral form into the target cell genome, the availability of stable producer cell lines, the lack of toxicity to target cells, and almost 10 years of experience in using them safely in clinical trials. Over the past several years,

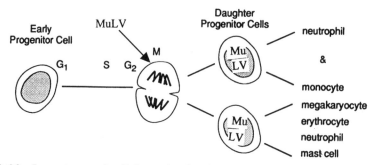

FIGURE 6.2 Importance of cellular activation by growth factors or cytokines to induce mitosis for transduction by Moloney murine leukemia virus (MuLV). Cells must pass through the mitotic phase of the cell cycle (M, middle frame) in order for the vector to gain access to the chromatin and integrate into the genome (right frame).

a number of modifications in the genetic sequences included in packaging cell lines has greatly decreased the risk of recombination events, and sensitive methods for detecting replication-competent virus have been established and are strictly utilized in all clinical trials. There have been no documented adverse events related to insertional mutagenesis in early human clinical studies or in preclinical animal studies using replication-defective viral vectors.

There appear to be two major limitations to the use of MuLV vectors for hematopoietic stem cell transduction. First, cells must pass through the mitotic phase of the cell cycle in order for the vector to gain access to the chromatin and integrate (Fig. 6.2). Most stem cells reside in the G_0 phase of the cell cycle, and manipulations that stimulate these cells to cycle ex vivo may result in irreversible lineage commitment or apoptosis. Second, the receptor for MuLV retroviral vectors (amphotropic vectors) on human and primate cells has been identified and appears to be broadly expressed in most human tissues. However, the low levels of this receptor on primitive HSCs may be limiting. To redirect receptor specificity, pseudotyping of vectors has been employed by replacement of MuLV envelope proteins with gibbon ape leukemia virus (GALV) envelope proteins. This technique improves transduction efficiency of mature lymphocytes and possibly hematopoietic stem cells. The vesicular stomatitis virus (VSV) envelope protein allows direct membrane fusion, circumventing the need for a specific cell surface receptor, but toxicity of the envelope protein to both producer cell lines and target cells hinders development of this approach.

Lentiviruses Recently, there has been an intensive focus on the development of vectors based on lentiviruses such as the human immunodeficiency virus (HIV)-1 or 2. Certain characteristics of HIV may overcome some of the limitations of the MuLV vectors. Pseudotyping of HIV-based vectors with VSV or amphotropic envelope proteins would allow transduction of hematopoietic progenitor and stem cells. Use of the HIV envelope gene would allow specific transduction of CD4+ targets. HIV and other lentiviruses transduce target cells without the need for cell division. The mechanism for this property is not fully understood. But, the dissection of the HIV genome and incorporation of the nuclear transport mechanism(s) into otherwise standard MuLV vectors for gene therapy has not been successful. Beyond these efforts, there are obviously major safety concerns that preclude clinical applications of HIV. Absolutely convincing preclinical data regarding efficacy and lack of replication-competent virus must be obtained prior to human use. Non-HIV-1 lentiviral vectors are also of great interest and are very early in development, as are vectors based on the human foamy virus (HFV), another retrovirus that appears to have little pathogenicity.

For adenoassociated virus (AAV), utility in hematopoietic stem cell gene transfer is unlikely. However, applications requiring only transient expression in lymphocytes or dendritic cells are attractive. Most recently, promising data has been obtained using AAV to transduce muscle cells in vivo, allowing prolonged production of soluble factors important in hematologic diseases such as factor IX for hemophilia or erythropoietin for anemia of chronic renal failure. AAV vectors package 5.2 kb of new genetic material precluding the transfer of large genes such as factor VIII.

Adenovirus (Ad) vectors have been explored primarily for in vivo gene delivery for the transfection of both dividing and nondividing cells. The immune response induced by Ad vectors, although a major disadvantage, is also being considered as a possible advantage for transduction of tumor cells with cytokines, co-stimulatory molecules, or other immune modulators in cancer vaccine protocols (see Chapter 13). These applications, thoroughly investigated in solid tumor animal models, are also being applied to hematologic malignancies such as leukemias and lymphomas. Normal primitive hematopoietic cells can be transduced by Ad, but only with very highly concentrated vector preparations that also result in significant toxicity. Transient expression in primitive cells may be of interest in manipulating homing after transplantation.

The simplest approach to gene transfer is to use naked plasmid deoxyribonucleic acid (DNA), with necessary control sequences and the transgene, as the vector. The advantages of nonviral vectors include the lack of any risk of generation of replication-competent infectious particles, independence from target cell cycling during transduction, and elimination of antivector immune response induced by viral proteins. There are few size constraints. However, transduction efficiency of primary cells is very low, and physical methods such as electroporation or chemical shock used to increase gene transfer efficiency of plasmids into cell lines are either inefficient or toxic. Encapsulation by lipsomes has been useful for some primary cell types, as has conjugation to molecular conjugates including polyamines and inactivated adenovirus. However, none of these nonviral methods has shown any promise in the transduction of hematopoietic stem or progenitor cells. Limited success has been reported transducing primary human lymphocytes with a device called the "gene gun," introducing plasmid DNA into cells using colloid gold particles. None of these vectors integrate, and expression levels are generally lower than reported with viral vectors.

HEMATOPOIETIC STEM AND PROGENITOR CELLS AS TARGETS FOR GENE THERAPY

The concept of genetic correction or modification of HSCs has been an ongoing primary focus of gene therapy research. The properties of both self-renewal and differentiation of HSC can provide for the continuous maintenance of the transgene in cells of hematopoetic origin, including red blood cells, platelets, neutrophils, and lymphocytes. Less obvious are the application to tissue macrophages, dendritic cells, and central nervous system microglial cells (Chapter 9). Lineage-specific control elements need to be included to allow for differential expression in the appropriate mature cell type; for example, the use of hemoglobin gene enhancers to target expression to red blood cells. The genetic correction of these cells offer a potential curative, one-time therapy for a wide variety of congenital disorders such as hemoglobinopathies, immunodeficiencies, or metabolic storage diseases. Gene therapy also allows consideration of novel approaches to malignancies and HIV infection such as differential chemoprotection and intracellular immunization (see Chapter 11).

The feasibility of harvesting transplantable stem cells from the bone marrow (BM) and the maintenance in short-term ex vivo cell culture were a crucial advantages in early animal studies. The discovery and isolation of hematopoietic cytokines in the mid-1980s allowed successful ex vivo culture and transduction, resulting in the first successful demonstration of efficient gene transfer into murine repopulating stem cells. More recently, the discovery of alternative sources of stem cells such as mobilized PB and umbilical cord blood (UCB) broadens the potential for HSC gene therapy to neonates or conditions requiring very high dose stem cell reinfusion.

However, several obstacles have limited progress toward efficient gene transfer into HSCs. Some are methodologic. No in vitro assays exist to identify and quantitate true human stem cells. Further, gene transfer strategies efficient in transduction of in vitro surrogates, such as day 14 colony forming units (CFU) or the primitive multipotential long-term culture initiating cells (LTCIC), have not resulted in similar high levels of transduction of actual repopulating cells in early clinical trials or large animal models. Thus, optimization of protocols and testing of new approaches has been hampered. An additional obstacle is the observation that the most primitive pluripotent hematopoietic cells appear to be predominantly in the quiescent G_0 phase of the cell cycle. These cells are thus resistant to transduction with MuLV retroviral vectors (Fig. 6.2). Attempts to increase cycling of primitive cells during transduction by prolonged culture in the presence of various combinations of hematopoietic cytokines has resulted in decreased engrafting ability. This is due to either loss of self-renewal properties, induction of apoptosis, or alteration in homing ability. Additionally, a characteristic of primitive hematopoietic stem and progenitor cells that inhibits efficient gene transfer is the low level of expression of receptors for a number of vectors including retroviruses and adenoassociated viruses. Lastly, many clinical applications are in nonmalignant disease where the use of high-dose ablative conditioning therapy prior to reinfusion of genetically corrected autologous stem cells is unacceptably toxic. Only with the use of high doses of stem cells can significant levels of engraftment occur without the use of high-dose conditioning chemotherapy or total body irradiation.

Preclinical Studies

Initial retroviral gene transfer into murine hematopoietic repopulating cells was achieved in 1984. The discovery, availability, and application of various hematopoietic growth factors improved the efficiency of ex vivo retroviral transduction of murine hematopoietic cells. Several different combinations of growth factors have been successfully used for supporting gene transfer into murine stem cells. These include the combination of interleukin 3 (IL-3), interleukin 6 (IL-6), and stem cell factor (SCF). Inclusion of recently discovered early acting growth factors such as *flt-3* ligand and megakaryocyte growth and development factor (MGDF)/thrombopoietin (TPO) have augmented the level of genetically modified cells. These cytokines and growth factors maintain primitive cell physiology ex vivo and potentially stimulate primitive cells to cycle without differentiation. They may also upregulate retroviral cell surface receptors. Other manipulations that have been found beneficial in the murine system include (1) treatment of animals with 5-fluorouracil before marrow harvest to stimulate cycling of primitive cells, (2) the co-culture of target cells directly on a layer of retroviral producer cells or other stromal support, (3) the use of high titer (greater than 10^5 viral particles per ml) vector and (4) co-localization of vector and target cells using fibronectin-coated dishes.

Under these enhanced conditions, retroviral gene transfer into murine BM hematopoietic cells is now achieved in vivo with long-term marking at 10 to 100% in all cell lineages. The persistence of vector sequences in short-lived granulocytes and in multiple-lineage hematopoietic cells from serially transplanted mice indicates that murine repopulating stem cells can be successfully modified with retroviral vectors. Other supportive data include retroviral integration site analysis documenting the common transduced clones from different lineages. The repopulation of murine stem cells in nonablative or partially ablative conditioning transplant models has been increased by pretreatment of recipient mice with G-CSF/SCF.

These results in the murine model have raised concerns about long-term expression of transgenes from integrated vectors. Studies have shown poor or decreasing in vivo expression of the transgene or transgenes, especially with serial transplants, despite persistence of vector sequences. A hypothesis for this down-regulation in expression is the methylation of specific sequences in the vector promoter and enhancer regions. To counter this down-regulation in gene expression, many modifications have been made in basic MuLV vectors. These include the exchange of control sequences in the long terminal repeats (LTRs) with sequences from other retroviruses with lineage specificity of expression and the mutagenesis of putative negative regulatory sequences. Data suggest that modified vectors show improved long-term in vivo expression, although, equivalent long-term expression from standard MuLV vectors has been acheived under certain circumstances.

Evaluation of ex vivo gene transfer protocols using human cells mainly relies on in vitro progenitor cell assays, including CFU (representing committed progenitors), and long-term culture initiating cell (LTCIC), a putative in vitro stem cell surrogate. Using similar optimized conditions to the murine model, 50% or more progenitor colonies were transduced by retroviral vectors. Equally high LTCIC transduction has also been observed. Although BM has been the traditional source for HSCs, optimized gene transfer into CFU or LTCIC indicates that mobilized PB and UCB can be sources for HSCs.

Purification for primitive cells by panning—the exposure of whole BM or mobilized PB to antibodies directed against cell surface antigens found only on primitive cells, such as CD34—followed by flow cytometric sorting or immunoabsorption results in the isolation of approximately 1 to 5% of total cells. These enriched progenitor cells have reconstituting properties in clinical transplantation protocols. Selection for CD34$^+$/CD38$^-$ or HLA-DR populations can further purify stem cells. Recent studies show that CD34- cell populations also possess repopulating activity, potentially arguing against the use of CD34-enriched cells for gene transfer and other applications. Use of purified target cells permits practical culture volumes and higher vector particle to target cell ratios (MOI) during transduction, thereby increasing gene transfer efficiency.

As data emerge suggesting that the use of in vitro surrogate assays do not predict levels of gene transfer seen in vivo in early human clinical trials, attention has refocused on studying in vivo repopulating cells. One approach is the use of large animal models since the stem cell dynamics, cytokine responsiveness, and retroviral receptor properties appear to be similar between humans and nonhuman primates. However, very few research centers have the facilities and resources to carry out such transplant studies, and thus current studies are feasible as small proof of principle experiments, with little ability to study the impact of changing multiple variables. Rhesus or cynamologous monkeys and baboons are currently used most extensively. The persistence of vector sequences was first observed in a rhesus monkey transplantation model in 1989. In this seminal study, the CD34-enriched marrow cells were transduced with a high titer vector producer cell line (greater than 10^{8-10} viral particles per ml) secreting both human IL-6 and gibbon IL-3. However, this high titer producer cell line also produced significant titers of replication-competent helper virus due to recombination between vector and helper sequences in the producer cell line. Thus, in vivo marking in these animals could not be interpreted. Moreover, high-grade T-cell lymphomas were found in some recipients several months posttransplantation because of insertional mutagenesis by the replication-competent contaminating virus. This complication resulted in wide agreement that it is absolutely necessary to use helper-free producer cell lines and vector stocks in any clinical application. As well, it is necessary to assess safety in large animals before human clinical use.

Subsequent studies have documented long-term genetic modification of multiple hematopoietic lineages in primates using a number of different helper-free retroviral vectors. These successful transductions have been performed in the presence of growth factors, using unpurified or CD34-enriched BM or mobilized PB cells. Lower levels of gene-modified circulating cells were reported when compared to the mouse model (generally less than 0.01 to 1%), although similar optimized transduction conditions were used in both systems. Improved marking levels of up to 1 to 4% have been reported by transducing growth factor-stimulated PB or BM hematopoietic cells in the presence of a cell line engineered to express a transmembrane form of human SCF. Recently, studies report further encouraging data when *flt-3* ligand is added to the transduction cytokine combination, either in the presence of a fibronectin support surface or autologous stroma. Marking levels of 10 to 20% in vivo for at least 20 weeks were confirmed by Southern blotting.

Some important results of retroviral transduction were obtained from the canine autologous transplantation model. For instance, effective transduction of G-CSF-mobilized peripheral blood repopulating cells was first observed in the dog. Partially or fully ablative conditioning was necessary to obtain detectable engraftment with transduced HSCs. Using this model, high levels (up to 10%) of transduced marrow CFU after transplantation have been reported using a 3-week long-term marrow culture for transduction and reinfusion without conditioning.

The expense and difficulty of transplanting large animals have resulted in the transplantation of gene-modified human hematopoietic cells in immunodeficient mice as an alternative model. The major obstacle of this method is the low-level engraftment with human cells. Improved results have been obtained by inclusion of co-transplantation of stromal cells secreting human IL-3, the use of more immunodeficient strains such as NOD/SCID, and transplantation into immunodeficient transgenic mice expressing human cytokines. Identical retroviral integration sites were documented in human myeloid and T-cell clones obtained from a mouse posttransplantation, suggesting that pluripotent human HSCs were transduced. Cord blood cells engraft with greater efficiency than adult BM or mobilized PB. Thus studies have employ CB to a greater extent and extrapolate the data to other cell sources for gene therapy. The predictive value of data derived from xenograft models remains to be proven through the direct comparison with results from human clinical studies, thereby tracking the same gene-modified cell population in both patients and immunodeficient mice.

Clinical Genetic Marking Studies

Genetic marking of cells with an integrating vector is a unique method for tracking autologous transplanted cells and their progeny in vivo. Early human clinical gene transfer trials used retroviral vectors carrying nontherapeutic marker genes to transduce a fraction of an autologous graft in patients undergoing autologous transplantation for an underlying malignancy. These studies provided proof of principle and safety data.

Several studies used retroviral marking to track whether reinfused tumor cells contribute to relapse after autologous transplantation. In two pediatric gene-marking studies, unpurged autologous marrow from children with acute myeloid leukemia or neuroblastoma was briefly exposed to a retroviral vector carrying the *Neo* gene. Genetically marked tumor cells were detected in several patients at relapse. This observation suggested that the reinfused marrow had contributed to progression and that purging was necessary. One adult marking study did not detect marked tumor cells in patients with acute leukemia at relapse, but overall transduction efficiencies in this study were lower. Marked relapses were demonstrated in chronic myelogenous leukemia: *bcr/abl*+ marrow CFU-C were shown to contain the marker gene. No marked relapses have been detected in adult patients with multiple myeloma and breast cancer transplanted with genetically marked bone marrow and peripheral blood cells. However, the marrow and blood cells were CD34-enriched before transduction, thus purging the starting population by at least 2 logs of tumor cells.

Another outcome of these marking studies was to assess in vivo gene transfer efficiency. In the pediatric study, a fraction of the bone marrow graft was briefly exposed to retroviral supernatant without growth factors or autologous stroma. As many as 5 to 20% of marrow CFU were shown to be neomycin-resistant between 6 and 18 months posttransplantation, suggesting effective transduction and ongoing transgene expression. This surprisingly high level of stable marked marrow progenitors may be explained in part by active cell cycle kinetics of the primitive HSCs from these children likely due to their young age. Additionally, the primitive HSCs may have been undergoing hematopoietic recovery from high-dose chemotherapy just before BM collection. However, only 0.1 to 1% of circulating mature cells were marked.

Treated adults have undergone autologous bone marrow and mobilized peripheral blood stem cell transplantation for multiple myeloma and breast cancer. Bone marrow and peripheral blood CD34-enriched cells were transduced with different retroviral vectors containing the *Neo* gene in order to assess the relative contribution to marking and engraftment of marrow and peripheral blood populations. Transduction was performed for 3 days in the presence of the cytokines IL-3, IL-6, and SCF. Circulating marked cells were detected after engraftment in all patients. Marked cells were also detected in three of nine recipients for over 18 months. Although granulocytes, B cells, and T cells were positive for the transgene, the gene transfer efficiency was lower than in the pediatric studies. Less than 0.1% of circulating cells were marked long term, and no high-level marking of marrow CFU-C was detected. Because both the bone marrow and peripheral blood grafts contributed to long-term marking, this study documented that mobilized peripheral blood grafts can produce multilineage engraftment. This study was also important evidence that allogeneic transplantation could be performed safely with this cell source. These investigators also tested the brief single transduction protocol that was effective in the pediatric study, but no persistent marking was detected in adult patients.

Clinical Studies Using Therapeutic Genes

A main objective of gene therapy is the replacement of defective or missing genes in congenital diseases. A number of single-gene disorders such as the hemoglobinopathies, Fanconi anemia, chronic granulomatous disease, and Gaucher disease have been the focus of clinical trials. The hematological deficiencies in these disorders can be successfully treated by allogeneic BMT, implying that normal stem cells can reverse the pathophysiology of the disorders. Despite the low level of gene transfer into long-term repopulating stem cells achieved in large animal models and early human marking studies, several clinical trials exploring potentially therapeutic genes have been reported or are ongoing (Table 6.2). Important information has been obtained on safety and feasibility of stem cell engraftment without ablation, and there are glimmers of hope regarding clinical benefit.

Severe combined immunodeficiency due to adenosine deaminase (ADA) mutations was the first disease involving gene therapy of hematopoietic cells for several reasons. The human ADA gene was cloned in the early 1980s and the small 1.5-kb (cDNA) could easily fit into a retroviral vector along a selectable marker gene such as *Neo*. Even a low level of gene transfer efficiency might be efficacious because ADA normal cells should have an in vivo survival and proliferative advantage. Thus, the correction of only 1 to 5% of target cells may have clinical benefit. Hematopoietic stem cells could be better gene correction targets than T cells in this and other immunodeficiency disorders because of the potential for permanent and complete reconstitution of the T-cell repertoire. However, it has been difficult to achieve stable long-term efficient transduction of HSCs, thus T cells were the initial targets chosen. To directly address this issue, two ADA-deficient children in Italy received both autologous bone marrow and mature T lymphocytes transduced with distinguishable retroviral vectors carrying both the ADA and *Neo* genes. The patients were then repeatedly reinfused with both cell products without conditioning. In the first year, vector-containing T cells originated from the transduced mature T cells; but, with time, there was a shift to vector-containing T cells originating from transduced bone marrow cells. A normalization of the immune repertoire and

TABLE 6.2 Published Clinical Trials of Gene Transfer into Hematopoietic Cells

Disease	Target Cell	Gene	Results
Melanoma	Tumor infiltrating lyphocytes	*Neo*	Detection of marked TILs in tumor
Acute leukemia	BM	*Neo*	Marked tumor at relapse; Persistence of marked normal CFU
Neuroblastoma	BM	*Neo*	Marked tumor at relapse; Persistence of marked normal CFU
Chronic myeloid leukemia	BM	*Neo*	Marked *bcr/abl* + CFU; Marked normal CFU
Breast cancer/multiple myeloma	BM and PB CD34+ cells	*Neo*	Persistence of marked cells of multiple lineages from PB and BM grafts
Severe combined immunodeficiency	UCB CD34+ cells	ADA, *Neo*	Gene-marked T cells, and low-level marking of other lineages
EBV-induced lymphoproliferative disorders (EBV-LPD) after BMT	EBV-specific cytotoxic lymphocytes	*Neo*	Transient detection of marked T cells, then in vivo expansion with EBV activation
Severe combined immunodeficiency	T lymphocytes	ADA, *Neo*	Persistence of gene-corrected T cells (1–30%)
Severe combined immunodeficiency	T lymphocytes BM	ADA, *Neo*	Gene-corrected T cells from both transduced lymphocytes and stem cells
Acute leukemia	BM	*Neo*	No marked tumor cells or persistence of marked hematopoietic cells
Fanconi anemia	PB CD34+ cells	FACC	Marking but no in vivo selection
Chronic granulomatous disease	PB CD34+ cells	*p47 phox*	Prolonged (6 months) production of gene-corrected granulocytes (0.004–0.05%)
EBV-LPD, relapsed leukemia and GVHD after BMT	Donor lymphocytes	HSV-TK, *Neo*, NGFR	Anti-EBV effect preserved, then elimination of GVHD by ganciclovir
Breast/ovarian/brain tumor	BM/PB CD34+ cells	MDR1	Transient or low-level gene transfer, no clear in vivo selection with chemotherapy

restoration of cellular and humoral immunity were documented after gene therapy. Data showed a surprisingly high number of marrow CFU resistant to neomycin. This was despite the lack of conditioning and the authors hypothesize an in vivo selective advantage for gene-corrected cells of all lineages.

In genetic disorders diagnosed in utero, an exciting alternative approach is the use of cord blood. These cells may contain relatively greater numbers of primitive repopulating cells more susceptible to retroviral transduction. Moreover, early treatment is crucial before disease progresses chronically to irreversible damage. The cord blood was collected at the time of delivery from three neonates diagnosed in utero with ADA deficiency. The cells were CD34-enriched and transduced with an ADA/*Neo* retroviral vector. The transformed cells were reinfused into the children without ablation. Vector sequences were detected in circulating mononuclear cells and in granulocytes of all three children for longer than 18 months but at low levels of less than 0.05%. However, when treatment with exogenous PEG-ADA was discontinued in one child, the proportion of vector-containing T cells increased to 10% or more. This was an unexpected finding that implied in vivo selection for corrected cells. Over time, however, the child's immune function declined and PEG-ADA therapy restarted. What can be concluded form the study is that the level of expression of ADA from the MuLV vectors remained low in unstimulated T cell in vivo. These cells were not fully functional despite a possible survival advantage in the development of the T cells from precursors.

Fanconi anemia (FA) is a hematopoietic genetic disorder that may be an excellent candidate for gene therapy. FA is a bone marrow failure syndrome, characterized by physical anomalies, and an increased susceptibility to malignancies. Cells from these patients are hypersensitive to DNA-damaging agents. FA can be functionally divided into at least five different complementation groups termed (A–E). Two different FA genes, FAC and FAA, have been identified from two different subsets of patients. Phenotypic correction of these abnormalities in cells from two patient groups was successful after transduction with retroviral vectors carrying the FAC or FAA gene. A possible in vivo survival advantage for gene-corrected primitive cells and their progeny has made FA an attractive candidate disease for stem cell gene therapy. A clinical trial has tested this hypothesis using G-CSF-mobilized peripheral blood CD34+ cells from three FAC patients as targets. The results of this trial suggest that gene complementation has at least transient positive effects on FA hematopoiesis as measured by progenitor growth and marrow cellularity. However, no clear clinical benefit or in vivo survival advantage for transduced cells has been demonstrated.

Chronic granulomatous disease (CGD) is a rare inherited immunodeficiency disorder of the NAPDH oxidase system and consequently of phagocytic cell function. It is characterized by recurrent bacterial and fungal infections that induce granuloma formation and threaten the life of patient. Four different genetic defects have been found to be responsible for this disease. Current clinical management of CGD patients includes administration of antibiotics, interferon-γ, or allogeneic BMT, but unsatisfactory clinical results make the development of gene therapy strategies highly desirable. Low levels of correction may have clinical impact as healthy X-linked CGD carrier females have been identified with only 1 to 10% of normal levels of NADPH function. In a clinical trial, five CGD patients with p47phox deficient have been reinfused with CD34+ peripheral blood stem cells transduced with a retroviral vector containing p47phox without conditioning. Genetically-modified granulocytes were detected by PCR and correction of neutrophil oxidase activity was documented during the first few months after infusion. But within 6 months these cells became undetectable. Similar results have been reported for a clinical trial carried out in patients with Gaucher disease. Without ablation, vector-containing cells were detected at low levels and only transiently after reinfusion.

LYMPHOCYTE GENE TRANSFER

Lymphocytes have characteristics that are advantageous for some gene therapy applications as compared to hematopoietic stem cells. Lymphocytes are easily harvested in large numbers and can be cultured ex vivo without major perturbation of phenotype, immune responsiveness, or proliferative potential. Lymphocytes may be repeatedly harvested and ablative conditioning is not necessary for persistence of infused cells. Both preclinical animal data and early clinical trials have reported encouraging results. However, they have also provided troublesome evidence of strong immune responses developing against exogenous genes expressed by lymphocytes.

Preclinical Studies

Stable, long-term ex vivo expression of transgenes has been achieved by using a retroviral vector containing *Neo* and human *ADA* genes in both murine and human T lymphocytes. Transduced murine lymphocytes could be selected by growth in G418 and subsequently expanded without changing their antigenic specificity. Infusion of these cells into *nude* mice has resulted in the persistence of *Neo*-resistant cells that continued to produce human ADA for several months.

Modified transduction protocols have been explored to further improve gene transfer to lymphocytes. Pseudotyping of MuLV vectors with a GALV envelope has increased lymphocyte transduction efficiency because lymphocytes appear to have more GALV receptors than amphotropic receptors. Other technical improvements during transduction have included centrifugation to increase the interaction between target cell and virus, phosphate depletion to up-regulate the amphotropic or GALV receptors, and low-temperature incubation to stabilize vector particles. Under these optimized conditions, up to 50% of lymphocytes can be transduced ex vivo without changes in viability, phenotype, or expansion capability. In an in vivo marking study, rhesus peripheral blood lymphocytes were transduced successfully with a vector encoding the *Neo* gene and HIV-1 *tat/rev* antisense sequences using these techniques. Following reinfusion, 3 to 30% of circulating CD4$^+$ cells contained the vector for at least several months, and lymph node sampling demonstrated that these cells could traffic normally.

Clinical Genetic Marking Studies

The initial controlled and monitored human gene transfer study used retroviral marking to monitor the fate of tumor-infiltrating lymphocytes (TIL) in vivo. Low-level marking was detected in tumor deposits. However, marking levels were too low to assess any preferential trafficking of TIL cells to residual tumor. In subsequent gene marking studies, behavior of transduced donor lymphocytes was studied in patients undergoing allogeneic transplantation. To control Epstein–Barr virus (EBV)-induced lymphoproliferative disorders (EBV-LPD) postallogeneic BMT, EBV-specific donor T cells were isolated, expanded, and gene marked in culture with EBV-transformed donor lymphoblasts as stimulators. After transplantation, the transduced T cells were reinfused, and two to three orders of magnitude expansion of marked cells were measured in vivo. EBV-specific cytotoxicity in the peripheral blood was greatly enhanced after the infusions. Although circulating marked cells became undetectable by 4 to 5 months after infusion, the persistence of memory cells from the infusion product was inferred in a patient with detectable marked lymphocytes in the blood after reactivation of latent EBV.

Suicide Gene Transfer

A similar approach has been utilized in patients with EBV-LPD with the modification of incorporating the *herpes simplex virus thymidine kinase* (*HSV-tk*) gene into the retroviral vector. This suicide gene converts the nontoxic prodrug ganciclovir to a toxic metabolite that kills the *tk*-expressing cell by inhibition of DNA synthesis. The inclusion of this gene in vectors allows elimination of transduced cells in vivo simply by ganciclovir administration postinfusion. For example, ganciclovir treatment could abrogate graft-versus-host disease (GVHD) in allogeneic BMT recipients if most of the allogeneic T cells contain the *tk* gene. This strategy depends on inclusion of a cell surface marker gene in the vector to allow positive selection of transduced cells before reinfusion. This would allow almost all infused cells to contain the *tk* gene and thus be sensitive to ganciclovir killing.

In allogeneic transplantation, donor lymphocytes play a therapeutic role in both graft-versus-leukemia (GVL) and immune reconstitution. However, their application is limited by the risk of severe GVHD. In a clinical trial, eight patients who relapsed or developed EBV-induced lymphoma after T-depleted BMT were treated with donor lymphocytes transduced with *HSV-tk* suicide gene. The transduced lymphocytes survived for up to 12 months, resulting in antitumor activity in five patients. Three patients developed GVHD, which could be effectively controlled by ganciclovir-induced elimination of the transduced cells. This study and other studies where patients with HIV disease received ex vivo expanded autologous lymphocytes transduced with a *tk*-hygromycin-resistant vector have reported troublesome evidence of an immune response developing against foreign gene products, such as herpes *tk* or drug-resistant genes. This immune response limits the persistence of transduced cells, as well as repeated infusions.

Therapeutic Genes

As noted earlier, the initial human gene therapy study used T lymphocytes as targets. Two children with severe combined immunodeficiency due to ADA deficiency received multiple infusions of autologous T cells transduced with a retroviral vector containing the human ADA gene. Both patients showed relative improvements in circulating T numbers and cellular and humoral immunity. In one child, the T-cell numbers rose to normal, lymphocyte ADA levels increased to roughly half that seen in heterozygote carriers of the disease, and the vector was detected in peripheral T lymphocytes at a concentration of approximately 1 copy per cell. In the second child, the T cell level rose temporarily during the infusions and then fell back. T-cell ADA activity did not increase, and only 0.1 to 1% of circulating T cells contained the vector even after multiple infusions. Both patients showed persistence of vector-containing cells for more than 2 years after the last T-cell infusion, which shows that transfused peripheral T cells can have a long life span. The expression level of ADA in these lymphocytes appears to be low, becoming significant with ex vivo activation. Thus, vector modifications may be needed to improve expression. Internationally, a similar study has been performed in one patient and the percentage of peripheral blood lymphocytes carrying the transduced ADA gene has remained stable at 10 to 20% during the 12 months since the fourth infusion. ADA enzyme activity in the patient's circulating T cells, which was only marginally detected before gene transfer, increased to levels comparable to those of a heterozygous carrier individual. This level was associated with increased T-lymphocyte counts and improvement of immune function.

CURRENT PROBLEMS AND FUTURE DIRECTIONS

In Vivo or Ex Vivo Selection

The observed low efficiency of gene transfer into hematopoietic stem and progenitor cells or other targets transduced ex vivo may be compensated by either positive selection of transduced cells before reinfusion or in vivo after engraftment. Rapid selection of transduced cells can be carried out using marker genes encoding proteins detectable by fluorescence-activated cell sorting (FACS) or other immunoselection techniques. The human cell surface protein CD24 or its murine analog, heat-stable antigen (HSA), has been tested as a selectable marker. Both small proteins (200 to 250 bp) take up little space in vector constructs, and noncrossreacting antibodies are available. Murine cells transduced with a vector containing human CD24 and selected before transplantation result in long-term reconstitution with a very high proportion of cells containing the vector. A vector expressing HSA allowed enrichment for transduced human progenitor cells. However, CD24 and HSA are glycosylphophatidylinositol-linked surface proteins. This class of proteins has been shown to be transferred from cell to cell both in vitro and in vivo, possibly complicating interpretation. A truncated, nonfunctional form of the human nerve growth factor receptor (NGFR) has also been developed as a selectable marker for hematopoietic cells, because hematopoietic cells do not express endogenous NGFR. Preclinical studies and early clinical trials have shown that transduction and sorting of lymphocytes using this marker is sensitive and specific. However, the introduction of new cell surface proteins has the theoretical disadvantage of altering trafficking or cell/cell interactions upon infusion of transduced cells. Alternative cytoplasmic markers such as jellyfish green fluorescent protein (GFP) are naturally fluorescent, avoiding the need for antibody staining. Reconstitution with enriched GFP$^+$ cells and long-term expression of GFP in multiple bone-marrow-derived cell lineages have been achieved in the murine model. However, a large animal study demonstrated that CD34-positive GFP-positive progenitor cells selected after 5-day culture in the presence of multiple cytokines are able to produce mature $CD13^+$ cells in the short-term. But these cells failed to engraft in the medium to long term.

In human studies, the positive selection of transduced lymphocytes using selection markers has already been achieved. The further expansion of transduced cells shows no changes in phenotype or in vivo function. However, it is still difficult to use ex vivo selection strategies on human hematopoietic stem cells posttransduction due to a low gene transfer efficiency. The major concern is that too few stem cells remain to allow safe and rapid hematopoietic reconstitution after enrichment of transduced cells, especially if ablative conditioning will be used. A potential solution to this problem is ex vivo expansion of selected transduced cells before reinfusion. It is unknown whether true long-term repopulating cells can be expanded or even maintained ex vivo using current culture conditions. Expanded cells have been documented to engraft lethally irradiated or stem-cell-deficient mice. However, a competitive disadvantage of ex vivo cultured cells against endogenous stem cells was shown in a nonablative model. In the ablative rhesus model, transduced $CD34^+$ cells expanded for 10 to 14 days ex vivo competed poorly against cells transduced and cultured for 4 days, despite 1 to 2 log expansion of total cells and CFU.

In vivo-selectable drug-resistant genes have been incorporated into retroviral vectors. There are at least two possible applications for this in vivo drug selection strategy: (1) induction of chemoprotection and (2) in vivo positive selection of genetically modified cells. Bone marrow suppression is one of the most common toxicities of chemotherapy regimens. One approach to increase the tolerated dose of chemotherapy is to introduce the human multidrug resistance 1 (*MDR1*) gene into bone marrow stem and progenitor cells. The protein product of this gene, called P-glycoprotein, can extrude many chemotherapy drugs out of cells, thereby resulting in a drug-resistant phenotype. These drugs include the anthracyclines, taxol, vinca alkaloids, and epipodophyllotoxins. Another potential application is to incorporate the gene into a vector with another gene of interest (e.g., glucocerebrosidase) to allow in vivo enrichment of the percentage of gene-modified cells into a therapeutically beneficial range by administration of MDR-effluxed drugs. Mice engrafted with *MDR1*-transduced marrow cells tolerate higher dose of MDR-effluxed drugs, and develop increasing percentages of circulating vector-containing cells. These cells are stable without further treatment suggesting selection at an early stem or progenitor cell level. The human clinical trials piloting this marrow-protective approach have been performed in patients undergoing autologous BMT for solid tumors such as ovarian and breast cancer. No clear evidence of chemoprotection or in vivo selection has been obtained. However, transductions in these trials were used in suboptimal protocols, and the level of marking was extremely low or undetectable. Thus, the results are not surprising. There has been a recent report of an aggressive myeloproliferative and eventually leukemic syndrome occurring in mice transplanted with *MDR1*-transduced marrow cells that were expanded ex vivo. This poor outcome possibly implicates the *MDR1* gene product in leukemogenesis and may terminate future clinical applications of *MDR1*.

Other drug-resistant genes have also been studied in vitro and in murine models. These include 06-alkylguanine-DNA-alkyltransferase or glutathione S-transferase, which confer protection against alkylating agents, and mutant dihydrofolate reductases (DHFRs) that confer resistance to trimetrexate as well as other antimetabolites. Each is very promising and may reach clinical trials in the near future. Issues with these strategies for chemoprotection are that nonhematologic toxicity may rapidly become limiting, and patients will not be protected from those side effects by engraftment with gene-modified, protected stem cells.

Alternative Vectors

The limitations of retroviral vectors has led to an intensive search for other viral vectors that can both transduce quiescent cells and integrate permanently into their genome. One type of candidate are vectors based on HIV. HIV vectors can transduce a high percentage of CD34$^+$ hematopoietic cells. In addition, G_0/G_1 primitive hematopoietic cells engrafting *NOD/SCID* mice can be transduced by lentiviral-based vectors and maintain their primitive phenotype, pluripotentiality, and transgene expression.

Although AAV has been investigated extensively for hematopoietic cell gene transfer, most current data argues against the use of AAV for these applications. This is because of inefficient AAV vector integration. Several laboratories have reported high transduction efficiency of both human and murine hematopoietic progenitors, as assayed by PCR or G418-resistant CFU-C, but primate studies indicate no advantage over retroviral vectors in gene transfer into repopulating stem cells.

Gene Correction

Current gene transfer strategies rely to a large extent on random insertion of a complete new copy of a defective gene or a corrective gene. A new copy is inserted even if the defect in the original gene is only a point mutation. Newer strategies aimed at repairing mutations in the endogenous gene are thus very attractive. One novel strategy is the correction of a mutation in the β-globin gene in EBV-transformed lymphocytes derived from patients with sickle cell anemia by use of chimeric RNA-DNA oligonucleotides. The analysis was only by PCR, with inherent potential for misinterpretation, and has not been reproduced. However, if this approach, or other similar methods, can reproducibly correct mutations in nondividing human hematopoietic stem cells, it will revolutionize the gene therapy field.

Immune Responses to Vectors and Transgenes

Immune responses against vector proteins or transgene-encoded proteins are clearly an obstacle to successful gene therapy. Repeated in vivo administration of complex vectors stimulates an active immune response to vector proteins. This results in clearance of subsequent vector before in vivo transduction as well as causing damage to transduced tissues. To overcome this problem, modified adenoviral vectors have been developed with minimal residual adenoviral genes. Non-human marker genes such as the *Neo* gene or suicide genes such as *tk* gene included in vectors for selection may also induce an immune response. The therapeutic gene itself may induce an immune response if the patient completely lacks the endogenous gene product.

In a murine allogeneic skin transplantation model, foreign genes expressed by hematopoietic stem cells and their progeny induce immune tolerance across MHC barriers. However, foreign transgene products expressed in lymphocytes, myocytes, and other non-stem cells clearly are capable of inducing an immune response. A dual strategy of engraftment of transduced stem cells and actual transduced target cells that need to be corrected (i.e., lymphocytes or muscle cells) may be necessary to induce tolerance.

Immune rejection of transduced cells has been controlled partially with immunosuppression using agents such as cyclosporine, cyclophosphamide, or IL-12. But these pharmacologic approaches are not desirable or practical for most gene therapy applications for hematological disorders. Thus, improved vector design and possible inclusion of anti-rejection mechanisms in the vector constructs are more desirable.

SUMMARY

The first genetic disease elucidated at the molecular level was sickle cell anemia. Gene therapy of disorders such as the hemoglobinopathies requires high-level correction and has been difficult to achieve. But it seems likely that some hematological disorders, such as severe combined immunodeficiency caused by ADA deficiency or chronic granulomatous disease, will become amenable to effective gene therapy. A better understanding of stem cell biology as well as the development of simple and reliable vectors are necessary for further progress. The wide variety of novel approaches for gene transfer currently being developed are certain to eventually achieve the promise of gene therapy first envisioned a decade ago.

KEY CONCEPTS

- Hematopoietic stem cells (HSCs) have the ability to self-renew and differentiate into all lineages of the hematopoietic system, including the reticuloendothelial system and central nervous system microglial cells. HSCs are easily collected from marrow, stimulated peripheral blood, and cord blood and can be cultured and transduced ex vivo before intravenous reinfusion. These features have made HSCs an ideal target for gene therapy of a wide variety of congenital disorders (immunodeficiencies, hemoglobinopathies, metabolic storage diseases) and acquired diseases (HIV infection and malignancies).
- Gene transfer into HSCs has been hampered by several biological obstacles, and early clinical trials have not shown clinically relevant levels of gene transfer in most instances. Problems include: (1) No in vitro assay to identify and quantitate true stem cell exists. (2) HSCs appear to be predominantly in G_0 phase of the cell cycle, making them resistant to proviral integration with the retroviral vectors in current clinical use. (3) The receptors for a number of vectors including retroviruses and adenoassociated virus are expressed at low levels on HSCs. (4) Chromosomal integration is necessary for delivery of the transgene to progeny cells. Non-DNA integrating delivery systems such as ade-

noviruses will result in only transient expression and thus correction. Improved gene transfer efficiency has been reported in many relevant preclinical studies, especially large animal models by the inclusion of new hematopoietic growth factors and fibronectin or stroma during transduction, pseudotyping of retroviral vectors, and application of lentiviral vectors.

· Lymphocytes have several features that make them more attractive than HSCs as targets for gene therapy. They are easily harvested, circulate in large numbers, and can be cultured ex vivo without changes of phenotype, immune responsiveness or proliferative potential. They may be repeatedly harvested, and ablative conditioning is not necessary for persistence of infused cells. They are used in the gene therapy for congenital and acquired immunodeficiencies, malignancies, and GVHD.

SUGGESTED READINGS

Gene Transfer

Kiem HP, Andrews RG, Morris J. Improved gene transfer into baboon marrow repopulating cells using recombinant human fibronectin fragment CH-296 in combination with interleukin-6, stem cell factor, FLT-3 ligand, and megakaryocyte growth and development factor. Blood 92:1878–1886, 1998.

Liu JM, Young NS, Walsh CE, et al. Retroviral mediated gene transfer of the Fanconi anemia complimentation group C gene to hematopoietic progenitors of group C patients. Hum Gene Therapy 8:1715–1730, 1997.

Lutzko C, Dube ID, Steward AK. Recent progress in gene transfer into hematopoietic stem cells. Crit Rev Onol Hematol 30:143–158, 1999.

Rosenberg SA, Aebersold P, Cornetta K. Gene transfer into humans—immunotherapy of patients with advanced melanoma, using tumor-infiltrating lymphocytes modified by retroviral gene transduction. N Engl J Med 323:570–578, 1990.

Gene Marking

Brenner MK, Rill DR, Holladay MS. Gene marking to determine whether autologous marrow infusion restores long-term haemopoiesis in cancer patients. Lancet 342:11347, 1993.

Brenner MK, Rill DR, Moen RC. Gene-marking to trace origin of relapse after autologous bone marrow transplantation. Lancet 341:85–86, 1993.

Deisseroth AB, Zu Z, Claxton D, et al. Genetic marking shows that Ph+ cell present in autologous transplants of chronic myelogenous leukemia (CML) contribute to relapse after autologous bone marrow transplantation in CML. Blood 83:3068–3076, 1994.

Dunbar CE, Cottler-Fox M, O'Shaughnessy J, et al. Retrovirally-marked CD34-enriched peripheral blood and bone marrow and cells contribute to long-term engraftment after autologous transplantation. Blood 85:3048–3057, 1995.

Tisdale JF, Hanazono Y, Sellers SE, et al. Ex vivo expansion of genetically marked rhesus peripheral blood progenitor cells results in diminished long-term repopulating ability. Blood 92:1131–1141, 1998.

Gene Therapy

Blaese RM, Culver KM, Miller AD, et al. T lymphocyte-directed gene therapy for ADA-SCID: Initial trial results after 4 years. Science 270:475–480, 1995.

Bordignon C, Notarangelo LD, Nobili N, et al. Gene therapy in peripheral blood lympho-cytes and bone marrow for ADA-immunodeficient patients. Science 270:470–475, 1995.

Hesdorffer C, Ayello J, Ward M, et al. Phase I trial of retroviral-mediated transfer of the human MDR1 gene as marrow chemoprotection in patients undergoing high-dose chemotherapy and autologous stem-cell transplantation. J Clin Oncol 16:165–172, 1998.

Heslop HE, Ng CYC, Li C, et al. Long-term restoration of immunity against Epstein-Barr virus infection by adoptive transfer of gene-modified virus-specific T lymphocytes. Nat Med 2:551–555, 1996.

Kohn DB, Weinberg KI, Nolta JA, et al. Engraftment of gene-modified umbilical cord blood cells in neonates with adenosine deaminase deficiency. Nat Med 1:1017–1023, 1995.

Malech HL, Maples PB, Whiting-Theobald N, et al. Prolonged production of NADPH oxidase-corrected granulocytes after gene therapy of chronic granulomatous disease. Proc Natl Acad Sci USA 94:12133–12138, 1997.

Riddell SR, Elliott MM, Lewinsohn DA, et al. T-cell mediated rejection of gene-modified HIV-specific cytotoxic T lymphocytes in HIV-infected patients. Nat Med 2:216–223, 1996.

Rill DR, Santana VM, Roberts WM, et al. Direct demonstration that autologous bone marrow transplantation for solid tumors can return a multiplicity of tumorigenic cells. Blood 84:380–383, 1994.

Rooney CM, Smith CA, Ng CYC, et al. Use of gene-modified virus-specific T lymphocytes to control Epstein-Barr-virus-related lymphoproliferation. Lancet 345:9–13, 1995.

Hematopoietic Stem Cell Gene Therapy

Dunbar CE. Gene transfer to hematopoietic stem cells: Implications for gene therapy of human disease. Annu Rev Med 47:11–20, 1996.

Kume A, Hanazono Y, Mizukami H, Urabe M, Ozawa K. Hematopoietic stem cell gene therapy: A current overview. Int J Hematol 69:227–233, 1999.

CHAPTER 7

Gene Therapy for Liver Disease

CHRISTY L. SCHILLING, MARTIN J. SCHUSTER, and GEORGE WU, M.D., PH.D.

BACKGROUND

The liver is a complex organ both in anatomy and function. These present challenges as well as provide opportunities for gene therapy of liver disease. Anatomically, the liver is a wedged-shaped, mutilobular, large organ. In adults, on the average, the liver comprises 1.8 to 3.1% of total body weight. In children, the ratio is even larger, up to 5.6% of body weight at birth. The liver receives blood from both the portal vein and the hepatic artery, thus providing systemic ports of entry for therapeutic approaches. The portal vein is the nutrient vessel carrying blood from the entire capillary system of the digestive tract, spleen, pancreas, and gallbladder. The hepatic artery provides an adequate supply of well-oxygenated blood to the liver. Innervation of the portal vein and hepatic artery alter the metabolic and hemodynamic functions of the liver. The functional unit of the liver is the acinus, which is a small parenchymal mass consisting of an arteriole, portal venule, bile ductule, and lymph vessels. A zonal relation exists between the cells of the acini and their blood supply. Different metabolic functions occur in the cells of each zone. For example, gluconeogenesis occurs in cells of zone 1, the area first to be supplied with fresh oxygenated blood. Cells of zone 3 actively metabolize alcohol and biotransform or detoxify drugs. Thus, different zones of liver tissue may need to be targeted for therapy of metabolic dysfunction. The recent discovery of hepatic stem cells and cellular lineages also has great implications to liver gene therapy. These discoveries indicate that cellular characteristics, phenotype, function, and metabolism are unique to a cellular level in the liver as well as based on zonal location. Thus, the liver exhibits both microheterogeneity and complexity at various levels that challenge the application of gene therapy to the organ.

INTRODUCTION

In the early years of gene therapy, the liver was not taken into consideration as a

---

An Introduction to Molecular Medicine and Gene Therapy, Edited by Thomas F. Kresina
ISBN 0-471-39188-3   © 2001 Wiley-Liss

major target organ. In contrast to bone marrow and peripheral blood cells, liver cells are not easily accessible and, in addition, there is no clearly separated pool of liver stem cells. Nevertheless, more recently, certain characteristics of the liver have drawn the attention of many researchers interested in gene therapy. The liver has the ability to synthesize large amounts of different proteins and performs many posttranslational modifications required for proper function of those proteins. It is also able to regenerate after partial injury. Many systemic inherited disorders such as hemophilia, familial hypercholesteremia, phenylketonuria, and other metabolic diseases could be treated by addressing the underlying genetic defect in liver cells. In addition, gene therapeutic strategies could theoretically be used to treat acquired diseases such as viral infections of the liver. Infections by hepatitis B and C viruses are major pulic health problems worldwide. For these reasons, the liver has become an important target organ for gene therapy.

At the same time, certain circumstances make the liver an especially challenging target for gene therapy. The liver is usually quiescent with respect to proliferation, that is, having few dividing cells, and, therefore, not an ideal target for gene vectors that require cell division. In addition, besides parenchymal hepatocytes, the liver contains a number of other different types of cells. These facts should be considered when choosing between different vectors and techniques of delivery of genes to liver cells. Accordingly, the first part of this chapter will discuss the basic tools, focusing on their application for hepatic gene delivery, while the second part will address the clinical applications attempted so far.

GENERAL PRINCIPLES FOR HEPATIC GENE THERAPY

There are two basic approaches for gene transfer into hepatocytes: ex vivo and in vivo strategies (Fig. 7.1). Ex vivo therapy requires the removal of a part of the liver. To obtain hepatocytes, the removed tissue is treated with collagenase, and hepatocytes are separated from nonparenchymal cells by density gradient centrifugation. Cells are then kept in culture and subjected to gene transfer by one of a variety of methods. The population of cells is selected for those successfully genetically engineered and finally reinfused via the portal vein into the patient's liver. However, hepatocytes are not readily cultured. They undergo a few rounds of cell division but not enough to substantially expand the population. Their viability is limited and culturing primary hepatocytes is hampered by some loss of differentiation. In addition, an already ill patient may not be able to undergo the harvesting procedure.

While hepatocytes are kept in culture, several methods can be used to introduce new genes. Deoxyribonucleic acid (DNA)-mediated techniques rely on commonly used transfection methods such as calcium phosphate co-precipitation with DNA and diethlyaminoethyl (DEAE) dextran complexed with DNA through electrostatic charges. These systems result in complexes that are taken up by the cell via endocytosis. Electroporation is another technique used to transfect cells. This involves the exposure of cells to electrical pulses that render the plasma membrane momentarily permeable. When performed in the presence of DNA, the membrane allows the nucleic acid to enter the cells. All three of these methods result in low levels of transfection efficiency and transient expression of the therapeutic gene. Alternatively, different viral vectors as well as liposomes can be used for ex vivo gene transfer.

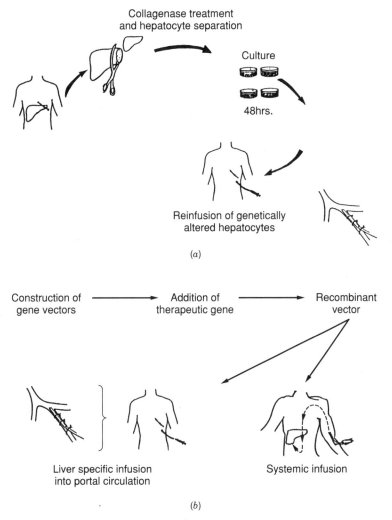

FIGURE 7.1 Two basic methods for the delivery of genes to the liver. (*a*) Shows the ex vivo approach. It requires the removal of part of the liver, usually the left lateral segment. The liver tissue is treated with collagenase and hepatocytes are separated from non-parenchymal cells by density gradient centrifugation. Hepatocytes are then propagated in culture and subjected to gene transfer. Finally successfully transformed cells are selected and reinfused via a catheter into the portal circulation of the patient's liver. (*b*) Shows the in vivo approach. A gene vector, suitable for the delivery of genes to the liver is constructed. The therapeutic gene is incorporated into this vector and the recombinant vector is infused into the patient. Systemic infusion over a peripheral vein is appropriate for vectors that selectively target the liver; direct infusion into the portal circulation is preferrable for vectors without liver targeting abilities.

For in vivo gene therapy, the therapeutic or normal gene is introduced directly into the host. On one hand, in vivo gene therapy circumvents the need for the invasive procedures of harvesting and reimplantation and eliminates the need to culture primary hepatocytes. On the other hand, it is necessary for any vehicle used for in vivo hepatic gene therapy to reach the liver efficiently. For systemic application, the gene vectors are ideally targeted to the liver, avoiding broad biodistribution and extrahepatic effects. Once inside the liver, a transgene has to pass through the fenestrations of endothelial cells to reach parenchymal liver cells, while simultaneously avoiding clearance through phagocytosis by Kupffer cells. In vivo gene therapy can also be mechanically directed to the liver by portal injection of the foreign gene construct. Presently several viral systems as well as liposomal preparations and protein–DNA conjugates have been used for in vivo gene therapy (Table 7.1).

Viral Vectors

Retrovirus   Retrovirus can infect many different types of mammalian cells including liver cells. One limitation to the use of prototype retroviruses in hepatic gene

TABLE 7.1   Advantages and Disadvantages of Vehicles Concerning Liver-Directed Gene Therapy

Vehicle	Advantages	Disadvantages
Retrovirus	No immune/inflammatory response	Requires dividing cells
	Absence of hepatic necrosis	Low expression in hepatic cells in vivo
	Integrates with stable expression	
Adenovirus	Targets hepatocytes specifically	Remains episomal
	Expressed in nondividing cells	Transient expression
		Inflammatory/immune response
		Injurious to hepatocytes
Adenoassociated virus	Expressed in nondividing cells	Small delivery capacity
	Integrates with stable expression	
	No inflammatory/immune response	
Liposomes	DNA protected from degradation	Uptake by nonparenchymal liver cells
	Large delivery capacity	Intracellular degradation in lysosomes
	No inflammatory/immune response	
Protein/DNA carriers	Liver specific	Intracellular degradation in lysosomes
	Large delivery capacity	Remains episomal
	No inflammatory/immune response	Transient expression

therapy is that only dividing cells are efficiently transduced. To circumvent this problem, researchers have performed partial hepatectomies before the administration of the retrovirus. Because the remaining liver tissue is induced to proliferate in response to this injury, the percentage of transduced cells could be increased.

Adenovirus In early adenoviral constructs, in addition to expression of the foreign gene, some viral genes were also expressed. The latter led to a virus-specific immune response manifested by development of hepatitis and destruction of the genetically altered hepatocytes. The expressed therapeutic protein usually became undetectable after a maximum period of 4 weeks. The formation of neutralizing antibodies by B lymphocytes against viral proteins make a periodic readministration less effective. This problem has been tackled by deleting additional viral genes to minimize the expression of viral proteins. It has been shown that the therapeutic gene expression level was increased in mouse liver while the immune response previously seen was decreased. Adenoviral constructs have recently been prepared in which all viral genes have been eliminated. Using a different approach, transient administration of an immunosuppressive drug resulted in the long-term expression of the adenoviral vector system. It has also been shown that it is possible to render rats immunotolerant to adenoviral antigens by intrathymic injections and oral administrations of adenoviral protein extracts or by neonatal administration of the virus in utero, thereby increasing long-term expression and allowing readministration of adenoviral vectors.

Adenoassociated Virus Adenoassociated virus (AAV) can infect dividing as well as nondividing cells making it a possible vector for use in organs such as the liver. The rate of transduction in nondividing cells, however, is lower than that of cells undergoing division. AAV transduces cells that are in S phase of the cell cycle. Treatments that interfere with DNA metabolism, such as hydroxyurea or aphidicolin and topoisomerase inhibitors, markedly increased the number of recombinant AAV transduced cells. γ-Irradiation has a similar effect on the efficiency of this system. After localized irradiation to the liver, hepatocyte transduction was increased up to 900-fold over hepatocytes of mice that were not irradiated. This is probably due to the fact that the irradiation is cytotoxic, thereby stimulating division of the surviving cells.

Nonviral Vectors

Liposomes Liposomes are microscopic vesicles consisting of one or multiple aqueous compartments. Liposome clearance from the circulation by the liver is dependent on the size and surface composition of liposomes. Because the fenestrations of the endothelial cells in the liver have a diameter of about 100 nm, particles larger than 250 kD cannot pass into the space of Disse and, therefore, do not interact significantly with hepatocytes (Fig. 7.2). For this reason, liposomes larger than 100 nm are cleared by phagocytosis through Kupffer and endothelial cells. Changing the size and lipid composition of the sphere can alter the biodistribution to the different cell populations within the liver. This allows for the targeting to either hepatocytes or Kupffer cells. One advantage of liposomes is the fact that DNA can simply be incorporated in the aqueous phase or associated with the

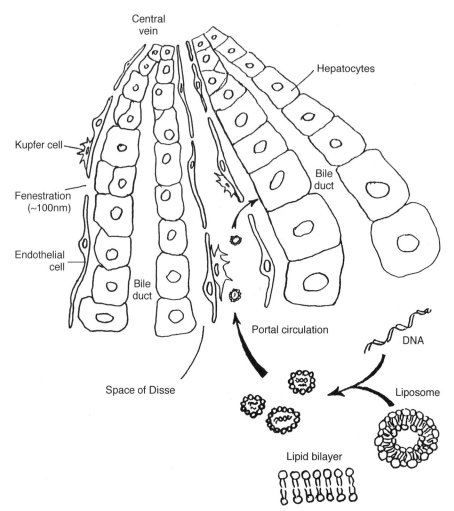

FIGURE 7.2   Liposomes are used as a device to deliver genes to hepatocytes. Liposomes are microscopic vesicles consisting of lipid bilayers enclosing one or multiple aqueous compartments. DNA is incorporated in the aqueous phase or associated with the lipid material after simply mixing with the lipid components. Liposomes enter the liver by the portal circulation. Their clearance from the circulation is largely dependent on their size and surface composition. Because the fenestrations of the endothelial cells in the liver have a diameter of about 100 nm, particles larger than 250 kD cannot pass into the space of Disse. Only small liposomes can escape uptake by Kupffer and endothelial cells and interact with parenchymal liver cells.

lipid material. In addition, the encapsulated gene is protected from enzymatic degradation.

Cationic liposomes have been used to form DNA complexes in which the DNA remains primarily on the outside of the microsphere. While this is an advantage because the DNA that can be trapped within the vesicle is limited, it may cause an aggregation of one or more liposomes and prevent uptake or promote phagocytosis by Kupffer cells. Liposomes are taken up by the cells via endocytosis and eventually enter lysosomes. In lysosomes, enzymatic degradation of the contents occurs and could decrease the efficiency of deliver of the therapeutic gene to the nucleus. To circumvent this problem, liposomes have been developed that are pH sensitive, avoiding fusion with the lysosomes. Following internalization, these liposomes change their properties when they are exposed to the low pH of endosomes. During endocytosis, they are able to destabilize the endosomal membrane or become fusogenic. In this way, the liposome may be able to deliver its contents into the cytoplasm before the liposome is delivered to lysosome.

Another means of improving the efficacy of liposomes to target parenchymal liver cells is the incorporation of various ligands recognized by receptors on the surface of hepatocytes. Examples of such targeting moieties are epidermal growth factor, lactosylceramide, asialofetuin, lactose mono-fatty acid esters, and β-galactoside. For many preparations, uptake by endothelial or Kupffer cells compared to parenchymal cells is still predominant, and there is no unanimity on the quantitative aspect of the differential uptake into different cell types in the liver. Liposomes with galactose residues are also recognized by Kupffer cells via the galactose-particle receptor, and the distribution between parenchymal and non-parenchymal liver cells is strongly size dependent, with only very small liposomes with limited loading capacity or vesicles containing lactosylceramide or lactose mono-fatty acid esters preferentially directed to parenchymal cells.

Protein–DNA Complexes Soluble conjugates between naturally occurring and recombinant proteins and DNA are attractive tools for gene therapy directed to the liver. An example of the use of targeted delivery of protein–DNA complexes is the use of asialoglycoprotein receptors. The asialoglycoprotein receptor is present in large numbers only on the plasma membrane of hepatocytes and binds galactose-terminated glycoproteins and neoglycoproteins with high affinity. Bound ligands are internalized by the cell via receptor-mediated endocytosis. Due to its specificity, the asialoglycoprotein receptor (AsGPr) has been exploited as a means to deliver drugs and DNA for therapeutic purposes, as well as diagnostic agents to hepatocytes.

A system, based on asialoglycoprotein-poly-L-lysine conjugates has been developed to target DNA to the liver via the AsGPr (Fig. 7.3). The α_1 acid glycoprotein, orosomucoid, was desialylated by treatment with neuraminidase to produce asialoorosomucoid (ASOR), a high-affinity ligand for the AsGPr. Poly-L-lysine (PL) was then covalently attached to the protein by carbodiimide-mediated amide bond formation. The resulting ASOR-PL conjugate bound the negatively charged DNA in a nondamaging electrostatic interaction and protected it from nuclease degradation. The complex was selectively and rapidly internalized into hepatocytes by receptor-mediated endocytosis, and foreign genes were expressed in vitro and in vivo. To further increase the persistence of foreign gene expression in vivo, a partial hepatectomy, leading to stimulated hepatocyte replication was performed. The underlying mechanism was shown to be the disruption of the microtubular network necessary for the translocation of endosomes to lysosomes, which could also be accomplished by colchicine administration.

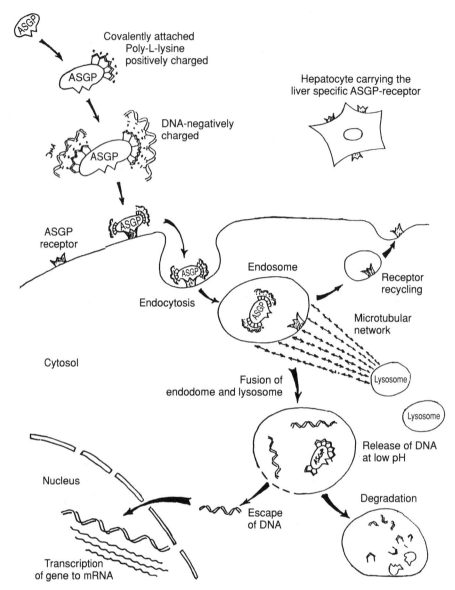

FIGURE 7.3   Use of asialoglycoprotein (ASGP) to target genes to the liver. The asialoglycoprotein receptor is present in large numbers only on the plasma membrane of hepatocytes and binds galactose-terminated glycoproteins with high affinity. Positively charged poly-L-lysine is covalently attached to ASGP by carbodiimide-mediated amide bond formation. The resulting ASOR-PL conjugate binds the negatively charged DNA in a nondamaging electrostatic interaction. The complex is internalized into hepatocytes by receptor-mediated endocytosis. After endocytosis the ligand dissociates from the receptor and the receptor recycles to the cell surface. The translocation of the endosome to the lysosome requires an intact microtubular network. After fusion of endosome and lysosome, the DNA is released from its carrier at low pH. Part of the DNA escapes the lysosome and reaches the nucleus where it can be transcribed into mRNA.

CLINICAL APPLICATIONS OF LIVER-DIRECTED GENE THERAPY

Familial Hypercholesterolemia

Familial hypercholesterolemia (FH) is an autosomal dominant disorder that affects one in every 500 people. It is caused by defects in the hepatic low-density lipoprotein (LDL) receptor gene. The reduced activity of the LDL receptor leads to an inefficient clearance of LDL particles by the liver and therefore, a limited metabolisim of LDL. Accordingly, this causes elevated serum LDL cholesterol levels, which leads to premature coronary artery disease. Heterozygotes for FH maintain only a portion of the normal LDL receptor function, and their serum LDL levels are almost double that of normal individuals. Homozygotes, having two mutant receptor genes, have only 0 to 20% of normal LDL receptor activity and show extremely elevated serum cholesterol levels. Without treatment, this usually leads to death by myocardial infarction before the age of 20.

The LDL receptor is, in fact, found on all cells. However, it is the hepatic expression of the receptor that plays the main role in regulating serum cholesterol levels. The liver is the only organ that is capable of converting cholesterol to bile acids and excreting them from the body. Pharmacological therapy for heterozygote FH patients, who express the LDL receptor at a low level involves upregulation of LDL receptor gene expression. Drugs, including 3-hydroxy-3-methylglutaryl coenzyme A reductase inhibitors and bile acid binders, act to reduce intracellular hepatic free cholesterol. This causes the LDL receptor gene expression to be influenced, accelerating LDL catabolism and, accordingly, reducing serum cholesterol. However, this treatment, combined with strict dietary reduction of cholesterol intake, is only feasible in the case of heterozygosity and does not reduce the serum cholesterol level into the normal range. For those patients that lack expression of a functional receptor due to homozygosity, or heterozygotes with an inefficient response to pharmacological therapy, weekly plasmapheresis or liver transplantation are the only alternatives. Both procedures are very expensive, and the latter is associated with morbidity and mortality and limited organ supply. For these reasons, hepatic gene therapy has been employed in an attempt to treat FH.

Early experiments in the Watanabe heritable hyperlipidemic (WHHL) rabbit, an animal model for FH, demonstrated the possibility of successful ex vivo gene therapy for FH. In these studies, hepatocytes were harvested, genetically modified ex vivo with retroviruses that contained an intact LDL receptor gene, and transplanted back into the animal. Control experiments with mock transfected hepatocytes demonstrated no cholesterol lowering effect, but showed a transient increase of the serum cholesterol levels probably due to the surgical procedure. Retroviral transduced hepatocytes were shown to become stably engrafted into the animal's liver with a subsequent lowered serum cholesterol level. The effect was observed for 6.5 months, the duration of the experiment. Subsequent experiments with dogs and baboons also rendered encouraging results. In the case of the baboon, 1.5 years after gene therapy, the transgene was still being expressed. The results of these early experiments provided support for the efficacy of this treatment and paved the way for human clinical trials.

A 28-year-old French Canadian woman was the first recipient of liver-directed gene therapy. She was homozygous for a mutation in the LDL receptor gene, resulting in the expression of a nonfunctional receptor. After suffering a myocardial infarction at the age of 16, she had a coronary artery bypass at the age of 26. Her baseline serum LDL concentration was 482 mg/dl (normal range 194 ± 34), and her dyslipidemia did not respond to conventional drug therapy. The left lateral segment of the patient's liver, comprising about 15% of total mass, was removed and the parenchymal liver cells were isolated. The cells were then transduced with a retroviral vector containing the full-length human LDL receptor gene under the control of a chicken β-actin promoter and a cytomegalovirus (CMV) enhancer. To select for successful transduction, cells were analyzed for the ability to uptake fluorescent labeled LDL. Only genetically altered hepatocytes were reinfused into the portal circulation (Fig. 7.4). The patient tolerated the procedures well without relevant side effects.

Immediately following infusion of the genetically altered cells, the patient's serum LDL dropped by 180 mg/dl. A new baseline was established that was 17% lower than before gene therapy. As her (LDL) decreased, her high-density lipoproteins (HDL) levels increased, improving her LDL/HDL ratio from 11 ± 0.4 to 7.9 ± 0.9. It is unclear as to why the HDL increased, although this same phenomenon has been observed in patients that underwent orthotopic liver transplantation. The patient also responded to the drug lovastatin, which prior to gene therapy had no effect. Lovastatin is thought to deplete intracellular cholesterol, thereby up-regulating expression of the LDL receptor. The recombinant receptor gene had no transcriptional elements that could respond to cholesterol-mediated regulation. This indicates that the response to lovastatin was related to posttranscriptional regulation, a mechanism demonstrated in previous studies. The response to lovastatin diminished the patient's serum LDL level further to 356 ± 22 mg/dl, and the effect was meanwhile stable over a period of 2.5 years.

There was no immune response to the recombinant receptor. The patient's sera contained no antibodies to the recombinant receptor when a western blot analysis was performed. Also, there was no evidence for autoimmune hepatitis following gene therapy. In an extension of this study, four other FH individuals, including two receptor-negative patients, were treated in a similar manner. Engraftment of successfully transduced hepatocytes as well as transgene expression was shown for all patients, without significant side effects. Two out of four patients experienced a significant improvement in their serum lipid profile, with a maximum reduction in serum LDL of 150 mg/dl in one of the receptor-negative patients. None of the patients developed an immune response to the transgene or to retroviral proteins. Although gene transfer was demonstrated in all patients, the clinical impact on the disease was low with serum cholesterol levels still exceedingly above the normal range. In summary, this first human clinical trial showed the feasibility of ex vivo gene therapy for FH but demonstrated the need for substantial modifications to improve the percentage of transduced hepatocytes and the level and duration of gene expression.

In an alternative approach, in vivo gene delivery was performed to treat WHHL rabbits. The human LDL receptor gene was placed under the control of transcriptional elements from the mouse albumin gene, conferring efficient expression in hepatocytes. The construct was conjugated via poly-L-lysine to ASOR, a high-affinity ligand for the ASOR receptor. Following systemic injection of this complex, analysis of WHHL rabbits revealed a rapid and liver-specific uptake of the

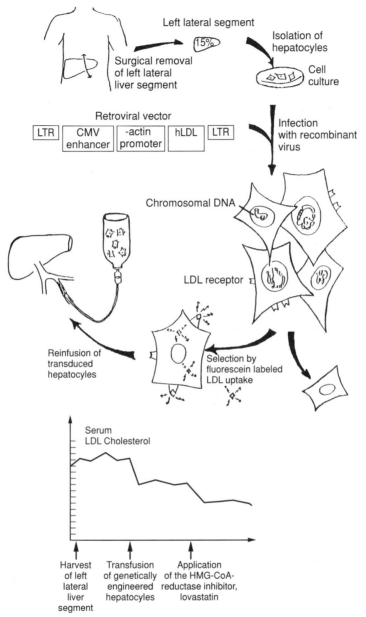

FIGURE 7.4 Gene therapy for LDL receptor deficiency. The left lateral liver segment of a patient homozygous for a mutation in the LDL receptor gene is removed and hepatocytes are isolated. The cells are transduced in culture with a retroviral vector containing the full-length human LDL receptor gene under the control of a chicken β-actin promoter and an cytomegalovirus (CMV) enhancer. The successfully transduced cells are selected by the use of fluorescein-labeled LDL. Only genetically altered hepatocytes are reinfused into the portal circulation of the patient. The patients baseline serum LDL concentration was 482 mg/dl. Immediately following infusion of the transduced hepatocytes, the patients serum LDL dropped by 180 mg/dl. In addition the patient now responded to lovastatin, a HMG-CoA reductase inhibitor, which prior to gene therapy had no effect. The observed reduction in the patients serum LDL level is meanwhile stable over a period of 2.5 years.

DNA–protein conjugate, followed by expression of the transgene. The animals experienced an immediate, but transient, decrease in total serum cholesterol by 153 ± 53 mg/dl. In control experiments, animal injected with a construct carrying the CAT (chloramphenicol acetyltransferase) reporter gene instead of LDL receptor gene showed CAT expression, but no diminuation of serum cholesterol levels. In this study the expression was only 2 to 4% of the endogenous level of LDL receptor expression, and the effect on the serum lipid profile lasted less than one week. These initial results were encouraging because of the specificity of the delivery. However, the low levels and short duration of recombinant gene expression were disappointing.

In recent animal studies, recombinant adenoviruses were used for in vivo liver-directed transfer of the LDL receptor gene. It was possible to restore LDL receptor expression in WHHL rabbits and LDL receptor knock-out mice, leading to substantial reductions in serum cholesterol levels. However, the expression of the recombinant receptor as well as the effect on the lipid profile has been only transient. This was due to the immune response that the host mounted against a low-level expression of viral proteins, with the subsequent destruction of the genetically altered cells. Especially in receptor-negative subjects, the expression of an LDL receptor could also trigger an immune response against the neoprotein, which would further reduce the expression of the transgene. To circumvent this problem, another group of researchers delivered the very low density lipoprotein (VLDL) receptor gene to the liver of LDL receptor knock-out mice using recombinant adenoviruses. Since the VLDL receptor is already expressed in extrahepatic tissue, there is no immune response to the receptor after hepatic expression. Also the VLDL receptor binds LDL with a low affinity. It mediates the uptake of VLDL, the precursor of LDL, and, therefore, results in a decrease of serum cholesterol.

Hemophilia B

Hemophilia B is an X-linked recessive coagulation disorder caused by a deficiency or functional defect of blood clotting factor IX. The condition can be life threatening without regular infusions of factor IX concentrates in patients with evidence of bleeding. Extensive testing of these products can eliminate impurities, but this form of therapy still bears the risk of transfusion-transmitted viruses such as hepatitis C and human immunodeficiency virus (HIV). In addition, the half life of factor IX is only 24 h and, therefore, makes repeated transfusions often necessary. The liver is the primary source for circulating factor IX and the prime target for a gene therapeutic approach to treat hemophilia B.

To date, attempts have been made in animal systems using the ex vivo approach. The problems with these therapies are similar to those that have been encountered with correcting other disorders: (1) the concentrations of circulating factor IX are low and (2) there is a loss of gene expression over time. The latter is due to loss of transduced cells or inactivation of the expression vectors.

There is a well-characterized canine model that has been used in preclinical trials for hemophilia B. These dogs have no detectable factor IX activity due to a missense mutation in the catalytic domain. A retrovirus vector that contained the canine factor IX gene under the control of retroviral promoter and enhancer elements was used for direct delivery to the dogs liver via infusion into the portal circulation.

Analysis by ELISA and a biological assay demonstrated that plasma levels of 2 to 10 ng/ml of factor IX were achieved. In a normal canine, the level is about 11.5 μg/ml. While the levels of circulating factor IX did not reach that of wild-type dogs, there was a dramatic improvement in the biochemical parameters of hemostasis. This was demonstrated by measuring the whole blood clotting time (WBCT), which in normal dogs is 6 to 8 min. In dogs that have hemophilia B, the WBCT was about 45 to 50 min. After undergoing gene therapy this time was reduced more than 50% with times in the range of 18 to 22 min. Although the concentration of factor IX was as little as 0.1% of normal values, there was a dramatic improvement in clotting times. Also, encouraging is the fact that this effect remained stable for over 9 months (Fig. 7.5).

Adenoviral vectors that express canine factor IX have also been used to treat hemophilia B dogs. Viral particles (2.4×10^{12}) were infused into the portal vasculature of the dogs. The animals produced 2 to 3 times the wild-type level of factor IX. However, the effect was only transient. The increase in factor IX concentration did normalize their clotting times, but the levels and clinical parameters returned to pretreatment levels in 2 months. While repeated administrations could be considered, it is possible that an immune response could develop with subsequent treatment.

Another group of researchers tried using adenoassociated viral (AAV) vectors to express human factor IX in mouse livers. They simply injected the mice in a tail vein with the recombinant vector after γ-irradiation was applied to the liver. As previously discussed, this treatment probably stimulates cells to divide, thereby improving the efficacy of adenoassociated viral gene therapy. The concentration of human factor IX in mice transduced with the AAV vector was between 0.1 and 1 ng/ml. This result is similar to that observed in the dog model. The normal values for human factor IX was 5 μg/ml, while levels of about 100 ng/ml would prevent chronic disease.

α₁-Antitrypsin Deficiency

α_1-Antitrypsin (AAT) is a serum glycoprotein, predominantly synthesized in the liver and secreted into the blood. It is a protease inhibitor whose function is essential in protecting the alveolar surface of the lung from destructive protease activity. Its major substrate, neutrophil elastase (NE), is released by neutrophils during phagocytosis, membrane perturbation, or cell lysis and cleaves connective tissue matrix proteins located in alveolar walls. In normal individuals the levels of AAT are sufficient to neutralize circulating NE. The different forms of AAT deficiency result in reduced plasma levels of the protease inhibitor and in the failure of NE to be neutralized. This is manifested in a high risk for the early development of pulmonary emphysema, due to proteolysis of the pulmonary extracellular matrix.

The normal gene for AAT is designated M and accounts for 95% of alleles in the caucasian American population. The most common mutants, called Z and S occur with an allelic frequency of 1 to 2% and 2 to 4%, respectively, in this population. In contrast Asians and African Americans are minimally affected. Homozygous individuals for the Z allele have only 10 to 15% circulating AAT levels bearing a certain risk for pulmonary emphysema. Homozygous individuals for the S allele and MS or MZ heterozygotes are phenotypically normal. However, some SZ heterozygotes

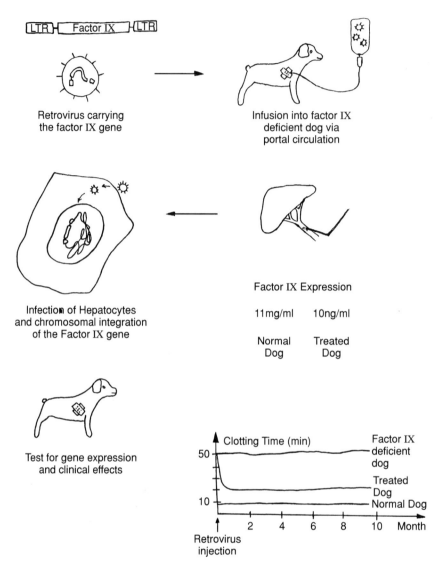

FIGURE 7.5 Gene therapy for factor IX deficiency. A recombinant retrovirus vector is constructed that contains the canine factor IX gene under the control of retroviral promoter and enhancer elements (LTR). This vector is infused into the portal circulation of dogs that have no detectable factor IX activity. The retrovirus is taken up by liver cells and the provirus DNA integrates into the chromosomal DNA. Analysis of the dogs' plasma by ELISA reveals plasma factor IX levels of 2 to 10 ng/ml. A normal canine has a plasma factor IX level of about 11.5 μg/ml. While the levels of circulating factor IX in the treated dog does not reach that of wild-type dogs, there was a dramatic improvement in the whole blood clotting time.

could display an increased risk for the manifestation of pulmonary emphysema. Homozygosity for the so-called null allele results in a complete lack of AAT in the plasma, and these patients are extremely likely to develop emphysema. The same is true for heterozygotes bearing an S or Z allele with the null allele. A number of different mutations are responsible for the null allele, ranging from point mutations to complete deletions. About 10% of individuals homozygous for the Z allele bear the additional risk of significant clinical liver injury, probably due to the accumulation of misfolded AAT in the ER of hepatocytes.

The current treatment for AAT deficiency consists of weekly intravenous applications or intratracheal inhalation of human AAT, produced from serum. Recombinant human AAT, synthesized in bacteria or yeast has the disadvantage of a shorter half-life and increased renal clearance due to improper posttranslational glycosylation. While the administration of human AAT has been shown to raise the serum AAT activities in patients, the response is only temporary, and a significant impact on the prevention of pulmonary damage has yet to be proven for the intravenous as well as the intratracheal application. α_1-Antitrypsin deficiency is another candidate disease for gene replacement therapy, whereby ideally, the correct gene could be delivered to hepatocytes and offer a long-term stable production of AAT. Attempts to correct this disorder have been studied on dogs where the introduction of the correct gene was performed in an ex vivo manner. After transplantation of retroviral transduced hepatocytes, the cells achieved peak production of human AAT in vivo at day 10 posttransplantation. However, these levels dropped and became undetectable around day 47.

Another group of investigators attempted an in vivo approach using small liposomes as the method of gene delivery. A plasmid containing the full-length human α_1-antitrypsin gene was encapsulated in small liposomes and was intravenously injected into mice. A single dose of liposomal-delivered plasmid induced the production of human AAT in mouse hepatocytes and resulted in measurable levels of human AAT in mouse plasma, still detectable after 11 days. In control experiments, the injection of free plasmid did not result in measurable AAT expression (Fig. 7.6).

Interestingly, there was no additive effect when additional doses of the liposome complex were delivered. However, partial hepatectomy performed 3 h after the intravenous application of the liposomal formulation increased human AAT plasma levels significantly. On day 11 after the intravenous (IV) injection, human AAT levels had increased 6.4 times compared to animals injected without the performance of partial hepatectomy. It is unclear why the repetitive application did not further increase the gene expression. Also, it is not completely understood why the stimulation of cell proliferation by partial hepatectomy increased gene expression. Presumably, this may be due to mechanisms that alter the compartmentalization of liposomal-delivered DNA within the cells, allowing escape from the lysosomal degradative pathway.

Crigler–Najjar Syndrome (Bilirubin UDP β-D Glucuronosyltransferase Deficiency)

Bilirubin is the principal degradation product of heme. The enzyme that catalyzes the coupling of bilirubin with glucuronic acid is bilirubin UDP-glucuronosyltransferase (B-UGT). The prototype of an inherited bilirubin conjugation disorder is Crigler–Najjar (CN) syndrome type I. Patients with this recessively inherited disease are characterized by high serum levels of unconjugated bilirubin, with little or no conjugated pigment in the bile. They do not respond to enzyme induction therapy with phenobarbitol and suffer a variety of neurological damages such as motor

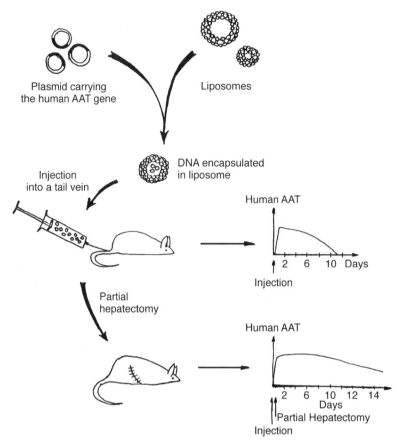

FIGURE 7.6 Gene therapy for α_1-antitrypsin (AAT) deficiency. A plasmid that contains the full-length human AAT gene is encapsulated in small liposomes. The liposomes are injected into the tail vein of a mouse. A single dose of liposomal-delivered plasmid induces the production of human AAT in mouse hepatocytes and results in measurable levels of human AAT in mouse plasma, lasting 11 days. If a partial hepatectomy is performed 3 h after the intravenous application of the liposomal formulation, AAT plasma levels are significantly higher.

abnormalities, hearing loss, kernicterus, and finally death. At present, the only definitive treatment for this disorder is liver transplantation. A similar defect exists in Gunn rats, which are homozygous for the mutation and, therefore, show no hepatic B-UGT activity. These rats exhibit lifelong hyperbilirubinemia and develop bilirubin encephalopathy. They provide a model system for studies on the efficacy of gene therapy for Crigler–Najjar syndrome type I.

An example of transient in vivo correction of this defect has been made by targeted delivery of the human *B-UGT* gene to the liver of Gunn rats using asialoglycoprotein poly-L-lysine DNA conjugates as previously described. As a strategy to prolong the duration of targeted gene expression, advantage was taken of the fact that the translocation of endosomes to lysosomes as part of the endocytotic degradative pathway requires an intact microtubular network. Colchicine, a microtubule disruptive agent, was administered 30 min prior to the injection of the ASOR–DNA complex to prevent the translocation of the endosomal vesicles containing the ligand to lysosomes. Targeted delivery of *B-UGT* under these conditions resulted in the persistence of the delivered DNA in the liver for 10 weeks. Bilirubin glucuronides were excreted in the bile and serum bilirubin levels decreased by 25 to 35% in 2 to 4 weeks and remained reduced for a period of 8 weeks. Without treatment with colchicine, the DNA remained in the liver for only 2 days and there was no effect on serum bilirubin levels.

These studies used concentrations of colchicine that would be toxic to humans. There are other drugs that could produce the same effect yet are safe for application in clinical human trials. Alternatively, to avoid side effects and broad biodistribution, colchicine could be delivered in a liver-specific manner. In this way, microtubular disruption provided a noninvasive method for prolonging the effect of this liver-specific method of gene therapy.

As discussed previously, recombinant adenoviruses are efficient in transferring foreign genes to quiescent, nondividing cells and high levels of gene expression can be achieved using this vector system. However, since they do not integrate their DNA into the host genome, subsequent administrations will be necessary. Therefore, the immune response, usually evoked after the initial injection has yet to be circumvented. Gunn rats were used to address this problem. Previously delivering the human *B-UGT* gene via recombinant adenovirus has proven to be effective for a short period. Treated animals showed excretion of bilirubin glucuronides and a 70% reduction of serum bilirubin levels. This effect was only transient due to the immune response mounted against adenoviral antigens, expressed by transduced hepatocytes. The same effect was not seen in subsequent applications to the same animals due to neutralizing antibodies. A group of researchers investigated whether the administration of recombinant adenovirus during the neonatal period could induce a tolerance to the recombinant adenovirus. Gunn rats (1 to 3 days old) were injected with 1×10^8 plaque forming units (pfu) of recombinant adenovirus carrying the human *B-UGT* gene. Subsequent injections were administered 56 and 112 days later. Control experiments were performed using recombinant adenovirus that contain the *Lacz* reporter gene. Animals that received the *B-UGT*, but not those that received *Lacz*, had a reduction of serum bilirubin levels by 70 to 76% as compared to untreated animals. There was a gradual increase of serum bilirubin levels by day 53, but the second and third injection of recombinant adenovirus had an additive effect on serum bilirubin levels. Analysis also showed that antibodies and cytotoxic lymphocyte activity to the recombinant adenovirus were not detectable. This demonstrates that injecting the recombinant adenovirus during the neonatal stage tolerized the animals and permitted long-term therapy with repeated administrations.

One concern with this treatment is the question if the induction of tolerance against the recombinant adenovirus could result in tolerance to wild-type virus as well. Adenoviral infections are common throughout the life span of a human being, usually manifested as self-limited, uncomplicated disease. The same group of researchers injected two doses of wild-type virus into Gunn rats previously tolerized with three doses of recombinant adenoviruses starting in the neonatal period. The animals elicited a cytotoxic T-lymphocyte immune response after the first injection of wild-type virus, which was further increased after the second injection. Interestingly, the animals continued to express the transferred *B-UGT* gene and did not experience an increase in unconjugated serum bilirubin levels.

Gene Therapy for Viral Infections

In contrast to many other gene therapeutic strategies, where replacement of a defective gene is the predominant goal, the therapy of viral infections by means of gene therapeutic technology is to inhibit viral replication, transcription, or translation of viral genes or assembly of viral particles. If the nucleic acid sequence of a viral gene is known, antisense oligonucleotides consisting of short single strands of DNA can be designed to bind the corresponding messenger ribonucleic acid (mRNA) (e.g., the sense strand) by complementary base pairing. This can result in direct inhibition of translation or cleavage of the RNA component of RNA–DNA hybrids by intracellular RNase H (Fig. 7.7). Antisense oligonucleotides are usually 15 to 20 bases long and made by the use of an automated DNA synthesizer.

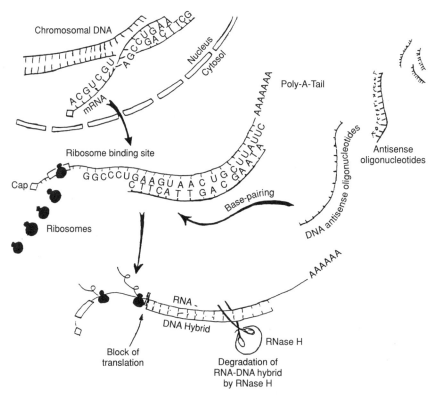

FIGURE 7.7 Antisense oligonucleotides. Chromosomal DNA is transcribed into messenger RNA (mRNA), containing a cap at the 5′ and a poly-A-tail at the 3′ terminus. Messenger RNA leaves the nucleus for the cytosol where translation into proteins takes place. Translation is performed by ribosome and requires a ribosome binding site. Antisense oligonucleotides consist of a short single strand of DNA. If the nucleic acid sequence of a viral gene is known, they can be designed to bind the viral mRNA by complementary base pairing. This results in direct inhibition of translation or cleavage of the RNA component of the RNA–DNA hybrid by RNase H. Replication of hepatitis B and C virus depends on an RNA intermediate. Therefore antisense oligonucleotides can interfere with viral replication.

A related strategy uses ribozymes to suppress viral replication or transcription of viral genes. Ribozymes are RNA molecules with a catalytic moiety capable of cleaving target RNA molecules surrounded by RNA arms able to bind to the target sequence by complementary base pairing similar to antisense oligonucleotides (Fig. 7.8). Theoretically, one ribozyme can cleave many target RNA molecules. Transfection of a vector containing the sequence of a ribozyme could result in the generation of many copies of therapeutic ribozyme molecules within target cells.

Another antiviral strategy consists of the use of dominant negative polypeptides, designed to interact with their native counterparts, thereby interrupting viral assembly or enzyme function.

Chronic Viral Hepatitis

There are at least five different viruses causing hepatitis in human. Hepatitis A virus and hepatitis E virus, contagious predominantly through a fecal-oral route, cause

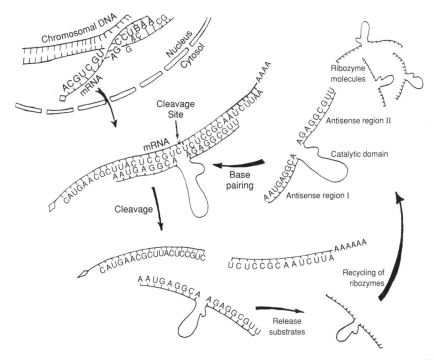

FIGURE 7.8 Ribozymes. Chromosomal DNA is transcribed into messenger RNA (mRNA), containing a cap at the 5′ and a poly-A-tail at the 3′ terminus. Messenger RNA leaves the nucleus for the cytosol where translation into proteins takes place. Ribozymes are RNA molecules with a catalytic moiety capable of cleaving target RNA molecules. The catalytic domain is surrounded by two RNA arms designated as antisense regions. The antisense regions are designed to bind the target sequence by complementary base pairing. After cleavage the substrate is released and the ribozyme recycles to cleave other target molecules. Ribozymes can cleave mRNA molecules as well as viral RNA involved in viral replication.

acute self-limited disease. Three other well-characterized viruses, hepatitis B virus (HBV), hepatitis C virus (HCV), and hepatitis D virus (HDV) are known to cause persistent infection and chronic disease of the liver.

Hepatitis B Virus HBV is a small DNA virus with a partially double-stranded circular DNA molecule of about 3200 base pairs. It belongs to a group of hepatotropic DNA viruses (hepadnaviruses) that includes the hepatitis virus of the woodchuck, ground squirrel, Pekin duck, and heron. The virus consists of an outer envelope and an internal core (nucleocapsid). The envelope is composed mainly of hepatitis B surface antigen (HBsAg). The nucleocapsid contains hepatitis core antigen (HBcAg), a DNA polymerase/reverse trancriptase, and the viral genome. Different from all other known mammalian DNA viruses, hepadnaviruses replicate via reverse transcription of an RNA intermediate, in a manner endogenous to the life cycle of RNA retroviruses (e.g., HIV). Based on this fundamental step in the replication of the virus, antiviral strategies aimed at the reverse transcription of HIV RNA or at HIV reverse transcriptase are also potentially useful against HBV infection.

A number of antisense sequences that are capable of inhibiting the replication of hepatitis B and hepatitis C viruses in vitro have been identified. Efficacy has also been observed with an antisense phosphorothioate DNA in vivo. However, because oligonucleotide uptake by cells is generally low, and susceptibility to degradation in plasma can be quite high, some form of targeting would be desirable for successful use of antisense strategies for therapy of viral hepatitis in vivo. A system, based on asialoglycoprotein-poly-L-lysine conjugates, was used to prepare ASOR-PL complexes with an 21-mer antisense oligonucleotide complementary to the sequence of the polyadenylation signal of the HBV genome. By using a radioactive end-labeled species, it was determined that the oligo alone was taken up with a rate of 0.05 pmol/h/million cells by two hepatoma cell lines, HepG2 (AsGPr positive) or SK Hep1 (AsGPr negative). However, the uptake of oligo conjugated to ASOR-PL was 10 times faster into HepG2 cells but was not changed in SK Hep1 cells. Co-incubation with an excess asialoorosomucoid blocked the uptake. To show whether the targeted antisense has antiviral activity, the HepG2 2.2.15 cell line was used. This cell line possesses AsGPrs, is stably transfected with the complete HBV genome, and secretes viral antigens as well as infectious virus particles. Administration of complexed antisense DNA blocked the expression of HBsAg in these cells, and reduced the replication of viral DNA by about 80% compared to untreated controls. A complexed oligonucleotide with random sequence had no effect, and the antisense oligo DNA alone decreased the expression of surface antigen and viral replication by only approximately 30%.

In a subsequent investigation, ASOR-PL complexed to a 21-mer phosphorothioate antisense oligonucleotide against the polyadenylation region and adjacent upstream sequences of WHV was used to treat WHV-infected woodchucks. Animals were injected intravenously with ASOR-PL complexes containing 0.4 mg antisense for 5 consecutive days (total dose 2 mg/animal, 0.1 mg/kg/day). Although there was no difference in the levels of surface antigen between treated and untreated animals, a significant decrease in viral burden was observed. Treated animals showed a 1 to 2 log decrease in circulating viral DNA, 25 days posttreatment. The decline lasted for approximately 2 weeks, after which there was a gradual rise in DNA levels.

Antisense alone or a complex containing a random oligo DNA of the same size and linkage failed to have any significant effect on viral DNA levels.

Targeted pretreatment of hepatocytes with the above antisense oligonucleotide complexed to ASOR-PL was used to prevent subsequent infection with HBV. Usually, it cannot be anticipated when an acute exposure to HBV will occur. However, after liver transplantation in patients infected with HBV, the grafts are invariably reinfected. Furthermore, there is an accelerated course in most cases. Protection of the graft by pretreatment could prevent reinfection. Pretreatment of Huh7 cells (AsGPr positive) with ASOR-PL antisense complexes before lipofection with an HBV plasmid (6.5 million copies of plasmid per cell) inhibited the amount of newly synthesized, core-associated viral DNA in Huh7 cells to undetectable levels, or less than 0.1 pg, as assessed by quantitative PCR. HBsAg, secreted by the cells into the medium, was inhibited in a dose-dependent manner by a maximum of 97%, and the inhibition lasted for 6 days. Pretreatment with unconjugated antisense or complexed random oligo showed no significant effects.

Very recently, a related targeting device, consisting of human adenovirus particles conjugated to *N*-acetyl-glucosamine-modified bovine serum albumin, streptavidin, and PL, was used to deliver phophorothioate-modified 16-mer antisense oligonucleotides to hepatocytes via the AsGPr. The oligonucleotide was directed against the encapsulation signal of the core gene. Chicken hepatoma cells (LHM) were transfected by complexed HBV–DNA. When the cells were treated with complexed oligonucleotide before and after treatment with complexed HBV–DNA, an approximately 80% inhibition of core-particle-associated HBV–DNA level was observed.

Another antiviral strategy consists of the use of dominant negative polypeptides, designed to interact with their native counterparts, thereby interrupting viral assembly or enzyme function. Mutants of HBV core protein were shown to inhibit wild-type viral replication by interference with nucleocapsid formation.

Hepatitis C Virus   HCV contains a single-stranded RNA genome of positive polarity and is about 9500 bp in length. Its replication requires a negative stranded RNA intermediate synthesized by the viral RNA dependent–RNA polymerase. The viral genome encodes a single polyprotein of 3010 to 3033 amino acids in length. Posttranslational processing results in the RNA binding nucleocapsid protein C, the envelope proteins E1 and E2, and the nonstructural proteins NS1 to NS5, including RNA-dependent RNA polymerase. At both termini of the RNA genome exist conserved sequences called noncoding regions (NCR), involved in RNA replication, translation initiation, and presumably RNA packaging.

Presently, animal models are limited to chimpanzees. For this reason, in vitro studies using artificial reporter constructs frequently are employed to investigate new treatment involving gene therapy for hepatitis C. In an early investigation, hepatitis C virus cDNA was cloned and used for screening highly conserved regions of the hepatitis C genome for potential target sequences in an antisense approach. After transcription with T7 RNA polymerase, HCV RNA was purified and mixed with a 10-fold molar excess with sense or antisense oligonucleotides. These mixtures were used for in vitro translation in a rabbit reticulocyte lysate in the presence of ^{35}S-methionine to synthesize HCV proteins. Sense oligonucleotides showed no significant inhibitory effect on HCV protein synthesis as measured by the incorporation of ^{35}S-methionine. In contrast, an antisense oligonucleotide directed against the 5′ NCR inhibited in vitro translation more than 50%. Another antisense oligonucleotide directed against the start codon of the HCV core gene inhibited in vitro translation up to 97%. Interestingly, antisense oligonucleotides directed against further downstream sequences had no inhibitory effect on translation, presumably due to the inefficiency blocking ribosomal translocation during translation. It is noteworthy that a molar ratio of oligonucleotide to HCV RNA of 10 to 1 was necessary to achieve the reported effects.

In subsequent studies, the ability of antisense oligonucleotides to inhibit translation in cell culture was investigated. Human hepatoma cell lines were transfected with plasmids carrying conserved HCV target regions either downstream of a CMV promoter or upstream of a luciferase reporter gene. Four different antisense oligonucleotides that were directed against the 5′ NCR were co-transfected with the reporter construct. At a concentration of 0.3 μM (~3 μg per 35 mm cell culture dish) two showed an inhibitory effect of 95% on luciferase activity. It is important to note that sense oligonucleotides also inhibited luciferase expression up to 30%.

Ribozymes have been shown to be effective against hepatitis B and hepatitis C viral RNA. Until now experiments using ribozyme technology have been demonstrated to cleave HBV RNA in vitro, but no suppression of HBV replication or HBV protein translation have been reported in cell systems or in vivo.

For HCV, suppression of viral gene expression in cells by ribozymes was successfully demonstrated. Again a plasmid carrying an HCV-luciferase reporter gene was constructed with the 5′ NCR and part of the core gene placed between a CMV promoter and the luciferase gene. Additionally, four vectors carrying the sequence for hammerhead ribozymes directed against the 5′ NCR or core region were used to synthesize ribozyme molecules for in vitro studies. After in vitro transcription of HCV-luciferase RNA, the different ribozyme molecules were investigated for their cleavage activity. The human hepatoma cell line Huh7 was then used to investigate the in vivo activity. Cells were co-transfected with ribozyme RNA and HCV-luciferase RNA at molar ratios of 0:1, 3:1, 10:1, and 30:1, the first ratio serving as the control. Two of the ribozymes, directed against the 5′ NCR and core region, respectively, suppressed luciferase activity by 73% (ribozyme:reporter gene ratio 10:1) and 55% (30:1), respectively. Control experiments with ribozymes harboring a mutation in their catalytic region did not show any inhibitory effect at the same molar ratio. Co-transfection of the HCV reporter plasmid and eukaryotic expression vectors encoding the two most promising ribozymes with a 20-fold molar excess of the ribozyme vector showed suppression of luciferase activity by approximately 50 and 40%. Control experiments with ribozymes not directed against HCV or co-transfection of a vector carrying the luciferase gene without upstream HCV sequences proved the specificity of the observed effect.

Finally, cell lines constitutively producing the two most promising ribozymes after stable transfection with the ribozyme carrying vectors were investigated. Ribozyme expressing cells were transiently transfected with the HCV-luciferase reporter plasmid and showed an inhibition of luciferase activity of 30 and 50% compared to parental cells transiently transfected with the reporter construct. When a conventional luciferase reporter plasmid was transiently transfected, ribozyme-expressing cell lines and parental cells showed no difference in luciferase activity.

Hepatocellular Carcinoma

Hepatocellular carcinoma (HCC) is one of the most common malignancies affecting man and causes an estimated one million deaths per year worldwide. Identified major risk factors are chronic infection with hepatitis B or C virus, liver cirrhosis, especially due to alcohol abuse or genetic hemochromatosis, and repeated exposure to aflatoxin. Surgery is the only curative therapy for HCC. However, due to the extent of the tumor and associated cirrhosis at the time of diagnosis, it is inappropriate in the majority of patients. The search for new therapies has not yet resulted in a significant improvement of the extremely poor prognosis of patients with unresectable HCC.

Compared to the above-mentioned disorders, gene therapy for HCC faces additional challenges. For example, it should be noted that tumors are diverse, and a single malignancy does not contain a homogenous population of cells. Tumor cells can be diverse in reference to cell surface receptors as well as cell turnover. Solid tumors contain rapidly dividing cells as well as quiescent cells. Perhaps the most difficult task is the fact that many HCC are multilocular or metastatic at the time of diagnosis, requiring systemic treatment. Until now gene therapeutic trials for HCC have been investigated in animal models and have not reached the state of clinical trials.

At the present time, most of the studies on gene therapy for HCC attempt to increase the immunogenicity of the tumor. This can be accomplished by transferring a gene that codes for a neoantigen into tumor cells or by amplifying or evoking an immune response against the malignant cells through the introduction of genes coding for a cytokine. Alternatively, the "suicide-gene" approach, in which a gene, coding for an enzyme, is introduced into tumor cells to convert a harmless prodrug into a cytotoxic agent inside of tumor cells making the tumor sensitive to exposure to prodrug.

In one of the first studies, recombinant retroviruses were constructed, carrying the varicella-zoster virus thymidine kinase (VZV-*tk*) gene transcriptionally regulated by either the hepatoma-associated α-fetoprotein or the liver-associated albumin promoter sequences. Cells expressing VZV-*tk* became selectively sensitive to the harmless prodrug araM which is converted to the cytotoxic araATP by VZV-*tk*, producing a cell-specific cytotoxic effect. With the inclusion of the α-fetoprotein promoter, the expression of the VZV-*tk* should only occur in HCC cells producing α-fetoprotein and not in normal α-fetoprotein negative hepatocytes (Fig. 7.9).

In subsequent studies HCC cells were transduced by the use of an adenoviral vector containing the herpes simplex virus thymidine kinase (HSV-*tk*) gene, rendering cells sensitive to the prodrug gancyclovir, which is also converted by the thymidine kinase into a toxic triphosphate form. After implantation of gene-transduced tumor cells into nude mice, complete regression of these tumors could be achieved by gancyclovir exposure. It was also possible to demonstrate an antitumor effect by the direct injection of the adenoviral vector into preestablished tumors. In addition, since the HSV-*tk* gene was under the control of an α-fetoprotein promoter, only tumors expressing α-fetoprotein could be successfully treated and, therefore, all other cells are spared. It was shown that the transduction of only a small number of tumor cells can result in almost a complete regression of the mass. The explanation for this observation is called the "bystander" effect and most likely due to

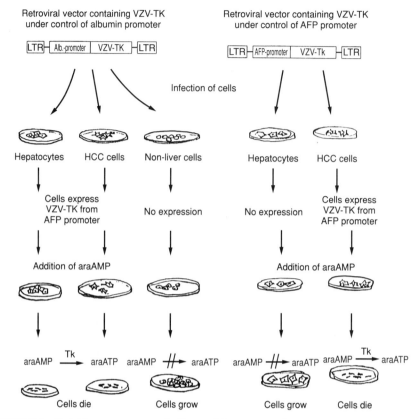

FIGURE 7.9 Suicide gene approach. Recombinant retroviruses are constructed, carrying the varicella-zoster virus thymidine kinase (VZV-*tk*) gene under control of either the albumin (alb, left part) or the α-fetoprotein promoter (right part). Hepatocytes or HCC cells express albumin and therefore express VZV-*tk* from the albumin promoter. Nonliver cells do not express VZV-*tk* from the albumin promoter (left part). HCC cells express α-fetoprotein and therefore express VZV-*tk* from the α-fetoprotein promoter. Hepatocytes do not express VZV-*tk* from the α-fetoprotein promoter (right part). Cells expressing VZV-*tk* become selectively sensitive to the harmless prodrug araM, which is converted to the cytotoxic araATP by VZV-*tk*.

immunological mechanisms evoked by the death of the transduced tumor cells or by the release of the cytotoxic triphosphate into the extracellular space.

In an alternative approach, a retrovirus vector expressing the TNF-α gene was used to transduce hepatocellular carcinoma cells. The use of albumin or α-fetoprotein regulatory elements results again in a liver cell or HCC cell specific gene expression. Neither the infection nor the expression of TNF-α had any cytotoxic effect on the proliferation or the viability of the cells in vitro, compared to the unmodified parental HCC cells. This was true for both of the TNF-α encoding retrovirus vectors, as well as for a control retrovirus vector, containing only the neomycin resistance gene. After subcutaneous injection of the transduced HCC cells into mice, only 1 of 20 animals developed a tumor, whereas 10 of 10 and 8 of 10 mice injected with the parental HCC cells or the control vector-infected HCC cells, respectively, developed tumors. The former group of 19 animals, which had not experienced any tumor growth after injection with TNF-α-transduced HCC cells, showed a partial resistance to the parental tumor cells. This was demonstrated by a rechallenge with the same number of parental HCC cells implanted in the vicinity of the previous injection site, which resulted in the development of subcutaneous tumors in only 4 of 19 animals. However, there is no unanimity about the involved immunological mechanisms: neither the prevention of chemotactic recruitment and migration of macrophages nor the depletion of CD4 or CD8 T lymphocytes nor a sublethal dose of whole-body radiation before the injection of the tumor cells prevented the effect of TNF-α. On the other hand, the method was shown to be effective in nude mice, and therefore, appeared to be independent of an intact T-lymphocyte function. The involvement of macrophages as well as T lymphocytes was demonstrated by immunohistochemical analysis. However, it remains unclear what mechanisms of the host response are critical to the rejection or growth of the transduced cells. It is reasonable to assume, that local production of TNF-α induces indirect immunological mechanisms leading to the rejection of parental tumor cells, and it would be of major interest if the same effect could be observed after a rechallenge of the resistant animals with tumor cells at a distant site.

In contrast to tumor models currently employed, the usual clinical situation requires the treatment of an established tumor. To address this problem other experiments went further in demonstrating that TNF-α-transduced HCC cells can prevent the tumor growth of previously implanted unmodified HCC cells. All animals given unmodified cells, or cells infected with the control vector at the second injection, developed tumors, but only 6 of 20 mice that received TNF-α-transduced HCC cells developed tumors at the site of the prior injection.

Most HCC are multilocular or metastatic at the time of diagnosis, requiring systemic treatment. The major limitation of many trials in gene therapy for the treatment of cancer is the lack of systemic effect of the applied strategy. The only study to date showing a regression of a disseminated intrahepatic tumor used the vascular delivery of retrovirus-producing cells encoding interleukin-2 or -4 by intrasplenic injection, and, thereby demonstrated the efficacy against multilocular but not systemic disease.

Alcoholic Liver Disease

Innovative approaches in gene therapy allow biomedical research investigations in behavioral-induced diseases. Alcoholic liver disease is such an example. The chronic consumption of alcohol in certain individuals leads to liver diseases resulting in liver failure. To date, therapy for alcoholic liver disease is the cessation of alcohol consumption and in the case of end-stage liver disease (liver failure) liver transplantation. Liver transplantation is a difficult option due to the shortage of donor organs. Thus, new options for therapy are needed. Recent studies have provided new insights in the pathogenic mechanisms of alcoholic liver disease. These studies have shown that two mediators are independently important for the induction of liver fibrosis due to ethanol (see Fig. 7.10). These mediators are TNF-α and TGF-β and are targets for gene therapy approaches to prevent liver fibrosis due to ethanol consumption.

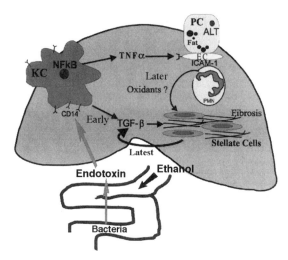

FIGURE 7.10   Possible biological mechanisms, as proposed by Thurman, for the generation of liver pathology—alcoholic hepatic steatohepatitis, inflammation, and fibrosis. Ethanol consumption induces "leaky gut syndrome," thereby altering intestinal permeability to Gram-negative bacteria colonizing the gut. Endotoxin, derived from the bacteria, increases in the blood and is transported to the liver. In the liver endotoxin binds to a plasma receptor (CD14) of Kupffer cells (liver macrophages). The Kupffer cells release tumor necrosis factor α (TNF-α), which in turn up-regulates the expression of intercellular adhesion molecules (ICAMs), which induce a neutrophil cellular infiltration. Subsequent stellate cell activation induces TNF-α, which augments extracellular matrix deposition in the liver (fibrosis).

Early events in alcoholic liver injury appear to be mediated by TNF-α, which is produced by Kupffer cells in the liver in response to gut-derived endotoxin. Gut-derived endotoxin is found in the liver due to increased permeability of the intestinal lining, the so-called leaky gut syndrome due to alcohol ingestion. Recent studies in the mouse have produced a gene knock-out mouse for the cellular receptor for TNF-α, TNF-R1. These animals are protected from alcoholic liver disease regardless of the level of alcohol consumed. Thus, targeted gene therapy approaches that inhibit or knock-out the expression of the TNF receptor, TNF-R1, are currently being investigated in animal models of alcoholic liver disease. In a similar approach, antisense techniques are being used to inhibit expression of TGF-β, a growth factor that induces fibrosis by increasing extracellular matrix deposition in the liver. Here the target is stellate cells in the liver that secrete large amounts of collagen when activated by TGF-β. Recent studies have shown TGF-β gene therapy approaches to be protective in animals. Here the infusion into the portal vein of a dominant negative receptor for TGF-β using an adenoviral vector-blocked fibrosis. Thus, inhibiting the expression of mediators of pathogensis is an important approach that can be utilized in the use of gene therapy of the liver.

SUMMARY

The liver synthesizes a large variety of proteins and, therefore, genetic defects in liver-specific genes can be responsible for many different inherited diseases. In addition the liver is the target for viral infections that can lead to acute and chronic disorders as well as hepatocellular carcinoma. While the liver is a challenging organ to deliver therapeutic genes, investigators have developed several methods that make the outlook of gene therapy for liver diseases promising.

The ex vivo approach has already been used with some success for the treatment of familial hypercholesterolemia in clinical trials but still requires modifications to improve the level of gene expression. Also encouraging is the result of the transfer of the human AAT gene into dogs using the ex vivo method. Ideally, however, gene therapy can be accomplished to correct genetic defects by in vivo methods.

Ideally, a vehicle for in vivo gene therapy for the treatment of liver disease must be liver specific, be able to pass through the endothelial lining to reach the parenchymal hepatocytes while avoiding clearance by Kupffer cells, and be effective in nondividing cells. It is also necessary for the therapeutic gene and the vehicle of delivery to avoid an immune response by the host. Several vehicles are under investigation to be used for the in vivo delivery of therapeutic genes. These include retroviral, adenoviral, and adenoassociated viral vectors as well as liposomes and protein–DNA complexes. All of these vehicles have advantages and disadvantages as shown in Table 7.1. Investigators are working to manipulate these systems to overcome the disadvantages so that the criteria needed for effective treatment can be met.

KEY CONCEPTS

- Many systemic inherited disorders such as hemophilia, familial hypercholesterolemia, phenylketonuria, and other metabolic diseases could be treated by addressing the underlying genetic defect in liver cells. In addition, gene therapeutic strategies could theoretically be used to treat acquired diseases such as viral infections of the liver.

- There are two basic approaches for gene transfer into hepatocytes: ex vivo and in vivo strategies.

- A vehicle for in vivo gene therapy for the treatment of liver disease should have optimum properties. They should (1) be liver specific, (2) pass through the endothelial lining to reach the parenchymal hepatocytes, (3) avoid clearance by Kupffer cells, (4) be effective in nondividing cells, and (5) avoid an immune response by the host.

- Soluble conjugates between naturally occurring and recombinant proteins and DNA are attractive tools for gene therapy directed to the liver. An example of the use of targeted delivery of protein–DNA complexes is the use of asialoglycoprotein receptors.

- The results of early gene therapy experiments using animal models of liver disease provided support for the efficacy of this treatment and paved the way for human clinical trials.

- Hepatitis B virus (HBV), hepatitis C virus (HCV), and hepatitis D virus (HDV) are known to cause persistent infection and chronic disease of the liver and are serious threats to public health. Gene therapy approaches to therapy are the use of targeted antisense molecules, ribozymes and dominant negative molecules with limited success in animal models of infection and disease.

SUGGESTED READINGS

General

Branch AD. A hitchhiker's guide to antisense and nonantisense biochemical pathways. Hepatology 24(6):1517–1529, 1996.

Grasso AW, Wu GY. Therapeutic implications of delivery and expression of foreign genes in hepatocytes. Adv Pharmacol 28:169–192, 1994.

Kormis KK, Wu GY. Prospects of therapy of liver diseases with foreign genes. Semin Liver Dis 15(3):257–267, 1995.

Nunes FA, Raper SE. Liver-directed gene therapy. Med Clin North Am 80(5):1201–1213, 1996.

Metabolic Diseases
Familial Hypercholesterolemia

Grossman M, Rader DJ, Muller WM, Kolansky DM, Kozarsky K, Clarke BJ, Stein EA, Lupien PJ, Brewer HB, Raper SE, Wilson JM. A pilot study of ex vivo gene therapy for homozygous familial hypercholesterolaemia. Nat Med 1(11):1148–1154, 1995.

Grossman M, Raper SE, Kozarsky K, Stein EA, Engelhardt JF, Muller WM, Lupien PJ, Wilson JM. Successful ex vivo gene therapy directed to liver in a patient with familial hypercholesterolaemia. Nat Genet 6:335–341, 1994.

Kozarsky KF, Jooss K, Donahee M, Strauss JF, Wilson JM. Effective treatment of familial hypercholesterolaemia in the mouse model using adenovirus-mediated transfer of the VLDL receptor gene. Nat Genet 13:54–62, 1996.

Hemophilia B (Factor IX Deficiency)

Kay MA. Hepatic gene therapy for hemophilia B. In: Aledort M, et al. (Eds.), Inhibitors to Coagulation Factors. Plenum, New York, 1995, pp. 229–234.

Koeberl DD, Alexander IE, Halbert CL, Russell DW, Miller AD. Persistent expression of human clotting factor IX from mouse liver after intravenous injection of adeno-associated virus vectors. Proc Natl Acad Sci USA 94:1426–1431, 1997.

Crigler–Najjar (Bilirubin UDPB-D Glucuronosyltransferase Deficiency)

Roy Chowdhury N, Hays RM, Bommineni VR, Franki N, Roy Chowdhury J, Wu CH, Wu GY. Microtubular disruption prolongs the expression of human bilirubin-uridinediphosphoglucuronate-glucuronosyltransferase-1 gene transferred into Gunn rat livers. J Biol Chem 271(4):2341–2346, 1996.

Takahashi M, Ilan Y, Roy Chowdhury N, Guida J, Horwitz M, Roy Chowdhury J. Long term correction of bilirubin-UDP-glucuronosyltransferase deficiency in Gunn rats by administration of a recombinant adenovirus during the neonatal period. J Biol Chem 271(43):26536–26542, 1996.

α_1-Antitrypsin Deficiency

Alino SF, Crespo J, Bobadilla M, Lejarreta M, Blaya C, Crespo A. Expression of human α_1-antitrypsin in mouse after in vivo gene transfer to hepatocytes by small liposomes. Biochem Biophys Res Commun 204(3):1023–1030, 1994.

Viral Hepatitis

Alt M, Renz R, Hofschneider PH, Paumgartner G, Caselmann WH. Specific inhibition of hepatitis C viral gene expression by antisense phosphorothioate oligodeoxynucleotides. Hepatology 22(3):707–717, 1995.

Bartholomew RM, Carmichael EP, Findeis MA, Wu CH, Wu GY. Targeted delivery of antisense DNA in woodchuck hepatitis virus-infected woodchucks. J Viral Hepatitis 2:273–278, 1995.

Nakazono K, Ito Y, Wu CH, Wu GY. Inhibition of hepatitis B virus replication by targeted pretreatment of complexed antisense DNA in vitro. Hepatology 23(6):1297–1303, 1996.

Sakamoto N, Wu CH, Wu GY. Intracellular cleavage of hepatitis C virus RNA and inhibition of viral protein translation by hammerhead ribozymes. J Clin Invest 98(12):2720–2728, 1996.

Schuster MJ, Wu GY. Targeted therapy for viral hepatitis. Drugs Today 32(8):653–661, 1996.

Hepatocellular Carcinoma

Cao G, Kuriyama S, Du P, Sakamoto T, Kong X, Masui K, Qi Z. Complete regression of established murine hepatocellular carcinoma by in vivo tumor necrosis factor α gene transfer. Gastroenterology 112:501–510, 1997.

Chen S-H, Kosai K-I, Xu B, Pham-Nguyen K, Contant C, Finegold MJ, Woo SLC. Combination suicide and cytokine gene therapy for hepatic metastases of colon carcinoma: Sustained antitumor immunity prolongs animal survival. Cancer Res 56:3758–3762, 1996.

Freeman SM, Ramesh R, Marrogi AJ. Immune system in suicide-gene therapy. Lancet 349:2–3, 1997.

Kaneko S, Hallenbeck P, Kotani T, Nakabayashi H, McGarrity G, Tamaoki T, Anderson WF, Chiang YL. Adenovirus-mediated gene therapy of hepatocellular carcinoma using cancer-specific gene expression. Cancer Res 55:5283–5287, 1995.

Schuster MJ, Wu GY. Gene therapy for hepatocellular carcinoma: Progress but many stones yet unturned! Gastroenterology 112(2):656–658, 1997.

Alcoholic Liver Disease

Tu GC, Cao QN, Zhou F, Israel Y. Tetranucleotide GGGA motif in primary RNA transcripts. Novel target site for antisense design. J Biol Chem 273:25125–25131, 1998.

Yin M, Ikejima K, Wheeler MD, Kono H, Bradford BU, Gallucci RM, Luster MI, Thurman RG. Essential role of tumor necrosis factor alpha in alcohol-induced liver injury in mice. Gastroenterology 117:942–952, 1999.

GENETIC MANIPULATION OF CARDIOVASCULAR TISSUE

Modulating Gene Expression in Cardiovascular Tissue

Gene therapy can be defined as any manipulation of gene expression that influences disease. This manipulation is generally achieved via the transfection of foreign deoxyribonucleic acid (DNA) into cells. Gene therapy can involve either the delivery of whole, active genes (gene transfer) or the blockade of native gene expression by the transfection of cells with short chains of nucleic acids known as oligonucleotides (Fig. 8.1).

The gene transfer approach allows for replacement of a missing or defective gene or for the overexpression of a native or foreign protein. The protein may be active only intracellularly, in which case very high gene transfer efficiency may be necessary to alter the overall function of an organ or tissue. Alternatively, proteins secreted by target cells may act on other cells in a paracrine or endocrine manner, in which case delivery to a small subpopulation of cells may yield a sufficient therapeutic result.

Gene blockade can be accomplished by transfection of cells with short chains of DNA known as antisense oligodeoxynucleotides (ODN). This approach attempts to alter cellular function by the inhibition of specific gene expression. Antisense ODN have a base sequence that is complementary to a segment of the target gene. This complimentary sequence allows the ODN to bind specifically to the corresponding segment of messenger ribonucleic acid (mRNA) that is transcribed from the gene, preventing the translation into protein. Another form of gene blockade is the use

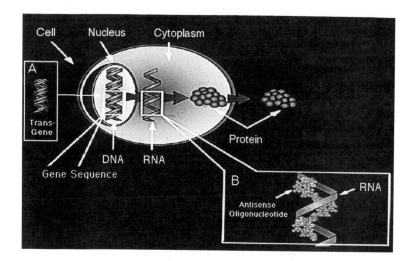

FIGURE 8.1   Gene therapy strategies. See color insert. (A) Gene transfer involves delivery of an entire gene, either by viral infection or by nonviral vectors, to the nucleus of a target cell. Expression of the gene via transcription into mRNA and translation into a protein gene product yields a functional protein that either achieves a therapeutic effect within a transduced cell or is secreted to act on other cells. (B) Gene blockade involves the introduction into the cell of short sequences of nucleic acids that block gene expression, such as antisense ODN that bind mRNA in a sequence-specific fashion and prevents translation into protein.

CHAPTER 8

Gene Therapy in Cardiovascular Dis

VICTOR J. DZAU, M.D., AFSHIN EHSAN, M.D., and MICHAEL J. MANN, M.I

INTRODUCTION

The explosive growth in understanding the changes in gene expression
with the onset and progression of acquired diseases has created a pi
revolutionizing the clinician's approach to common disorders. Noting
graphics of cardiovascular diseases in the population of the United State
is the medical revolution more likely to impact a significant population
than in the arena of cardiovascular disease. Gene therapy offers the p
alter, or even reverse, pathobiology at its roots. As researchers learn mor
genetic blueprints of disease, the scope of applicability of this exciting
peutic approach will continue to expand.

The therapeutic manipulation of genetic processes has come to embra
introduction of functional genetic material into living cells as well as the
specific blockade of certain active genes. As a better understanding oj
tic contribution to disease has evolved, so has the breadth of gene m;
technology. Therapeutic targets have been identified in an effort to im
ventional cardiovascular therapies, such as balloon angioplasty or bypa
Entirely novel approaches toward the treatment of acquired diseases, s
induction of angiogenesis in ischernic tissues, are also being developed. /
asm grows for these new experimental strategies, it is important for clini
aware of their limitations as well as their strengths, and for careful proces
uation to pave the possible integration of these therapies into routine pra
the basic principles of gene manipulation and its applicability to the tr
cardiovascular disease are presented as well as a review of the use of ge
in animal models and in clinical trials.

An Introduction to Molecular Medicine and Gene Therapy, Edited by Thomas F. Kresina
ISBN 0-471-39188-3 © 2001 Wiley-Liss

of ribozymes, segments of RNA that can act like enzymes to destroy only specific sequences of target mRNA. A third type of gene inhibition involves the blockade of transcription factors. Double-stranded ODN can be designed to mimic the transcription factor binding sites and act as decoys, preventing the transcription factor from activating target genes.

Cardiovascular DNA Delivery Vector

Plasmids are circular chains of DNA that were originally discovered as a natural means of gene transfer between bacteria. Naked plasmids can also be used to transfer DNA into mammalian cells. The direct injection of plasmid DNA into tissues in vivo can result in transgene expression. Plasmid uptake and expression, however, has generally been achieved at reasonable levels only in skeletal and myocardial muscle. The "ideal" cardiovascular DNA delivery vector would be capable of safe and highly efficient delivery to all cell types, both proliferating and quiescent, with the opportunity to select either short-term or indefinite gene expression. This ideal vector would also have the flexibility to accommodate genes of all sizes, incorporate control of the temporal pattern and degree of gene expression, and to recognize specific cell types for tailored delivery or expression. While progress is being made on each of these fronts individually, gene therapy remains far from possessing a single vector with all of the desired characteristics. Instead, a spectrum of vectors has evolved, each of which may find a niche in different early clinical gene therapy strategies.

Recombinant, *replication-deficient retroviral vectors* have been used extensively for gene transfer in cultured cardiovascular cells in vitro, where cell proliferation can be manipulated easily. Their use in vivo has been more limited due to low transduction efficiencies, particularly in the cardiovascular system where most cells remain quiescent. The random integration of traditional retroviral vectors such as molorey murine leukemia virus (MMLV) into chromosomal DNA involve a potential hazard of oncogene activation and neoplastic cell growth. While the risk may be low, safety monitoring will be an important aspect of clinical trials using viral vectors. Recent improvements in packaging systems (particularly the development of "pseudotyped" retroviral vectors incorporate vesicular stomatitis virus G-protein) have improved the stability of retroviral particles and facilitated their use in a wider spectrum of target cells.

Recombinant *adenoviruses* have become the most widely used viral vectors for experimental in vivo cardiovascular gene transfer. Adenoviruses infect nondividing cells and generally do not integrate into the host genome. These vectors can therefore achieve relatively efficient gene transfer in some quiescent cardiovascular cell types, but transgenes are generally lost when cells are stimulated into rounds of cell division. The immune response to adenoviral antigens represents the greatest limitation to their use in gene therapy. Conventional vectors have generally achieved gene expression for only 1 to 2 weeks after infection. It is not certain to what extent the destruction of infected cells contributes to the termination of transgene expression given that the suppression of episomal transgene promoters appears to occur as well. In the vasculature, physical barriers such as the internal elastic lamina apparently limits infection to the endothelium, with gene transfer to the media and adventitia only occurring after injury has disrupted the vessel architecture. Although gene delivery to 30 to 60% of cells after balloon injury has been reported with adenoviral vectors carrying reporter genes, the fact that atherosclerotic disease has also been found to limit the efficiency of adenoviral transduction may pose a significant problem for the treatment of human disease.

Adenoassociated virus (AAV) can infect a wide range of target cells and can establish a latent infection by integration into the genome of the cell, thereby yielding stable gene transfer as in the case of retroviral vectors. Although AAV vectors transduce replicating cells at a more rapid rate, they possess the ability to infect nonreplicating cells both in vitro and in vivo. The efficiency of AAV-mediated gene transfer to vascular cells, and the potential use of AAV vectors for in vivo vascular gene therapy, remains to be determined. However, a number of studies have reported successful transduction of myocardial cells after direct injection of AAV suspensions into heart tissue, and these infections have yielded relatively stable expression for greater than 60 days.

The development of effective methods of nonviral transfection in vivo has posed a significant challenge to cardiovascular and other clinical researchers. Lipid-based gene transfer methods are easier to prepare and have greater flexibility in terms of substituting transgene constructs than the relatively complex recombinant viral vector processes. A growing variety of cationic *liposomes* have been used extensively during the last 5 to 10 years for in vivo and in vitro delivery of plasmid DNA and antisense oligonucleotides. Other substances, such as lipopolyamines and cationic polypeptides, are also being investigated as potential vehicles for enhanced DNA delivery both for gene transfer and gene blockade strategies. In vivo DNA transfer efficiency with any of these methods, however, continues to be very low. The addition of inactivated Sendai viral particles to liposome preparations has been shown to enhance the fusigenic properties of the lipids and may be a means of improving DNA delivery. In addition, the controlled application of a pressurized environment to vascular tissue in a nondistended manner has recently been found to enhance oligonucleotide uptake and nuclear localization. This method may be particularly useful for ex vivo applications such as vein grafting or transplantation and may represent a means of enhancing plasmid gene delivery.

Controlling Gene Expression in Cardiovascular Tissue

In addition to effective gene delivery, many therapeutic settings will demand some degree of control over the duration, location, and degree of transgene expression. To this end, researchers have developed early gene promoter systems that allow the clinician to regulate the spatial or temporal pattern of gene expression. These systems include tissue-specific promoters that have been isolated from genetic sequences encoding proteins with natural restriction to the target tissue, such as the von Willebrand factor promoter in endothelial cells and the α-myosin heavy-chain promoter in myocarium. Promoters have also been isolated from nonmammalian systems that can either promote or inhibit downstream gene expression in the presence of a pharmacologic agent such as tetracycline, zinc, or steroids. In addition, regulation of transgene expression may even be relegated to the physiologic conditions, with the incorporation of promoters, enhancers, or other regulatory elements that respond to developmental stages or specific conditions such as hypoxia or increased oxidative stress.

GENE THERAPY OF RESTENOSIS

Pathophysiology

Recurrent narrowing of arteries following percutaneous angioplasty, atherectomy, or other disobliterative techniques is a common clinical problem. It severely limits the durability of these procedures for patients with atherosclerotic occlusive diseases. In the case of balloon angioplasty, restenosis occurs in approximately 30 to 40% of treated coronary lesions and 30 to 50% of superficial femoral artery lesions within the first year. Intravascular stents reduce the restenosis rates in some settings, however, the incidence remains significant and long-term data are limited. Despite impressive technological advances in the development of minimally invasive and endovascular approaches to treat arterial occlusions, the full benefit of these gains awaits the resolution of this fundamental biologic problem.

The pathophysiology of restenosis is comprised of a contraction and fibrosis of the vessel wall known as remodeling, and an active growth of a fibrocellular lesion composed primarily of vascular smooth muscle cells (VSMC) and extracellular matrix. The latter process, known as neointimal hyperplasia, involves the stimulation of the normally "quiescent" VSMC in the arterial media into the "activated" state characterized by rapid proliferation and migration. A number of growth factors are believed to play a role in the stimulation of VSMC during neointimal hyperplasia, including platelet-derived growth factor (PDGF), basic fibroblast growth factor (bFGF), transforming growth factor beta (TGF-β), and angiotensin II. Activated VSMC has also been found to produce a variety of enzymes, cytokines, adhesion molecules, and other proteins that not only enhance the inflammatory response within the vessel wall but also stimulate further vascular cell abnormality. Although it is now thought that remodeling may account for the majority of late lumen loss after balloon dilation of atherosclerotic vessels, proliferation has been the predominant target of experimental genetic interventions.

Cytostatic and Cytotoxic Approaches

There have been two general approaches—cytostatic, in which cells are prevented from progressing through the cell cycle to mitosis, and cytotoxic, in which cell death is induced. A group of molecules known as cell cycle regulatory proteins act at different points along the cell cycle (see Chapter 10), mediating progression toward division. It has been hypothesized that by blocking expression of the genes for one or more of the regulatory gene products, progression of VSMC through the cell cycle could be prevented. As well, neointimal hyperplasia could be inhibited. To support this hypothesis, near complete inhibition of neointimal hyperplasia after carotid balloon injury has been demonstrated. This has been via hemagglutinating virus of Japan (HVJ)–liposome-mediated transfection of the vessel wall with a combination of antisense ODN against cell cycle regulatory genes. Arrest of the cell cycle via antisense blockade of either of two proto-oncogenes, c-*myb* or c-*myc*, has been found to inhibit neointimal hyperplasia in models of arterial balloon injury. However, the specific antisense mechanism of the ODN used in these studies has subsequently been questioned.

In addition to transfection of cells with antisense ODN, cell cycle arrest can also be achieved through manipulation of transcription factor activity. The activity of a number of cell cycle regulatory genes is influenced by a single transcription factor known as E2F. In quiescent cells, E2F is bound to a complex of other proteins, including a protein known as the retinoblastoma (*Rb*) gene product. *Rb* prevents E2F's interaction with chromosomal DNA and stimulation of gene activity. In proliferating cells, E2F is released, resulting in cell cycle gene activation. A transcription factor decoy bearing the consensus binding sequence recognized by E2F can be employed as a means to inhibit cellular proliferation. The use of this strategy to prevent VSMC proliferation and neointimat hyperplasia after rat carotid balloon injury has been demonstrated. Alternatively, the approach of localized arterial infection with a replication-defective adenovirus encoding a nonphosphorylatable, constitutively active form of *Rb* at the time of balloon angioplasty has been studied. This approach significantly reduces smooth muscle cell proliferation and neointima formation in both the rat carotid and porcine femoral artery models of restenosis. Similar results were also obtained by adenovirus-mediated overexpression, a natural inhibitor of cell cycle progression, the cyclin-dependent kinase inhibitor, p2l. Here, p21 likely prevents hyperphosphorylation of *Rb* in vivo. In addition to blockade of cell cycle gene expression, interruption of mitogenic signal transduction has been achieved in experimental models as well. For example, Ras proteins are key transducers of mitogenic signals from membrane to nucleus in many cell types. The local delivery of DNA vectors expressing Ras-dominant negative mutants, which interfere with Ras function, reduced neointimal lesion formation in a rat carotid artery balloon injury model.

Nitric oxide mediates a number of biologic processes that are thought to mitigate neointima formation in the vessel wall. These include inhibition of VSMC proliferation, reduction of platelet adherence, vasorelaxation, promotion of endothelial cell survival, and possible reduction of oxidative stress. In vivo transfer of plasmid DNA coding for endothelial cell nitric oxide synthase (ecNOS) has been investigated as a potential paracrine strategy to block neointimal disease. EcNOS complementary DNA (cDNA) driven by a β-actin promoter and CMV enhancer was transfected into the VSMC of rat carotid arteries after balloon injury. This model is known to have no significant regrowth of endothelial cells within 2 to 3 weeks after injury and therefore capable of loss of endogenous ecNOS expression. Results revealed expression of the transgene in the vessel wall, along with improved vasomotor reactivity and a 70% inhibition of neointima formation (Fig. 8.2).

A direct cytotoxic approach to the prevention of neointima formation can involve the transfer of a suicide gene such as the herpes simplex virus thymidine kinase (HSV-*tk*) gene into VSMC. Using an adenoviral vector, HSV-*tk* was introduced into the VSMC of porcine arteries rendering the smooth muscle cells sensitive to the nucleoside analog gancyclovir given immediately after balloon injury. After one course of gancyclovir treatment, neointimal hyperplasia decreased by about 50% in that model system. More recently, studies induced endogenous machinery for VSMC suicide, in a strategy designed to inhibit the growth or achieve regression of neointimal lesions. This strategy involved antisense ODN blockade of a survival gene, known as *Bcl-x*, that helps protect cells from activation of programmed cell death, or apoptosis.

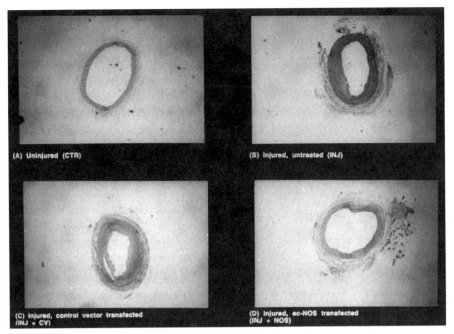

FIGURE 8.2 Inhibition of neointimal hyperplasia by in vivo gene transfer of endothelial cell–nitric oxide synthase (ecNOS) in balloon-injured rat carotid arteries. See color insert.

GENE THERAPY FOR ANGIOGENESIS

Angiogenesis and Angiogenic Factors

The identification and characterization of angiogenic growth factors has created an opportunity to attempt the therapeutic neovascularization of tissue rendered ischemic by occlusive disease in the native arterial bed. It has been clearly established, in a number of animal models, that angiogenic factors can stimulate the growth of capillary networks in vivo. But, it is less certain that these molecules can induce the development of larger, more complex vessels in adult tissues needed for carrying significantly increased bulk blood flow. Nevertheless, the possibility of an improvement, even of just the microvascular collateralization as a biological approach to the treatment of tissue ischemia, has sparked the beginning of human clinical trials in neovascularization therapy.

The intial description of the angiogenic effect of fibroblast growth factors (FGFs) prompted the discovery of an abundance of proangiogenic factors. These factors either stimulated endothelial cell proliferation or enhanced endothelial cell migration. In some cases both activities were observed. The list of angiogenic factors includes such diverse molecules as insulinlike growth factor, hepatocyte growth factor, angiopoeitin, and platelet-derived endothelial growth factor. The molecules that have received the most attention as potential therapeutic agents for neovascularization, however, are vascular endothelial growth factor (VEGF) and two members of the FGF family, acidic FGF (FGF-1) and basic FGF (FGF-2). All angiogenic factors share some ability to stimulate capillary growth in classical models such as the chick aflantoic membrane. However, much debate persists regarding the optimum agent and the optimum route of delivery for angiogenic therapy in the ischemic human myocardium or lower extremity. VEGF may be the most selective agent for stimulating endothelial cell proliferation, although VEGF receptors are also expressed on a number of inflammatory cells including members of the monocyte-macrophage lineage. This selectivity has been viewed as an advantage since the unwanted stimulation of fibroblasts and VSMC in native arteries might exacerbate the growth of neointimal or atherosclerotic lesions. The FGFs are believed to be potent stimulators of endothelial cell proliferation, but, as their name implies, they are much less selective in their pro-proliferative action.

Angiogenic Gene Therapy

Preclinical studies of angiogenic gene therapy have utilized a number of models of chronic ischemia. An increase in capillary density was reported in an ischemic rabbit hind limb model after VEGF administration. This result did not differ significantly regardless of whether VF-GF was delivered as a single intra-arterial bolus of protein, as plasmid DNA applied to surface of an upstream arterial wall, or via direct injection of the plasmid into the ischemic limb. Direct injection of an adenoviral vector encoding VEGF also succeeded in improving regional myocardial perfusion and ventricular fractional wall thickening at stress. These results were shown in a pig model of chronic myocardial ischemia induced via placement of a slowly occluding Ameroid constrictor around the circumflex coronary artery.

Unlike VEGF, FGF-1 and -2 do not possess signal sequences that facilitate secretion of the protein. Thus, the transfer of these genetic sequences is less likely to yield an adequate supply of growth factor to target endothelial cells. To overcome this limitation, a plasmid was devised encoding a modified FGF-I molecule onto which a hydrophobic leader sequence had been added to enhance secretion. Delivery of this plasmid to the femoral artery wall, even at low transfection efficiencies, was found to improve capillary density and reduce vascular resistance in the ischemic rabbit hind limb. Applying a similar strategy, 10^{11} viral particles of an adenoviral vector encoding human FGF-5, containing a secretary signal sequence at its amino terminus, were injected via intracoronary infusion. This protocol resulted in enhanced wall thickening with stress and a higher number of capillary structures per myocardial muscle fiber 2 weeks after gene transfer.

Another novel approach to molecular neovascularization has been the combination of growth factor gene transfer with a potentially synergistic method of angiogenic stimulation: transmyocardial laser therapy. The formation of transmural laser channels is not yet fully established as an effective means of generating increased collateral flow. But documented clinical success in reducing angina scores and improving myocardial perfusion in otherwise untreatable patients has been observed. In a porcine Ameroid model, direct injection of plasmid DNA encoding VEGF in the region surrounding laser channel formation yielded better normalization of myocardial function than therapy alone. This therapeutic strategy can now be delivered either through minimally invasive thoracotomy or a percutaneous catheter-based approach (Fig. 8.3).

A number of phase I safety studies have been reported in which angiogenic

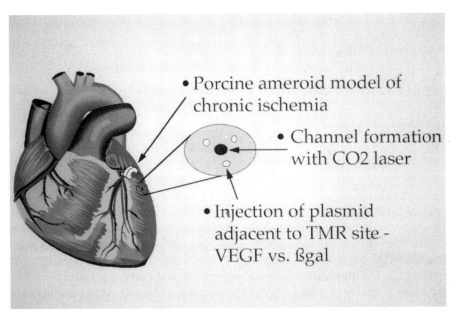

FIGURE 8.3 Combined gene transfer and transmyocardial laser revascularization (TMR). See color insert. Schematic representation of chronic ischemia induced by placement of Ameroid constrictor around the circumflex coronary artery in pigs. Ischemic hearts that underwent TN4R followed by injection of plasmid encoding VEGF demonstrated better normalization of myocardial function than either therapy alone.

factors or the genes encoding these factors have been administered to a small number of patients. These studies have involved either the use of angiogenic factors with peripheral vascular or coronary artery disease in patients who were not candidates for conventional revascularization therapies or the application of proangiogenic factors as an adjunct to conventional revascularization. The modest doses of either protein factors or genetic material delivered in these studies were not associated with any acute toxicities. Concerns remain, however, regarding the safety of potential systemic exposure to molecules known to enhance the growth of possible occult neoplasms or that can enhance diabetic retinopathy and potentially even occlusive arterial disease itself. Despite early enthusiasm, there is little experience with the administration of live viral vectors to a large number of patients. Thus, it is uncertain whether potential biological hazards of reversion to replication-competent states or mutation and recombination will eventually become manifest.

In addition, it is also unclear whether the clinical success of conventional revascularization, which has involved the resumption of lost bulk blood flow through larger conduits, will be reproduced via biological strategies that primarily increase microscopic collateral networks. It must also be remembered that neovascularization is itself a naturally occurring process. The addition of a single factor may not overcome conditions that have resulted in an inadequate endogenous neovascularization response in patients suffering from myocardial and lower limb ischemia. Despite these limitations, angiogenic gene therapy may provide an alternative not currently available to a significant number of patients suffering from untreatable disease. In addition, angiogenic gene therapy may offer an adjunct to traditional therapies that improves long-term outcomes.

GENE THERAPY OF VASCULAR GRAFTS

Modification of Vein Graft Biology

The long-term success of surgical revascularization in the lower extremity and coronary circulations has been limited by significant rates of autologous vein graft failure. A pharmacologic approach has not been successful at preventing long-term graft diseases such as neointimal hyperplasia or graft atherosclerosis. Gene therapy offers a new avenue for the modification of vein graft biology that might lead to a reduction in clinical morbidity from graft failures. Intraoperative transfection of the vein graft also offers an opportunity to combine intact tissue DNA transfer techniques with the increased safety of ex vivo transfection. A number of studies have documented the feasibility of ex vivo gene transfer into vein grafts using viral vectors.

The vast majority of vein graft failures that have been linked to the neointimal disease is part of graft remodeling after surgery. Although neointimal hyperplasia contributes to the reduction of wall stress in vein grafts after bypass, this process can also lead to luminal narrowing of the graft conduit during the first years after the operation. Furthermore, the abnormal neointimal layer, producing proinflammatory proteins, is the basis for an accelerated form of atherosclerosis that causes late graft failure.

As in the arterial balloon injury model, a combination of antisense ODN inhibiting expression of at least two cell cycle regulatory genes could significantly block neointimal hyperplasia in vein grafts. Additionally, E2F decoy ODN yield similar efficacy in the vein graft when compared to the arterial injury model. In contrast to arterial balloon injury, however, vein grafts are not only subjected to a single injury at the time of operation, but they are also exposed to chronic hemodynamic stimuli for remodeling. Despite these chronic stimuli, a single, intraoperative decoy ODN treatment of vein grafts resulted in a resistance to neointimal hyperplasia that lasted for at least 6 months in the rabbit model. During that time period, the grafts treated with cell cylce blockage were able to adapt to arterial conditions via hypertrophy of the medial layer. Furthermore, these genetically engineered conduits proved resistant to diet-induced graft atherosclerosis (Fig. 8.4). They were also associated with preserved endothelial function.

An initial prospective, randomized double-blind clinical trials of human vein graft treatment with E2F decoy ODN has recently been undertaken. Efficient delivery of the ODN is accomplished within 15 min during the operation by placement of the graft after harvest in a device that exposes the vessel to ODN in physiologic solution. This device creates a nondistending pressurized environment of 300 mmHg (Fig. 8.5). Preliminary findings indicated ODN delivery to greater than 80% of graft cells and effective blockade of targeted gene expression. This study will measure the effect of cell cycle gene blockade on primary graft failure rates and represents one of the first attempts to definitively determine the feasibility of clinical genetic manipulation in the treatment of a common cardiovascular disorder.

With the development of viral-mediated gene delivery methods, some investiga-

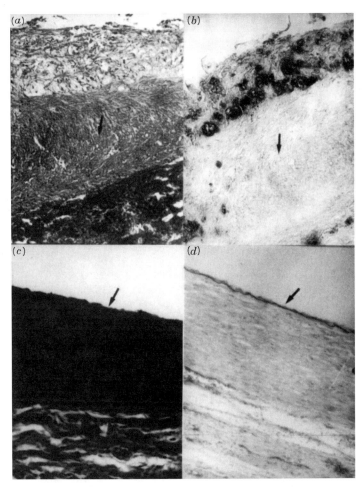

FIGURE 8.4 Control oligonucleotide-treated (A and B) and antisense oligonucleotide (against c and 2 kinase/PCNA)-treated vein grafts (C and D) in hypercholesterolernic rabbits, 6 weeks after surgery (×7O). See color insert. Sections were stained with hematoxylin/van Gieson (A and C) and a monoclonal antibody against rabbit macrophages (B and D). Arrows indicate the location of the internal elastic lamina.

tors have begun to explore the possibility of using these systems ex vivo in autologous vein grafts. Studies have demonstrated the expression of the marker gene β-galactosidase along the luminal surface and in the adventitia of 3-day porcine vein grafts infected with a replication-deficient adenoviral vector for 2 h at the time of surgery. Other studies have explored the use of a novel adenovirus-based transduction system in which adenoviral particles are linked to plasmid DNA via biotin/streptavidin-transferrin/polylysine complexes. β-Galactosidase expression was documented 3 and 7 days after surgery in rabbit vein grafts incubated for 1 h with complexes prior to grafting. Expression was greatest on the luminal surfaces of the grafts. The presence of transfected cells in the medial and adventitial layers was also reported.

The feasibility of gene transfer in vein grafts has subsequently lead to the inves-

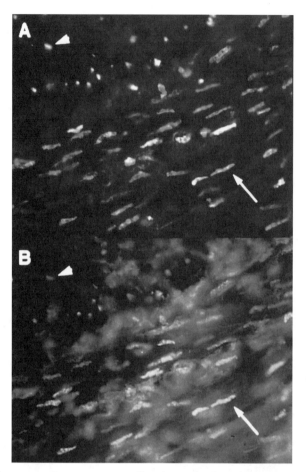

FIGURE 8.5 Intraoperative pressure-mediated transfection of fluorescent-labeled ODN to saphenous vein graft cells. See color insert. (A) Hoechst 33,342 nuclear chromatin staining of vein graft in cross section, illustrating location of nuclei within the graft wall (100×). (B) Same section of saphenous vein viewed under FITC-epifluoreseence at 100×. Note the pattern of enhanced green fluorescence in the nuclei of cells within the graft wall, indicating nuclear localization of labeled ODN.

tigation of potential therapeutic endpoints such as neointima formation. Studies using a replication-deficient adenovirus expressing tissue inhibitor of metalloproteinase-2 (TIMP-2) demonstrate a decrease in neointimal formation in a saphenous vein organ culture model. Other studies using intraoperative transfection of the senescent cell-derived inhibitor (*sdi, I*) gene, a downstream mediator of the tumor suppresser gene *p53* and the HVJ–liposome system, demonstrated a reduction in neointima formation.

Bioengineering and Gene Therapy

The use of gene transfer in vein grafts may go beyond the treatment of the graft itself. The thrombogenicity of prosthetic materials, such as poly(tetrafluoroethylene)

(PTFE) or Dacron, has limited their use as small caliber arterial substitutes. A combined bioengineering, cell-based gene therapy strategy may decrease this thrombogenicity. Successful isolation of autologous endothelial cells and their seeding onto prosthetic grafts in animal models have been well characterized. Furthermore, it has been hypothesized that one can enhance the function of these endothelial cells via the transfer of genes prior to seeding of the cells on the graft surface. The initial report of the use of this strategy achieved successful endothelialization of a prosthetic vascular graft with autologous endothelial cells transduced with a recombinant retrovirus encoding the *lacz* gene. Successful clinical applications of these concepts, however, have not been reported. In an attempt to decrease graft thrombogenicity, 4-mm Dacron grafts were seeded with retroviral transduced endothelial cells encoding the gene for human tissue plasminogen activator (TPA). The grafts were subsequently implanted into the femoral and carotid circulation of sheep. The proteolytic action of TPA resulted in a decrease in seeded endothelial cell adherence, with no improvement in surface thrombogenicity.

GENE THERAPY FOR THE HEART

The myocardium has been shown to be receptive to the introduction of foreign genes. As seen in noncardiac muscle, measurable levels of gene activity has been found after direct injection of plasmids into myocardial tissue in vivo. Although limited to a few millimeters surrounding the injection site, these observations have laid the basis for consideration of gene transfer as a therapeutic approach to cardiac disease. Additionally, both adenoviral and adenoassociated viral vectors can be delivered to the myocardial and coronary vascular cells via either direct injection or intracoronary infusion of concentrated preparations in rabbits and porcine models respectively. Gene transfer into the myocardium has also been achieved via either the direct injection or intracoronary infusion of myoblast cells that have been genetically engineered in cell culture.

Congestive Heart Failure

The β-adrenergic receptor (β-AR) is known to be a critical player in mediating the ionotropic state of the heart. This receptor has received significant attention as a target for genetic therapeutic intervention in congestive heart failure. Transgenic mice were generated expressing the β2-AR under the control of the cardiac major histocompatibility complex (X-MHC) promoter. These animals demonstrated an approximately 200-fold increase in the level of β2-AR along with highly enhanced contractility and increased heart rates in the absence of exogamous β-agonists. This genetic manipulation of the myocardium has generated considerable interest in the use of gene transfer of the β-AR gene into the ailing myocardium as a means of therapeutic intervention. To date, attempts at exploring this exciting possibility have been primarily limited to cell culture systems. However, recent studies have move this technology into animal studies. For example, adenoviral-mediated gene transfer of the human β2-AR successfully demonstrated improved contractility in rabbit ventricular myocytes that were chronically paced to produce hemodynamic failure. An enhanced chronotropic effect resulting from the injection of a β2-AR plasmid construct into the right atrium of mice has been performed. But no evaluation of enhanced contractility by transfer of this gene into the ventricle has been reported. These results demonstrate the feasibility of using the βP-adrenergic pathway and its regulators as a means by which to treat the endpoint effect of the variety of cardiac insults.

There has also been recent interest in the enhancement of contractility through the manipulation of intracellular calcium levels. Sarcoplasmic reticulum Ca^{2+}-ATPase (SERCA2a) transporting enzyme, which regulates Ca^{2+} sequestration into the sarcoplasmic reticulum (SR), has been shown to be decreased in a variety of human and experimental cardiomyopathies. Over expression of the SERCA2a protein in neonatal rat cardiomyocytes using adenoviral-mediated gene transfer has been achieved. This leads to an increase in the peak (Ca^{2+} li) release, a decrease in resting (Ca^{2+} li) levels, and more importantly to enhanced contraction of the myocardial cells as detected by shortening measurements. The success of this approach in improving myocardial contractility has yet to be documented in vivo. But once again, gene therapy approaches provide a novel and potentially exciting means by which to treat the failed heart.

Myocardial Infarction

Myocardial infarction (MI) is the most common cause of heart failure. At the cellular level MI results in the formation of scar that is composed of cardiac fibroblasts. Given the terminal differentiation of cardiomyocytes, loss of cell mass due to infarction does not result in the regeneration of myocytes to repopulate the wound. Researchers have, therefore, pursued the possibility of genetically converting cardiac fibroblasts into functional cardiomyocytes. The feasibility of this notion gained support from gene transfer studies. These studies used retroviral-mediated gene transfer for the in vitro conversion of cardiac fibroblasts into cells resembling skeletal myocytes via the forced expression of a skeletal muscle lineage-determining gene, *MyoD*. Fibroblasts expressing the *MyoD* gene were observed to develop multinucleated myotubes similar to striated muscle that expressed MHC and myocyte-specific enhancer factor 2. Additional studies have shown that the tranfection of rat hearts injured by freeze–thaw with adenovirus containing the *MyoD* gene resulted in the expression of myogenin and embryonic skeletal MHC. At this time, however, functional cardiomyocytes have not yet been identified in regions of myocardial scarring treated with in vivo gene transfer.

Ischemia and Reperfusion

Coronary artery atherosclerosis, and resulting myocardial ischemia, is a leading cause of death in developed countries. Reperfusion injury has been linked to significant cellular damage and progression of the ischemic insult. In addition to stimulating therapeutic neovascularization, genetic manipulation may be used as a means to limit the degree of injury sustained by the myocardium after ischemia and reperfusion. The process of tissue damage resulting from ischemia and reperfusion has been well characterized.

Briefly, the period of ischemia leads to an accumulation of adenosine monophosphate that then leads to increased levels of hypoxanthine within and around cells in the affected area. Additionally, increased conversion of xanthine dehydrogenase into xanthine oxidase takes place. Upon exposure to oxygen during the period of reperfusion, hypoxanthine is converted to xanthine. This conversion results in the cytotoxic oxygen radical, superoxide anion (O_2^-). This free radical goes on to form hydrogen peroxide (H_2O_2), another oxygen radical species. Ferrous iron (Fe^{2+}) accumulates during ischemia and reacts with H_2O_2, forming the potent oxygen radical, hydroxyl anion (OH^-). These free radical species result in cellular injury via lipid peroxidation of the plasma membrane, oxidation of sulfhydryl groups of intracellular and membrane proteins, nucleic acid injury, and breakdown of components of the extracellular matrix such as collagen and hyaluronic acid. Natural oxygen radical scavengers, such as superoxide dismutase (SOD), catalase, glutathione peroxidase, and hemoxygenase (HO) function through various mechanisms to remove oxygen radicals produced in normal and injured tissues.

The level of oxygen radical formation after ischemia–reperfusion injury in the heart can overwhelm the natural scavenger systems. Thus, overexpression of either extracellular SOD (ecSOD) or manganese SOD (MnSOD) in transgenic mice has improved postischemic cardiac function and decreased cardiomyocyte mitochondrial injury in adriamycin-treated mice, respectively. These findings suggest a role for gene transfer of natural scavengers as a means to protect the myocardium in the event of an ischemia–reperfusion event. Substantial protection has been observed against myocardial stunning, using intra-arterial injection of an adenovirus containing the gene for Cu/Zn SOD (the cytoplasmic isoform) into rabbits. However, no studies have investigated the direct antioxidant effect and ensuing improvement in myocardial function of this treatment after ischernia and reperfusion injury. This application of gene therapy technology may offer a novel and exciting approach for prophylaxis against myocardial ischemic injury when incorporated into a system of long-term, regulated transgene expression.

In addition to the overexpression of antioxidant genes, some researchers have proposed intervening in the program of gene expression within the myocardium that lead to the downstream deleterious effects of ischemia reperfusion. For example, the transfection of rat myocardium with decoy oligonulceotides, blocking the activity of the oxidation-sensitive transcription factor NFκ-B, may be a useful approach. NFκ-B is linked to the expression of a number of proinflammatory genes. It inhibition succeeded in reducing infarct size after coronary artery ligation.

Genetic manipulation of donor tissues offers the opportunity to design organ-specific immunosuppression during cardiac transplantation. Although transgenic animals are being explored as potential sources for immunologically protected xenografts, the delivery of genes for immunosuppressive proteins, or the blockade of certain genes in human donor grafts, may allow site-specific, localized immuno-suppression. Alternatively, these approaches could result in a reduction or elimination of the need for toxic systemic immunosuppressive regimens. Gene activity has been documented in transplanted mouse hearts for at least 2 weeks after intraoperative injection of the tissue with either plasmid DNA or retroviral or adenoviral vectors. The transfer of a gene for either TGF-β or interleukin-10 in a small area of the heart via direct injection, succeeded in promoting immunosuppression of graft reject. Cell-mediated immunity was inhibited and acute rejection was delayed. In another study, the systemic administration of antisense ODN directed against intercellular adhesion molecules (ICAM-1) also prolonged graft survival and induced long-term graft tolerance when combined with a monoclonal antibody against the ligand for ICAM-1, the leukocyte function antigen.

SUMMARY

The field of gene therapy is evolving from the realm of laboratory science into a clinically relevant therapeutic option. The current state of this technology has provided us with an exciting glimpse of its therapeutic potential. Routine application, however, will require improvement of existing techniques along with the development of novel methods for gene transfer. More importantly, no one method of gene transfer will serve as the defining approach. Rather, it will be the use of all available techniques, either individually or in combination, that will shape the application of this therapy. Over the past two decades, as scientists have begun to unlock the genetic code, more insight into the pathogenesis of disease has been gained. With the use of gene manipulation technology, this new information can be used to further improve the understanding and treatment of complex acquired and congenital diseases previously unresponsive to traditional surgical and pharmacologic therapy.

KEY CONCEPTS

- The ideal cardiovascular DNA delivery vector would be capable of safe and highly efficient delivery to all cell types, both proliferating and quiescent, with the opportunity to select either short-term or indefinite gene expression. This ideal vector would also have the flexibility to accommodate genes of all sizes, incorporate control of the temporal pattern and degree of gene expression, and to recognize specific cell types for tailored delivery or expression.
- Recombinant, replication-deficient retroviral vectors have been used extensively for gene transfer in cultured cardiovascular cells in vitro, where cell proliferation can be manipulated easily. Recombinant adenoviruses have become the most widely used viral vectors for experimental in vivo cardiovascular gene transfer. Adenoassociated virus has successfully transduced myocardial cells after direct injection of viral suspensions into heart tissue; and these infections have yielded relatively stable expression for greater than 60 days. For nonviral gene delivery, the controlled application of a pressurized environment to vascular tissue in a nondistended manner has recently been found to enhance oligonucleotide uptake and nuclear localization. This method may be particularly useful for ex vivo applications, such as vein grafting or transplantation, and may represent a means of enhancing plasmid gene delivery.
- Gene therapy approaches using either cytostatic, in which cells are prevented from progressing through the cell cycle to mitosis, or cytotoxic, in which cell death is induced, may inhibit neointimal hyperplasia of restenosis.
- Gene therapy for therapeutic neovascularization targets angiogenic growth factors.

- Gene therapy offers a new avenue for the modification of vein graft biology that might lead to a reduction in clinical morbidity from graft failures. Intraoperative transfection of the vein graft offers an opportunity to combine intact tissue DNA transfer techniques with the increased safety of ex vivo transfection.
- For gene therapy of the heart, genetic manipulation of the myocardium has generated considerable interest in the use of gene transfer of the β-adrenergic recepter gene into the ailing myocardium as a means of therapeutic intervention. For myocardial infarction, gene therapy offers the ability to genetically convert cardiac fibroblasts into functional cardiomyocytes. Genetic manipulation may be used to limit the degree of injury sustained by the myocardium after ischemia and reperfusion through the transfer of natural scavengers of oxidative tissue injury.

SUGGESTED READINGS

Cardiovascular Gene Therapy

Allen, MD. Myocardial protection: Is there a role for gene therapy. Ann Thorac Surg 68:1924–1928, 1999.

Amant C, Berthou L, Walsh K. Angiogenesis and gene therapy in man: Dream or reality. Drugs 59(Spec No 33–36), 1999.

Ponder KP. Systemic gene therapy for cardiovascular disease. Trends Cardiovasc Med 9:158–162, 1999.

Zoldhelyi P, Eichstaedt H, Jax T, McNatt JM, Chen ZQ, Shelat HS, Rose H, Willerson JT. The emerging clinical potential of cardiovascular gene therapy. Semin Interv Cardiol 4:151–65, 1999.

Vascular/Smooth Muscle Gene Therapy

Chang MW, Barr E, Lu MM. Adenovirus-mediated over-expression of the cyclin/cyclin dependent kinase inhibitor, p2l inhibits vascular smooth muscle cell proliferation and neointima formation in the rat carotid artery model of balloon angioplasty. J Clin Invest 96:2260–2268, 1995.

Chang MW, Barr E, Seltzer J. Cytostatic gene therapy for vascular proliferative disorders with a constitutively active form of the retinoblastoma gene product. Science 267:518–522, 1995.

Dunn PF, Newman KD, Jones M. Seeding of vascular grafts with genetically modified endothelial cells. Secretion of recombinant TPA results in decreased seeded cell retention in vitro and in vivo. Circulation 93:1439–1446, 1996.

George SJ, Baker AH, Angelini GD. Gene transfer of tissue inhibitor of metalloproteinase-2 inhibits metalloproteinase activity and neointima formation in human saphenous veins. Gene Therapy 5:1552–1560, 1998.

Gibbons GH, Dzau VJ. The emerging concept of vascular remodeling. N Engl J Med 330:1431–1438, 1994.

Houston P, White BP, Campbell CJ, Braddock M. Delivery and expression of fluid shear stress-inducible promoters to the vessel wall: Applications for cardiovascular gene therapy. Hum Gene Therapy 10:3031–3044, 1999.

Mann MJ, Gibbons GH, Tsao PS. Cell cycle inhibition preserves endothelial function in genetically engineered rabbit vein grafts. J Clin Invest 99:1295–1301, 1997.

Mann MJ, Whittemore AD, Donaldson MC. Preliminary clinical experience with genetic engineering of human vein grafts: Evidence for target gene inhibition. Circulation 96:14–18, 1997.

Morishita R, Gibbons GH, Horiuchi M. A novel molecular strategy using cis element "decoy" of E2F binding site inhibits smooth muscle proliferation in vivo. Proc Natl Acad Sci USA 92:5855–5859, 1995.

Morishita R, Gibbons GH, Kaneda Y. Pharmacokinetics of antisense oligodeoxynucleotides (cyclin B I and c&2 kinase) in the vessel wall in vivo: Enhanced therapeutic utility for restenosis by HVJ-liposome delivery. Gene 149:13–19, 1994.

Ohno T, Gordon D, San H, Pompili VJ. Gene therapy for vascular smooth muscle cell proliferation after arterial injury. Science 265:781–784, 1994.

Poliman MJ, Hall JL, Mann MJ. Inhibition of neointimal cell bcl-x expression induces apoptosis and regression of vascular disease. Nat Med 4:222–227, 1998.

Simons M, Edelman ER, DeKeyser JL. Antisense c-myb oligonucleotides inhibit intimal arterial smooth muscle cell accumulation in vivo. Nature 359:67–70, 1992.

Tabata H, Silver M, Isner JM. Arterial gene transfer of acidic fibroblast growth factor for therapeutic angiogenesis in vivo: Critical role of secretion signal in use of naked DNA. Cardiovasc Res 35:470–479, 1997.

Cardiac Gene Therapy

Akhter SA, Skaer CA, Kypson AP. Restoration of beta-adrenergic signaling in failing cardiac ventricular myocytes via adenoviral-mediated gene transfer. Proc Natl Acad Sci USA 94:12100–12105, 1997.

Barr E, Carroll J, Kalynych AM, Tripathy SK. Efficient catheter-mediated gene transfer into the heart using replication-defective adenovirus. Gene Therapy 1:51–58, 1994.

Edelberg JM, Aird WC, Rosenberg RD. Enhancement of murine cardiac chronotropy by the molecular transfer of human beta$_2$ adrenergic receptor DNA. J Clin Invest 101:337–343, 1998.

Giordano FJ, Ping P, McKiman MD. Intracoronary gene transfer of fibroblast growth factor-5 increases blood flow and contractile function in an ischaemic region of the heart. Nat Med 2:534–539, 1996.

Kaptitt MG, Xiao X, Samulski RJ. Long-term gene transfer in porcine myocardium after coronary infusion of an adeno-associated virus vector. Ann Thorac Surg 62:1669–1676, 1996.

Li Q, Bolli R, Qiu Y. Gene therapy with extracellular superoxide dismutase attenuates myocardial stunning in conscious rabbits. Circulation 98:1438–1448, 1998.

Lin H, Parmacek MS, Leiden JM. Expression of recombinant genes in myocardium in vivo after direct injection of DNA. Ciruclation 82:2217–2221, 1990.

Losordo DW, Vale PR, Symes JF. Gene therapy for myocardial angiogenesis: Initial clinical results with direct myocardial injection of phVEGF165 as sole therapy for myocardial ischemia. Circulation 98:2800–2804, 1998.

Mack CA, Patel SR, Schwarz EA. Biologic bypass with the use of adenovirus-mediated gene transfer of the complementary deoxyribonucleic acid for vascular endothelial growth factor 121 improves myocardial perfusion and function in the ischernic porcine heart. J Thorac Cardiovasc Surg 1 15:168–176, 1998.

Morishita R, Sugimoto T, Aoki M. In vivo transfection of cis element "decoy" against nuclear factor-kappab binding site prevents myocardial infarction. Nat Med 3:894–899, 1997.

Murry CE, Kay MA, Bartosek T. Muscle differentiation during repair of myocardial necrosis in rats via gene transfer with MyoD. J Clin Invest 98:2209–2217, 1996.

Poston RS, Mann MJ, Rode S. Ex vivo gene therapy and LFA—I monoclonal antibody combine to yield long-term tolerance to cardiac allografts. J Heart Lung Transp 16:41, 1997.

Qin L, Chavin KD, Ding Y. Retrovirus-mediated transfer of viral IL-10 gene prolongs murine cardiac allograft survival. J Immunol 156:2316–2323, 1996.

Sayeed-Shah U, Mann MJ, Martin J. Complete reversal of ischemic wall motion abnormalities by combined use of gene therapy with transmyocardial laser revascularization. J Thorac Cardiovasc Surg 1 16:763–769, 1998.

Schumacher B, Pecher P, von Specht BU. Induction of neoangiogenesis in ischemic myocardium by human growth factors: First clinical results of a new treatment of coronary heart disease. Circulation 97:645–650, 1998.

Tam SK, Gu W, Nadal-Ginard B. Molecular cardiomyoplasty: Potential cardiac gene therapy for chronic heart failure. J Thorac Cardiovasc Surg 109:918–924, 1995.

Yu Z, Redfern CS, Fishman GI. Conditional transgene expression in the heart. Circ Res 79:691–697, 1996.

CHAPTER 9

Components of Cell and Gene Therapy for Neurological Disorders

LAURIE C. DOERING, PH.D.

INTRODUCTION

The complexity of the nervous system poses several challenging problems for scientists and clinicians who seek to apply gene therapy to neurological disorders. In addition to the standard problems associated with gene therapy (discussed in Chapter 3), we deal with very delicate, complex networks of cells and face the issue of accessibility (Fig. 9.1) and targeting the desired cell type(s) when considering gene therapy strategies in the central nervous system. Unlike other organs in the body such as the liver or lungs where large proportions of the organs can be damaged with minimal or no functional consequences, damage to extremely small areas of the brain can be devastating. Therapeutic targeting to selective areas or cell types will be difficult to achieve in the central nervous system (CNS).

Excluding the identified genetic causes of neurodegenerative diseases, the etiology underlying the primary neurological disorders is unknown. While the principle cell types affected in disorders such as Parkinson's and Alzheimer's have been identified, the exact contributing factors or conditions that trigger relentless neuronal degeneration are presently unknown. Therefore, at this time, gene products that help to reduce the effects of neural dysfunction, offset neuronal death, inhibit apoptosis, or encourage cell survival form the basis of gene therapy in the nervous system. As gene therapy approaches are developed and refined, the outcome of gene therapy in the nervous system could be extremely effective.

In this chapter, the key aspects of neural dysfunction associated with the prominent nervous system disorders are explained. Promising advances with gene transfer to the CNS have been made with different families of virus vectors. A focus on the vectors and the cells used for gene delivery in animal models is provided. Important features of the clinical trials using genetically modified cells and trophic fac-

An Introduction to Molecular Medicine and Gene Therapy, Edited by Thomas F. Kresina
ISBN 0-471-39188-3 © 2001 Wiley-Liss

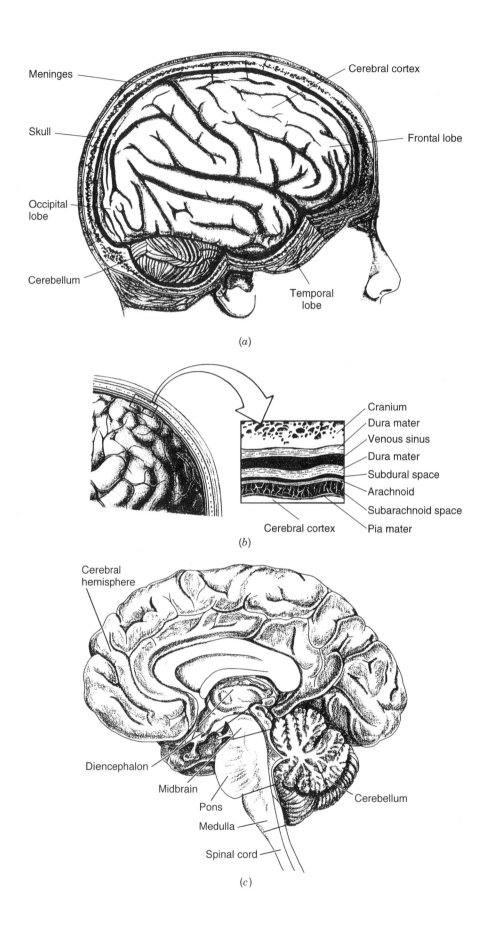

Meninges

Cerebral cortex

Skull

Frontal lobe

Occipital lobe

Cerebellum

Temporal lobe

(a)

Cranium
Dura mater
Venous sinus
Dura mater
Subdural space
Arachnoid
Subarachnoid space
Cerebral cortex
Pia mater

(b)

Cerebral hemisphere

Diencephalon

Midbrain

Pons

Medulla

Spinal cord

Cerebellum

(c)

tors for neurodegeneration are described, and we will illustrate how neuroscience research in combination with genetics and molecular biology is guiding the future of gene therapy applications in the nervous system.

SORTING OUT THE COMPLEXITY OF THE NERVOUS SYSTEM

The nervous system is divided into two main parts: (1) the central nervous system consisting of the brain and spinal cord and (2) the peripheral nervous system (PNS) composed of the nervous tissue in the form of nerves that emerge bilaterally from the brain and spinal cord that serve to keep the other tissues of the body in communication with the CNS (Fig. 9.2). Numerous types of neurons specialized to receive, process, and transmit information via electrical impulses are primarily responsible for the functional characteristics of the nervous system (Fig. 9.3). Neurons can be identified by their size, shape, development, and organization within the brain. Neurons work in networks and secrete neurotransmitters and other chemical messengers at sites of functional contact called synapses. At each synapse a region of the cell membrane in the presynaptic neuron is specialized for rapid secretion of one or more types of neurotransmitters. This area is closely apposed to a specialized region on the postsynaptic cell that contains the receptors for the neurotransmitter or other ligands. The binding of the neurotransmitter to the receptors triggers an electrical signal, the synaptic potential, in the postsynaptic cell (Fig. 9.4). Information in the nervous system is thereby transmitted and processed by elaborate networks that generate a spectrum of electrical and chemical signals.

Glial cells, often referred to as specialized support cells of the CNS, represent the second major class of cells that perform important functions that are key to the normal operation of the nervous system (Fig. 9.3). There are four main types of glial cells. Astrocytes act in a general supportive capacity and help to maintain the extracellular environment in the CNS. The astrocyte processes are intimately associated with the neuronal cell bodies, dendrites, and nerve terminals. They serve to insulate and isolate pathways and neuronal tracts from one another. Oligodendrocytes and Schwann cells form the myelin sheaths around axons in the CNS and PNS, respectively. The myelin is wrapped around segments of axons and serves to accelerate the conduction of the electrical signals. In the CNS, each oligodendrocyte may form and maintain myelin sheaths for approximately 60 axons. In the PNS, there is only one Schwann cell for each segment of one axon. Microglial cells in the CNS are analogous to macrophages and can be activated by a number of conditions, including inflammation and trauma.

◀―――

FIGURE 9.1 External view of the cerebral hemisphere. (*a*) Brain and spinal cord are protected by many layers including the skin, bone, and special connective tissue layers referred to as the meninges. (*b*) Schematic diagram of the protective layers that cover the brain. (*c*) Major divisions of the human brain as seen from a midsaggital view.

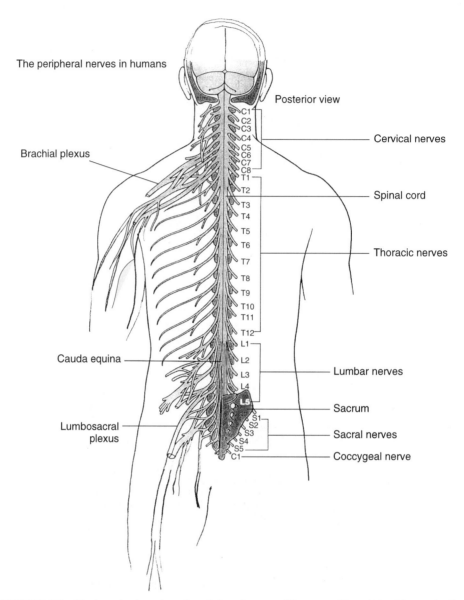

FIGURE 9.2   Brain, spinal cord, and peripheral nerves. There are 31 vertebral bones in the spinal column that house and protect the spinal cord. Between the vertebrae, spinal (peripheral) nerves emerge bilaterally. The individual nerves are made of sensory and motor fibers that interface the peripheral parts of the body with the central nervous system (brain and spinal cord).

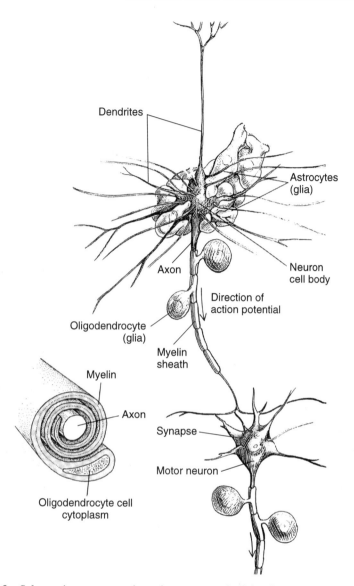

Dendrites

Astrocytes
(glia)

Axon

Neuron
cell body

Direction of
action potential

Oligodendrocyte
(glia)

Myelin
sheath

Myelin

Axon

Synapse

Motor neuron

Oligodendrocyte cell
cytoplasm

FIGURE 9.3 Schematic representation of neurons and glial cells. Neurons are surrounded by astrocytes that fill the interstices between neuronal cell bodies. Glia outnumber neurons by at least 10 to 1. Oligodendrocytes wrap around the axon and produce the myelin sheath. Inset shows how the myelin wraps around segments of the axon.

WHAT GOES WRONG IN NEUROLOGICAL DISORDERS?

Given the vast number and types of neurons and glial cells in the nervous system, one quickly realizes the potential for several neurological dysfunctions, depending on the cell type(s) affected. Neuronal degeneration can occur in selected areas of the brain or neurodegenerative events may affect the entire brain (global neu-

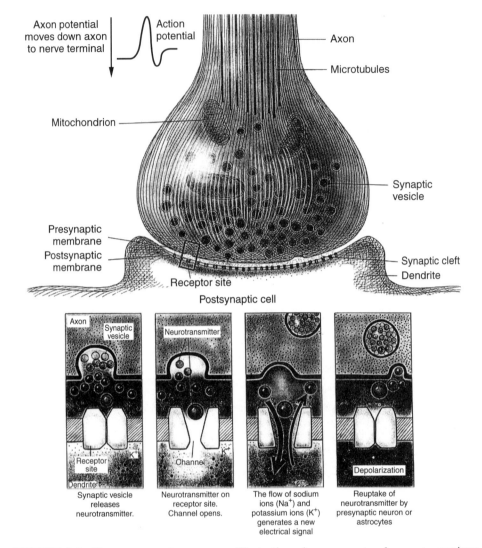

FIGURE 9.4 Components of a synapse. Illustration shows aspects of neurotransmitter release, receptor interaction, and generation of the electrical signal. All electrical signals arise from the action of various combinations of ion channel proteins that form aqueous pores through which ions traverse the membranes. When ion channels are open, ions move through the channels down their electrochemical gradients. Their net movement across the membrane constitutes a current that changes the membrane potential and generates an electrical signal.

rodegenerative conditions) as in the case of the neurogenetic lysosomal storage diseases (LSD) associated with single-gene mutations.

For the majority of neurological disorders, specific classes of neurons in the brain or spinal cord show selective vulnerability. Depending on the type of neuron/ neurotransmitter affected, changes will occur in behavior, memory, or movement. In Parkinson's, neurons located in the substantia nigra of the midbrain that contain the neurotransmitter dopamine undergo accelerated cell death. Loss of these neurons influences the normal function of the extrapyramidal system in the brain and results in rigidity and tremor of the limbs. Alzheimer's isolates the hippocampus and regions of the cerebral cortex due to death of acetylcholine-rich neurons, causes dementia, and prevents the formation of new memory. Amyotrophic lateral sclerosis (ALS) damages the motor neurons in the CNS and causes weakness and spasticity. Alternatively, when oligodendrocytes in the central nervous system are affected, problems develop with routine motor functions, and sensory deficits become noticeable in individuals with multiple sclerosis.

The LSD are genetic disorders resulting from mutations in genes that code for proteins involved with the degradation of normal body compounds that include lipids, proteins, and carbohydrates. Although most lysosomal disorders result from defects in genes that code for lysosomal enzymes, some are caused by genes coding for transport proteins, protective proteins, or enzymes that process the lysosomal enzymes. Individually, the LSD occur infrequently, but collectively they occur approximately in 1/5000 births. The accumulation of enzyme substrates in cells of the CNS characterizes disorders like the mucopolysaccharidoses or GM_1 gangliosidosis.

What triggers selected cell death in the nervous system? In some cases, genetic causes have been associated with neuronal degeneration. In Huntington's disease, a mutation (triplet repeat mutations) in chromosome 4 is linked with the death of neurons in a region of the brain called the caudate/putamen, a complex of interconnected structures tuned to modulate motor activities. The identification of unstable triplet repeat mutations represents one of the great discoveries of human neurogenetics. Genetic linkages discussed later in this chapter have also been determined for a small percentage of individuals with Alzheimer's and Parkinson's.

We have identified various types of cytological and molecular changes in neurons that are associated with the death of neurons. Research has identified numerous, specific changes in neurons at risk associated with the prevalent CNS disorders and also with the aging process. Abnormal accumulations of filaments and altered proteins are recognized as primary features of neurons targeted in neurological dysfunction. The accumulations may occur in the cytoplasm of the neuron or in the extracellular environment. In certain instances, the pattern of neuronal loss is dictated by how the neurons are connected to one another. Alzheimer's is an excellent example of this point. Virtually all the subgroups of neurons lost in Alzheimer's are found to be connected to regions of the cerebral cortex that show high levels of neuritic plaque formation—foci of degenerating processes and twisted arrays of cytoskeletal elements in the neurons referred to as neurofibrillary tangles.

What sets off the initial changes in neurons that lead to a cascade of cell death in specific areas and pathways of the nervous system? A number of molecular mechanisms at different levels of neuronal function have been proposed. Changes to the cytoskeleton, oxidative injury, deoxyribonucleic acid (DNA) modifications, changes in ribonucleic acid (RNA)/protein synthesis, abnormal protein accumulation, toxic-free radicals, reduced axonal transport, and programmed cell death have been identified as possible reasons for neurological disease. Several animal models are used to generate these molecular changes, and, in turn, they help define the possible etiology of neurodegeneration and provide a way to test gene therapy strategies for CNS disorders, injury, or aging.

NEUROTROPHIC FACTORS AND GENE THERAPY

Neurotrophic Factors

There are a variety of molecules in the nervous system that are important to the survival, differentiation, and maintenance of neurons in both the PNS and CNS. These molecules, referred to as neurotrophic factors (Table 9.1), induce pattern and synapse formation and create highly specialized neural circuits in the brain. The factors are secreted from the target innervated by the neurons, taken up at the nerve terminals, and then transported over long distances to the cell body where they act to regulate neuronal functioning by a variety of signaling mechanisms (Fig. 9.5). We now realize that neurotrophic factors bind to cell surface receptor proteins on the nerve terminals, become internalized (receptor-mediated endocytosis), and then move toward the cell body by the mechanism of retrograde axonal transport. Advances in the understanding of the structure of the receptors for neurotrophic factors indicate that they are similar to the receptors used by traditional growth factors and cytokines. The expression of the receptors for the neurotrophic factors is exclusively or predominantly in the nervous system, and, when activated, the factors display distinctive molecular actions.

Nerve growth factor (NGF) is the prototype member of the neurotrophins, a family of proteins that have common structural features. It was discovered and characterized in the 1950s by Rita Levi-Montalcini, Stanley Cohen, and Viktor Hamburger and was the first molecule to show potent nerve growth promoting activity on explants of neural tissue maintained in tissue culture. Since the discovery of NGF, a number of molecules have been identified and added to the expanding list of substances grouped under the broad umbrella of neurotrophic factors. Common, well-studied factors are listed in Table 9.1. Responses to the neurotrophins are mediated through receptor tyrosine kinases that belong to the *trk* family of protoonco-

TABLE 9.1 A Listing of Common Neurotrophic Factors

Class	Members	Receptor	Responsive Neurons
Neurotrophins	NGF	TrkA	Forebrain cholinergic neurons
	NT-3	TrkC	Corticospinal neurons
	NT4/5	TrkB	Caudate/putamen
	BDNF	TrkB	Substantia nigra
Transforming growth factor β	GDNF	Ret	Substantia nigra neurons
	TGF-β		Motor neurons
Cytokines	CNTF	CNTFα	Spinal cord motor neurons
	LIF	gp130/JAK LIFRβ/TYK	Spinal cord motor neurons
Insulinlike growth factors	IGF-1	IGF receptor	Forebrain cholinergic neurons Forebrain cholinergic neurons
	IGF-2		
Fibroblast growth factors	bFGF	FGF receptor	Forebrain cholinergic neurons Spinal cord motor neurons
	aFGF		

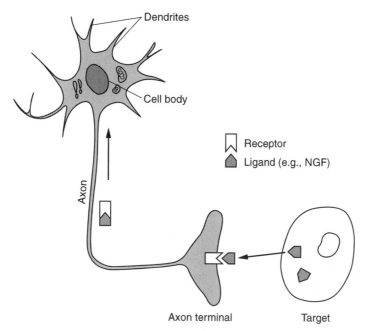

FIGURE 9.5   Retrograde signaling by neurotrophic factors. The neurotrophic factor ligand (supplied by a target tissue) binds to the receptor on the surface of the axon terminal. This receptor–ligand complex is then transported along the axon to the cell body. Retrograde trophic signals have been shown to modulate neuronal growth, survival, death, and the expression of neurotransmitters.

genes. It is now clear that neurotrophic factors can be provided by a number of sources including glial cells, afferent processes of neurons, muscle, and even by the extracellular matrix. Numerous biological events including neuronal growth, phenotype (neurotransmitter) expression, and programmed cell death have been linked with retrograde neurotrophic factor signaling. Hence, there are many possible lines of study to explore the effects of neurotrophic factor gene therapy in relation to basic neural cell survival and function for the treatment of neurodegenerative disorders.

From basic research, we have learned that if the brain is injured, these molecules can be released to play a significant role in the recovery process. In addition to limiting the loss of neurons, neurotrophic factors can stimulate new outgrowth from the axons and dendrites, regulate axon branching, modulate neurotransmitter synthesis, and influence synapse formation. This inherit property of structural and functional change in neurons in response to environmental cues (like the release of neurotrophic factors) is referred to as plasticity. Many factors have been shown to have overlapping effects (primarily on development and survival) on subsets of neurons in the central and peripheral nervous system. It is now very clear that any given type of central or peripheral neuron needs a combination of factors, rather than a single neurotrophic factor to optimize survival and function. Therefore, decisions must be made regarding the most effective combinations of factors for the neurons/neurological disorder in question. As discussed later in this chapter, the logic of combined neurotrophic factor therapy must, however, be balanced against the increased risk of adverse effects that have surfaced from many clinical trials.

The identification and characterization of each neurotrophic molecule has been followed by the establishment of transgenic (knock-out) mice that do not produce that factor or the associated receptor components to help unravel the physiological function of these molecules and to assess their contribution to the survival of different neuronal types. It should be pointed out, however, that we do not know if neurotrophic gene defects in humans are associated with any aspect of neurological dysfunction.

Extensive research has focused on the beneficial effects of delivering neurotrophic factors in the animal models of neurodegeneration and this research has set the foundation for a number of clinical trials (discussed later). The extent of the nervous system damage, the available concentration of neurotrophic factors, and the time at which the factor is released are key parameters in relation to the effectiveness of these molecules to rescue neurons from death. It should be realized that the precise roles of neurotrophic factors and their therapeutic potential in degeneration disorders remains to be elucidated.

Gene Therapy in Animal Models of Neural Degeneration

At the present time CNS gene therapy initiatives follow in vivo and ex vivo approaches. Gene transfer by viral vectors is currently the most common and preferred method of gene delivery to cells of the CNS. The in vivo method involves direct administration of the virus to the nervous system. For this approach, viral vectors are injected into specified locations of the brain or spinal cord. In the case of ex vivo gene transfer, new genes are first introduced into cells in a tissue culture environment, and then the cells are stereotaxically transplanted into desired regions of the nervous system.

As gene therapy efforts continue, the list of viral systems continues to grow. The types of viruses and cells that have been used for gene delivery in the nervous system are shown in Figure 9.6. Now, viral vectors and cells are used together and certain combinations show real promise and benefits over the gene and cell replacement procedures used just a few years ago. As each neurotrophic factor is identified, cells are genetically modified to secrete the factor and then tested in animal models for effects on neuronal survival and animal behavior (Table 9.2). Some of the gene therapy models are highlighted here with a special focus on the promising vectors and the cells used to transfer genes with therapeutic value in the CNS. The purpose of this section is to provide some examples of the streams of gene therapy used in the animal models for the neurodegenerative disorders described in this chapter.

To model Alzheimer's, animals are used that show cholinergic neuron loss, the formation of neurofibrillary tangles plaques, or the generation of the amyloid precursor protein. In mammals, transection of the fimbria-fornix pathway (connection between the hippocampus and medial septum) produces significant death (approximately 50%) of cholinergic neurons in the medial septum, paralleled by a loss of cholinergic inputs to the hippocampal formation. If a neurotrophin (e.g., NGF) is administered, the transection-induced neuronal loss in the medial septum/forebrain

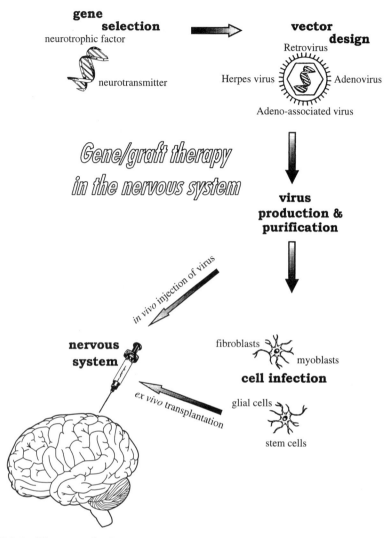

FIGURE 9.6 Viruses and cell types used for experimental gene/graft therapy in the nervous system.

region can be minimized. Infusions of NGF in animal models of age-related memory impairments will also improve the memory-associated tasks.

The possibility of supplying a neurotrophic factor to the brain via genetically engineered cells was first demonstrated by Fred Gage and co-workers in 1988. The investigators used a rat fibroblast cell line (208F) that had been modified with a retrovirus designed to synthesize and secrete NGF. The fibroblasts were implanted into the brains of rats with fimbria-fornix lesions. The engineered fibroblasts produced enough active NGF to rescue more than 90% of the cholinergic neurons from cell death. This work indicated that this approach to ex vivo gene therapy is feasible in the CNS. Similar neuroprotective effects on medial septal cholinergic neurons

TABLE 9.2 Rodent Models Used to Study Neurological Disorders

Disorder	Model	Principal Cell Type Affected	Related Survival Trophic Factor	Transgenic Mouse Model
Parkinson's	6-OHDA injection	Dopamine neurons	BDNF, GDNF	NURR 1
Alzheimer's	Transection of fimbria-fornix pathway	Cholinergic neurons	NGF, NT4/5	APP
Huntington's	Excitotoxin injection (e.g., kainic acid)	GABA neurons	BDNF, NT4/5, CNTF	CAG repeat
ALS	Injection of IDPN	Motor neurons	BDNF, CNTF	SOD1
MS	EAE	Oligodendrocytes	CNTF, IL-6	2–5 MBP

have been shown with primary fibroblasts, baby hamster kidney (BHK) cells, and neuroblastoma cells all modified to produce NGF.

In addition to gene therapy with neurotrophic factors, strategies that use regulatory proteins of cell death have been examined. Antiapoptotic factors like Bcl-xL is one of three isoforms of Bcl-x that protects cells from the damaging effect of reactive oxygen molecules. These antiapoptotic factors are being evaluated by gene therapy in animal models of neural degeneration (see section on programmed cell death and neurodegeneration).

The most popular animal model of Parkinson's is the rat model. Involving intracerebral injections of the catecholamine neurotoxin 6-hydroxydopamine (6-OHDA), this neurotoxin destroys the dopamine fibers that project from the substantia nigra to the striatum. This treatment results in a loss of dopamine and causes a circling behavior in the animals when they are given a dopamine agonist (e.g., amphetamine or apomorphine) to activate the dopamine receptors. The circling tendencies can be reduced when the enzyme tyrosine hydroxylase (rate-limiting enzyme for dopamine production) is made available to neurons in the striatum. Initial ex vivo gene therapy experiments in consideration of Parkinson's used cell lines of fibroblasts genetically modified in culture to express the gene for tyrosine hydroxylase. In this case, the function of the implanted fibroblasts was monitored by observing reductions in the circling behavior of the recipient host rats. In addition to fibroblasts, primary myoblasts and a variety of other cell lines have been modified to synthesize tyrosine hydroxylase and have shown to reduce the behavioral impairments in the 6-OHDA-lesioned rat model. It should also be pointed out that fibroblasts as well as other non-neuronal cell types do not make connections with the host brain circuitry but still produce strong functional effects when producing the transgene product. A primary drawback when using fibroblast cell lines has been the continued expansion of the fibroblast cell mass within the brain. To prevent tumor formation by these cell lines, the cells can be encapsulated by materials that allow for the exchange of the transgene product between the cells and the host tissue. Important advances that use primary cells, stem cells, and cell lines that withdraw from the cell cycle are now the focus of attention when considering the transplantation of cells into the nervous system.

Although we do not know why neurons that contain dopamine preferentially die in Parkinson's, neurotrophic factors that enhance the survival and function of these dopamine neurons are the center of attention for gene therapy possibilities with the hope of preventing the death of these neurons. Promising factors include brain-derived neurotrophic factor (BDNF), fibroblast growth factor (FGF), and glial-cell-line-derived neurotrophic factor (GDNF). These three factors show significant protection of dopaminergic neurons. Primary fibroblasts and fibroblast cell lines engineered to deliver BDNF by retroviral infection can prevent the degeneration of dopamine neurons when the fibroblasts are transplanted into the striatum of animals that model Parkinson's. In this situation, BDNF is taken up by the nerve terminals of the dopamine neurons and moved back to the cell body by retrograde transport. In the cell body, the BDNF activates a cascade of molecular signals that prevents neuronal death.

GDNF is a member of the transforming growth factor β (TGF-β) family, a large group of cytokines that play roles in the control of cell proliferation, migration, and morphogenesis. This molecule, discovered in the culture supernatants of a glial cell line by Leu-Fen Lin in the laboratory of Frank Collins in 1993 was shown to have potent effects on the survival of dopamine neurons. Replication-defective aden-ovirus vectors that encode for GDNF are able to reduce experimentally induced rotational behavior when injected into the 6-OHDA rat model of Parkinson's. These Ad vectors using the Rous sarcoma virus (RSV) promoter to control the GDNF transgene, however, showed significant reductions in transgene expression levels after 1 month. Host immune reactions to adenovirus and down-regulation of the viral promoters are common problems observed with adenoviral injections in the brain. Next generation Ad vectors will be designed to minimize the immune reactions and extend gene expression. Like other neurotrophic factors, GDNF now appears to have pharmacological effects on a wide variety of neurons. It is a potent survival factor for motor neurons in the spinal cord and for Purkinje neurons in the cerebellum.

Another technique to prevent neuronal degeneration has been to transplant support cells with fetal neurons. In this situation, referred to as a co-grafting strategy, the support cells assist with the survival of the transplanted neurons. Fibroblasts modified to produce a local supply of FGF helps maintain grafts of fetal dopamine neurons. The fibroblasts not only help to maintain the population of transplanted neurons but also help to reduce the need for large numbers of fetal cells when dissected from embryonic brains.

In consideration of Huntington's, encapsulated human fibroblasts made to secrete ciliary neurotrophic factor (CNTF) can prevent behavioral deficits and striatal degeneration in the rodent model of Huntington's disease. Experimental gene therapy in a monkey model of Huntington's has been evaluated. Monkeys given an injection of quinolinic acid show features of neurodegeneration that are characteristic of Huntington's disease. Researchers at CytoTherapeutics in Rhode Island engineered baby hamster kidney fibroblasts to secrete CNTF and then enclosed the cells in polymer capsules before implantation into the striatum. When the capsules containing the modified fibroblasts were grafted into the monkeys that model Huntington's, the production of CNTF protected several populations of cells including GABAergic and cholinergic neurons from death.

It should be noted that the vectors are designed to eliminate viral gene expression to avoid cytotoxic and immunological effects. The exclusion of these genes, however, often reduces the efficiency and length of transgene expression. Control of the gene product will be a critical aspect of successful gene therapy in the CNS. There are intense efforts to develop gene regulatory elements that offer cell-specific (spatial) expression and/or drug-dependent (temporal) expression of the desired therapeutic gene. Potential transgene promoter/regulatory elements to guide neuronal expression include the light neurofilament subunit, α-tubulin, neuron-specific enolase, and tyrosine hydroxylase. Promoters for glial fibrillary acidic protein and myelin basic protein have been constructed to drive transgene expression in astrocytes and oligodendrocytes, respectively. A common inducible (temporal) transgene system uses tetracycline or tetracycline derivatives as controlled promoters. Transcriptional control of tyrosine hydroxylase, various reporter genes, and CNTF has been achieved with the inducible tetracycline system in neural progenitors and in cell lines. The ability to control the genetic elements and the level of the new transgene via a pharmacological effector such as tetracycline will be very important in consideration of CNS gene therapy protocols that focus on the delivery of neurotrophic factors and neurotransmitters.

Exploiting the Properties of HIV for Gene Delivery in the CNS

The power and potential of molecular biology techniques is exemplified through the creation of very useful gene delivery vectors that are based on potentially harmful viruses such as the human immunodeficiency virus type 1 (HIV-1). Neurons in the nervous system reside in a nondividing state and therefore potential virus vectors for gene therapy must be capable of infecting postmitotic cells. A method developed by Inder Verma, Luigi Naldini, and Didier Trono at the Salk Institute in La Jolla, California, took advantage of HIV genome elements to generate recombinant viruses capable of infecting nondividing cells, including neurons. The HIV virus is a well-characterized lentivirus that belongs to the retrovirus family. Lentiviruses (from the Latin word *lentus* meaning slow) cause slow chronic and progressive degenerative diseases of the nervous, hematopoietic, musculoskeletal, and immune systems.

The lentiviruses have powerful gene regulatory systems and the HIV-1 *tat*-LTR (long terminal repeats) transactivator–promotor combination is one of the strongest known. These viruses are the only retroviruses able to integrate into the chromosomes of cells that are not mitotically active. This virus was stripped of its ability to reproduce but used the HIV nuclear import components to guide the integration of new genes into the nuclei of infected cells. The HIV genetic sequences that control integration into the target cells plus the elements from two other viral plasmids were used to produce highly efficient virus vectors that directed long-term, stable, novel gene expression in neurons. The efficiency of gene transfer is high and reports indicate that lentiviral vectors injected into the adult rat brain stably transduce terminally differentiated cells in vivo, without a decrease in transgene expression or toxicity for at least 6 months in vivo. Furthermore, the injection of HIV-derived vectors into the nervous system does not set off significant inflammatory or immune responses. The ability to construct HIV-based viral vectors for efficient and stable gene delivery into nondividing cells is an important step to increase the applicability of retroviral vectors in human gene therapy.

Programmed Cell Death and Neurodegeneration

Programmed cell death (PCD), also referred to as apoptosis, occurs during the development of all animals and is the process where cells activate an intrinsic death program. Recent attention has been focused on the observations of increased PCD rates in the major neurological disorders discussed in this chapter. While there is no definitive evidence that PCD is the key problem in neurological disorders, there is a rapidly growing body of evidence that PCD is involved with the death of neurons and glial cells. There are numerous genes that modulate PCD. These genes and their products show homology throughout the animal kingdom from the nematode to the primates. The products of the *Bcl-2* family of protooncogenes have been extensively characterized as proteins that regulate cell death. A possible therapeutic approach to preventing neuronal degeneration may be via the modulation of apoptosis by members of the *Bcl-2* family, including *bcl-xl* and *bax*. In Alzheimer's, levels of *Bcl-2* protein are significantly higher than aged-matched adult brain, and this protein is predominantly localized to activated astrocytes rather than neurons.

Overexpression of *bcl-2* in the superoxide dismutase (SOD) transgenic mouse model of ALS delays the onset of the motor neuron disorder but does affect the duration of the condition. *Bcl-2* has strong antioxidant properties. Thus, overexpression of *Bcl-2* may prevent the degeneration of motor neurons by inhibiting free radical mediated damage. Studies of this type suggest the possibility of *Bcl-2* gene therapy for ALS. However, these experiments indicate that potential treatment should begin before the clinical symptoms of ALS are apparent.

Poor survival of grafted neurons has been a major issue in neural transplantation. Attempts to increase the survival of grafted neurons have been made by expressing the *Bcl-2* gene in cells before transplantation. This concept has been tested with a cell line generated from the substantia nigra. When this cell line over-expresses the *Bcl-2* protein in the striatum of 6-OHDA treated rats, enhanced behavioral improvements are observed in the rat (i.e., reductions in apomorphine-induced rotation).

In the rodent fimbria-fornix lesion model of cholinergic neuron degeneration, neuroprotective effects have been demonstrated by the *Bcl-xL* gene. Expression of *Bcl-xL* by lentiviral vectors in this model significantly increases cholinergic neuron survival in the septal region subsequent to axotomy of the pathway. Studies of this nature provide evidence that overexpression of antiapoptotic factors via gene transfer in vivo is sufficient to rescue neuronal populations after axotomy.

A new family of anti-apoptotic proteins called inhibitors of apoptosis (IAP) has recently been discovered. Human IAP proteins include XIAP, HIAP1, HIAP2, NAIP, BRUCE, and Survivin. The neuronal apoptosis inhibitory protein (NAIP) is expressed in neuronal cells. The administration of NAIP with adenoviral vectors has been shown to reduce the death of hippocampal neurons in cases of ischemia and rescue motor neurons in laboratory axotomy models.

TABLE 9.3 Variables that Encourage Survival of Cells Grafted in the Brain

Fetal cells: survive better than adult cells
Young hosts: are more receptive to grafts
Trophic factor(s): improve cell survival and enhance process growth
Target access: is a key in long-term survival
Vascular supply: is essential for survival
Immune compatibility: reduces the risk of rejections

NEURAL TRANSPLANTS AND STEM CELLS

Experimental Transplantation to Clinical Application

In the 1970s as the concept of neural transplantation grew, the parameters to maximize the survival and function of grafted cells were established. One of the key pioneers in the field of transplantation research, Anders Björklund at the University of Lund, Sweden, has been instrumental in the refinement of cell grafting in the CNS. Within a span of approximately 20 years, the transplantation of cells into the brain evolved from the laboratory setting to clinical trials for severe Parkinson's. Table 9.3 lists several variables that have been identified as essential to maximize the survival of cells when grafted into the CNS.

Throughout the 1980s and 1990s, tissues were grafted into the brain to study aspects of neural cell development and to identify the function of different brain areas. Animal models of CNS degeneration and injury were refined; the survival and restorative effects of various neurons and glial cells in the animals were studied from cellular, molecular and behavioral perspectives. Now, we are in a new era of establishing the most appropriate cell grafting technologies for application in the clinic. Unfortunately, the dramatic restorative functional changes seen in certain animal models of neurological disease were not seen with the transfer of the grafting techniques to the human situation. Case in point—transplants of fetal substantia nigra neurons stereotaxically injected into the striatum of Parkinson's patients. While the 6-OHDA-lesioned rats (the rodent model of Parkinson's) with implants of substantia nigra showed significant and remarkable recovery of some behavioral impairments, the outcome for individuals with Parkinson's who received stereotaxic injections of fetal neurons was not as favorable as the laboratory findings. The results from the initial clinical trials were partly encouraging in that there were no major side effects from this type of operation. Some of the transplanted cells in the human striatum show extended survival for years, and for some patients there was a therapeutically significant reduction in the motor symptoms (rigidity and bradykinesia). In fact, survival of grafted fetal neurons up to 8 years has been reported. The modest to moderate improvement seen in some patients does, however, gradually disappear. There has been considerable variability in the outcome from patient to patient. To date we cannot predict with certainty that Parkinson's patients who are ideal candidates for a transplant will benefit from this grafting procedure.

One of the primary problems with transplanting neurons into the lab animal and human brain has been the issue of poor graft survival. In humans only about 5% of the fetal dopamine neurons survive using the current transplantation protocols.

However, in animals, poor cell survival has been correlated with surprisingly significant restoration of behavior. This raises the issue of just how representative are the animal models of human neurological disorders. Although fetal neurons have shown the greatest potential in terms of graft survival and clinical efficacy for Parkinson's, there are serious concerns associated with the use of human fetal neurons, namely tissue availability, quality control, and ethics. To circumvent some aspects of these problems, research has examined neural xenografts for Parkinson's and the use of stem or neuronal cells grown in culture. It is now possible to isolate subpopulations of stem or neuronal progenitor cells from the developing or adult nervous system, expand the cells in culture, and then use the cells for transplantation or as vehicles for gene delivery to selected sites of the nervous system. These cells survive in vitro in media enriched with growth factors and with passage express a neuronal phenotype. A major advantage of using progenitor cells for transplantation is that they have not been transformed or immortalized and exist naturally in the brain. Continued collaborative efforts between the basic and the clinical research sectors using stem or progenitor cells for ex vivo transgene delivery will be critical to the progression of effective therapy for Parkinson's and other neurodegenerative conditions.

As previously described, a variety of non-neuronal primary cells and cell lines have been used largely as a way to deliver an active substance that promotes survival or growth of neurons. Cells of non-neural origin (e.g., fibroblasts, myoblasts) do not integrate into the host brain tissue and therefore remain as isolated tissue masses. These types of cells are foreign to the brain and we do not know the long-term consequences of these foreign cells within the CNS. The ideal cells used for cell replacement should be derived from the CNS. Research centered on cell replacement strategies now focus predominantly on the use of neural stem cells. Cells that can fully differentiate and integrate in the CNS provide excellent prospects for therapy and also for the delivery of gene products.

Stem Cells in the Adult Brain

Until just a few years ago, it was generally assumed and believed that the adult brain was incapable of generating new neurons. Research on a number of fronts has established that the adult mammalian brain contains stem cells that can give rise to the full spectrum of neurons and glial cells. In particular, the subventricular zone, an important layer that forms during development and persists into adulthood retains the capacity to generate both neurons and glial cells (Fig. 9.7). Stem cells by strict definition over the lifetime of the animal must be able to proliferate, show self-renewal, produce progeny with multilineage characteristics, and divide when injured. Progenitor cells refer to cells with a more restricted potential than stem cells, and precursor cells refer to cells within a given developmental pathway. The presence of neural stem cells in the adult brain has established the possibility for using the mature brain as a source of precursor cells for transplantation and helps to establish new therapy directions for neurological injury and disease. In fact, as our understanding of stem cell neurobiology grows, it may be possible to control the proliferation and migration of such cells into areas of the nervous system affected by the diseases discussed in this chapter. The notion of self-repair in the brain is now visible at the basic research level. With eloquent neuroanatomical tech-

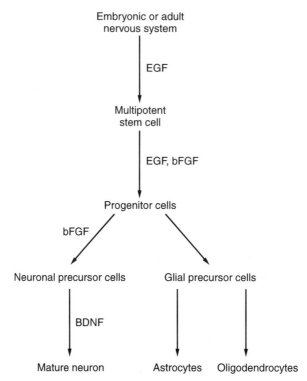

FIGURE 9.7 Theoretical model for the generation of neurons and glial cells from stem cells in the brain. The potential growth factors governing the commitment and differentiation of the neuronal lineage are indicated.

niques, Sanjay Magavi, Blair Leavitt, and Jeffrey Macklis of the Children's Hospital/Harvard Medical School have shown that stem cells in the adult mouse brain can migrate and replace neurons that undergo apoptosis in the neocortex. Moreover, these newly generated neurons had also made connections to their appropriate target.

Multipotent stem cell proliferation and differentiation can be regulated by neurotrophic factors. For example, epidermal growth factor (EGF) can induce the proliferation of stem cells from embryonic and adult CNS tissue in vitro. When growth factors are added in sequence to neural stem cells, they regulate whether the cells will acquire neuronal or glial characteristics. The addition of basic fibroblast growth factor to progenitor cells derived from EGF responsive stem cells produces neuronal progenitors.

One sector of gene therapy research focuses on a neural-stem-cell-based strategy. There is hope that progenitor or stem cells will play the critical role in effective CNS gene therapy. With the capability of differentiating along multiple cell lineages, stem cells may be very effective for the delivery of therapeutic gene products throughout the brain or spinal cord. The potential of combining progenitor cells with CNS gene therapy was demonstrated by Evan Snyder, Rosanne Taylor, and John Wolfe in 1995. They demonstrated that neural stem cells, engineered to secrete the enzyme β-glucuronidase (GUSβ) could deliver therapeutic levels of GUSβ sufficient to enhance the life span of mice modeled for a neurogenetic LSD—

mucopolysacchaidoses type VII (MPSVII). The enzyme deficiency in this mouse model causes lysosomal accumulations of undegraded glycosominoglycans in the brain and other tissues that results in fatal degenerative changes. Fibroblasts transduced by a retrovirus encoding GUSβ have also been successful in clearing the lysosomal lesions in this model. The ability to clear the lysosomal distentions from neurons and glial cells by gene therapy is an important advance because most patients are not diagnosed with LSD until the lesions are advanced enough to affect phenotype or developmental milestones. Similar therapeutic paradigms are also being evaluated for other inherited neurogenetic diseases that are characterized by an absence of discrete gene products. Engineered cells and progenitors are also being grafted into mouse models of hexosaminadase deficiencies causing Tay-Sachs and Sandhoff disease.

Oncogene Transfer to Neural Cells

A variety of methods have been developed to generate cell lines from primary cells and developmental neurobiologists have used specially constructed retrovirus vectors to establish cell lines from the developing CNS. Clones of stem cells or progenitor cells are used extensively to study aspects of differentiation along neuronal and glial lineages. These types of progenitor cell lines have been useful in the identification of molecules and neurotrophic factors that initiate and modulate differentiation at specific developmental time points. Stage-specific lines of neurons or glial cells have been established with retrovirus vectors containing oncogenes such as the simian virus 40 (SV40) large tumor T antigen, *neu*, and the *myc* family.

The *myc* family of protooncogenes consist of a number of well-characterized members including c-*myc*, N-*myc*, and L-*myc*. The *myc* gene was originally identified as the oncogene of the MC29 avian leukemia virus. This retrovirus induces a number of carcinomas in addition to the leukemic disorder *my*elocytomatosis (*myc*) in birds and can transform primary cells in tissue culture.

The transformation of cells from the developing nervous system with a retrovirus expressing v-*myc* have revealed extraordinary characteristics. In culture, progenitor cells immortalized with the v-*myc* oncogene divide continuously. However, when removed from the culture environment and transplanted back into the nervous system of laboratory animals, these v-*myc*-immortalized cells withdraw from the cell cycle and undergo terminal differentiation. In addition, certain neural progenitor cells generated with v-*myc* not only stop dividing in the animals' brain, but the cells also undergo site-specific differentiation. A well-characterized clonal cell line (termed C17.2) with stem cell features will acquire glial characteristics or neuronal features when situated in the white matter or gray matter, respectively. The C17.2 cells will also differentiate into the appropriate neuronal phenotype and express the neurotransmitter specific to the transplant region. Several hundred grafts of neural cells carrying the v-*myc* gene have been studied in laboratory animals in numerous regions of the central and peripheral nervous system, and not a single graft has shown continued proliferation (tumor growth). Hence, the cells with this oncogene fall into a special category with highly desired characteristics in consideration of cell replacement strategies for therapeutic restoration of nervous system function. At this time, the precise mechanism(s) that override the expression of the v-*myc* oncogene product and pull the cells from mitotic cycling are not known.

CLINICAL NEURODEGENERATIVE CONDITIONS

Alzheimer's

In the strictest sense, the conditions of Alzheimer's and also Parkinson's should be defined as disorders rather than diseases, since no etiological agents have been identified at this time. Alzheimer's represents the single greatest cause of mental deterioration in older people, affecting approximately 4 million in the United States and 300,000 in Canada. Men and women are affected almost equally. The German physician Alois Alzheimer first described this condition in 1907 as a case presentation of a 51-year-old woman whose symptoms included depression, hallucinations, dementia, and, upon postmortem examination, a "paucity of cells in the cerebral cortex . . . and clumps of filaments between the nerve cells."

Alzheimer's is a progressive, degenerative condition of the brain, usually associated with advancing age. Although the majority of individuals are in their sixties, Alzheimer's can develop at a younger age. No matter when a person is affected, the condition is always progressive and degenerative. Formerly self-reliant people eventually become dependent upon others for routine daily activities.

The first indication of Alzheimer's are subtle changes in behavior. Difficulty with short-term memory then becomes apparent. Adjustments to new places or situations may prove to be stressful. Learning, making decisions, or executing tasks becomes problematic. Eventually, emotional control becomes more and more difficult.

Although there are a number of promising clues, the definitive cause of Alzheimer's has not been determined. Scientists recognize that there are two forms of Alzheimer's—familial and sporadic. The familial (sometimes referred to as early-onset Alzheimer's) stream is known to be entirely inherited. These autosomal-dominant inheritance patterns are linked to specific mutations in the genes encoding presenilin 1 (PS1), presenilin 2 (PS2), and the amyloid precursor protein (APP). Mutations at all three of these loci lead to increased production of the amyloid polypeptide Aβ42. This peptide is derived from APP and spans the transmembrane region of cells. Abnormal phosphorylation events lead to the deposition of Aβ42 in the neuropil and blood vessel walls and may be the initiating factor in Alzheimer's. It is estimated that 10 to 20% of cases belong to the familial group. It progresses faster than the sporadic, late-onset form of the disorder, which generally develops after age 65. The late-onset forms have been associated with the presence of APOE-ζ4 alleles. APOE is a serum protein that mediates cholesterol storage, transport, and metabolism. It appears that the APOE allele type does not predict risk of Alzheimer's but influences the age at which the disease is likely to occur.

In Alzheimer's, axons and dendrites in the brain neurophil degenerate and disrupt the normal passage of signals between cells. These focal areas of degeneration (senile plaques) have specific cytological characteristics. The plaques are composed of degenerating neuronal processes associated with extracellular deposits of amyloid peptides. These foci tend to recruit astrocytes and microglia. In addition, changes also occur inside the neurons, leading to cytoskeletal disruption and the accumulation of abnormal filament proteins in twisted arrays called neurofibrillary tangles. Tangles consist predominantly of abnormal phosphorylated forms of tau—a protein that binds to microtubules as part of the neuronal cytoskeleton.

The severity of mental deterioration has been correlated with a high density of neuritic plaques and neurofibrillary tangles in the cortical areas of the brain. Acetylcholine and somatostatin are the principal neurotransmitters that are depleted in Alzheimer's.

There is strong evidence implicating cholinergic neurons as the mediators of memory loss in Alzheimer's. The illness results from selective damage of specific neuronal circuits in the neocortex, hippocampus, and basal forebrain cholinergic system. In fact, the extent of the cholinergic deficit correlates with the degree of memory impairment and the loss of cholinergic function appears to be one of the earliest changes. Nerve growth factor has a potent influence on the survival of cholinergic neurons, and NGF administration prevents cholinergic neuron atrophy during normal aging and in cases of experimental injury. These observations have provided part of the rationale for NGF therapy of Alzheimer's. This chapter describes experiments applying gene therapy to the animal models of Alzheimer's and Parkinson's as well as related clinical trials.

Parkinson's

In 1817, the British physician James Parkinson published a study entitled *An Essay on the Shaking Palsy*. In this work, he outlined the major symptoms of the disorder that would later bear his name.

Parkinson's runs a lifetime incidence of about 2% and an estimated one million people in the United States have this neurodegenerative disorder. It generally affects men and women 40 years of age or older. Symptoms appear slowly and in no particular order. In fact, many years may pass before early symptoms progress to the point where they interfere with normal activities. The four major hallmarks or symptoms are debilitating rigidity, resting tremor, bradykinesia or akinesia (slowness or lack of movement), and postural instability demonstrated by poor balance.

Parkinson's is caused by the progressive deterioration of a small area in the midbrain called the substantia nigra. This region contains neurons that produce the neurotransmitter dopamine. Dopamine is transported through the axons that terminate in the striatum—a large structure consisting of the caudate nucleus and the putamen. This structure is part of the basal nuclei and is involved in complex muscular activities such as postural adjustments, locomotion, and balance. The striatum may also be viewed as responsible for inhibiting unwanted movements and permitting selected actions. As neurons in the substantia nigra die, less dopamine is transported to the striatum. Other groups of neurons connected with the striatum may also die. Eventually a low threshold level of dopamine leads to the neurological symptoms (Fig. 9.8). There is muscle stiffness and difficulty with bending the extremities. Walking patterns change and the gait will often assume a shuffling pattern. There is freezing of movement when the movement is stopped and often the inability to resume motion. The finger-thumb rubbing (pill-rolling tremor) may be present. Changes in facial expression are described as a "masklike" appearance. Speech becomes slow and very low, with a monotone quality. There is also a loss of fine motor skills and hand writing takes on distinctive features.

A pattern of familial aggregation for the autosomal dominance and inheritance of early-onset Parkinsons' has been established, and a susceptible gene associated with this group has been located on the long arm (q) of chromosome 4 at band 21

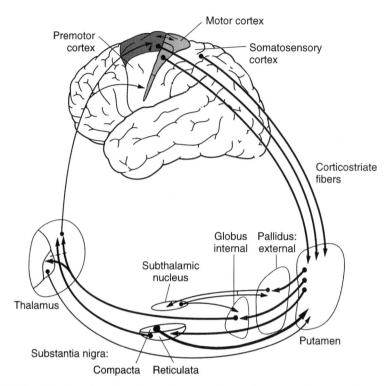

FIGURE 9.8 Circuits of the basal ganglia. A variety of reciprocal connections are made between neurons joining the substantia nigra with the striatum (putamen). Dopamine made in the substantia nigra is transported to the putamen (arrow). Death of substantia nigra neurons results in reduced levels of dopamine transported to the putamen and causes the neurological symptoms of Parkinson's.

(4q21). A mutation in the α-synuclein gene (a substitution of alanine to threonine at position 53), which codes for a presynaptic nerve terminal protein, was identified to be at fault in a large Italian family in 1997 by Mihael Polymeropoulos and co-workers at the National Human Genome Research Institute in Bethesda, Maryland. A number of additional defective genes including *Parkin*, *PARK3*, *UCH-LI*, and *2p13* have also been identified in certain family pedigrees.

Current treatment for Parkinson's is aimed at controlling the symptoms. The primary pharmacological therapy is based on increasing dopamine levels in the brain by supplying the precursor L-DOPA and disabling the side effects by the co-administration of a peripheral DOPA-decarboxylase inhibitor. Combined L-DOPA/carbidopa medication is the primary method to alleviate akinesia and rigidity in the early to middle stages of Parkinson's. Basic research and gene therapy initiatives are directed at preventing the loss of neurons that synthesize dopamine (possibly by supplying a neurotrophic factor) or by engineering cells to increase the dopamine concentration in the striatum.

Modern imaging techniques and an improved understanding of basal ganglia function and organization has revitalized the surgical treatments for Parkinson's. Magnetic resonance imaging and electrophysiologically monitoring during surgery permits detailed localization within the brain. Common procedures include the pallidotomy and thalamic deep brain stimulation. The presence of high-frequency stimulation through electrodes placed deep in the brain appears to produce a functional lesion in the desired target area (deep brain stimulation). One of the main applications of neurosurgery is the control of L-DOPA induced dyskinesia by electrical ablation of the posterior ventral globus pallidus (pallidotomy).

Huntington's

In 1872, George Huntington described a disease that he, his father, and his grandfather had observed in several generations of their patients. Huntington's disease (HD) is a hereditary neurodegenerative condition that results in a pattern of cumulative damage to the basal ganglia. HD is expressed in a dominant manner and affects about 5 in every 100,000 individuals. It is estimated that 30,000 persons have HD in the United States. However, 150,000 individuals are at a 50% risk of inheriting the disease from an affected parent. It usually develops in a subtle fashion in the fourth to fifth decade of life and gradually worsens over a course of 10 to 20 years until death. The hallmark feature is distinctive choreic (dancelike) movements. The motor symptoms develop gradually, initially characterized by involuntary movements. Uncontrolled movements increase until the patient is confined to a bed or wheelchair. Aspects of cognitive loss and psychiatric disturbances also surface. The movement symptoms appear in the form of clumsiness, stiffness, and trouble with walking. Aspects of dementia include a decline in memory, concentration, and problem solving. If psychiatric symptoms appear, there are episodes of depression, instability, and even personality changes associated with mood swings. At the neuropathological level, there is a selective loss of neurons that is most aggressive in the striatum (caudate and putamen regions). Specific sets of cholinergic, GABA, and substance P neurons die and leave the dopamine afferent terminals in the striatum relatively intact. Nerve cell death (up to 90%) in the striatum is thought to cause the chorea. Areas of astroglial propliferation are also evident. The marked atrophy of the striatum and enlargement of the ventricles is readily visible by computed axial tomography (CAT) scans and nuclear magnetic resonance (NMR) imaging. There is no specific therapy or treatment for this disease.

Although the genetic defect causing Huntington's was assigned to chromosome 4 in 1983, it took 10 additional years of intense research to identify the gene in question. This gene produces the protein termed *huntingtin*. The Huntington's Disease Collaborative Research Group showed that a section of the gene contains CAG nucleotides that repeat several times causing an elongated polyglutamine tract in the mutant huntingtin protein. There is an inverse relationship between the increased number of CAG repeats in the gene and the age of onset of the clinical symptoms. More than 50 CAG repeats are associated with the most extreme forms of juvenile Huntington's. Individuals with more than 40 repeats will develop Huntington's. No one with fewer than 30 repeats will develop Huntington's. The function of this trinucleotide sequence has not been identified. Despite the selective neuronal cell death, the transcripts for the mutated gene are widely expressed in brain and non-nervous system tissues. The gene has been implicated as a transcription factor to regulate the expression of other genes. Because HD is dominant, most HD patients carry one copy of the expanded triplet gene and one normal copy of the gene. Therefore, each of their children has a 50/50 chance of receiving the gene and a 50/50 chance of inheriting the condition.

Amyotrophic Lateral Sclerosis

Amyotrophic lateral sclerosis (ALS) is also called motor neuron disease. Since the 1930s, this disease has been widely referred to as Lou Gehrig's disease. The incidence of ALS in the United States is 1 to 3 per 100,000. In this condition, there is a system degeneration of the upper and lower motor neurons in the brain and spinal cord. Lower motor neurons constitute the large neurons in the anterior horn of the spinal cord that connects with the skeletal (voluntary) muscles of the body. The upper motor neurons refer to the pyramidal neurons in the cerebral cortex that interact and modulate the activity of the lower motor neurons. Neurons affected usually show accumulations of phosphorylated neurofilaments in swollen proximal regions of axons and in cell bodies. There are signs of axonal degeneration leading to a reduction in the number of motor neurons in the spinal cord and brain stem nuclei. A loss in the number of pyramidal neurons in the brain motor cortex is associated with degeneration of the corticospinal pathways (responsible for voluntary movement). This condition is very progressive, resulting in muscle weakness and an atrophy of muscle mass due to the degenerating neurons.

ALS occurs sporadically in 90% of the cases. In 10% of patients, a family history link can be found. Mutations of the copper–zinc superoxide dismutase (SOD1) gene, mapped to chromosome 21, have been associated with ALS in approximately 20% of the patients with the familial links. The SOD1 are a group of enzymes that catalyze the conversion of the radical $\cdot O_2$ to hydrogen peroxide and oxygen. These enzymes provide cellular defense against the radical $\cdot O_2$ and its toxic derivatives. The cause of ALS is not known and there is no known cure. Life expectancy from the time of diagnosis is about 2 to 5 years, but there is a wide range because some patients have prolonged survival. ALS is recognized and classified on clinical grounds since no definitive diagnostic test is currently available.

This condition presents in different ways, depending on the muscles initially affected. Symptoms may include stumbling, a loss of dexterity and strength in the hands, or difficulty in swallowing. With progression, muscle twitching and cramping become frequent. The degeneration of the neuromuscular components may be present for some time before the symptoms cause real concern. In the majority of cases, all voluntary muscles become affected, leaving the patient completely paralyzed.

Multiple Sclerosis

Multiple sclerosis (MS) is a chronic disorder of the CNS involving decreased nerve functioning. About 350,000 Americans have MS, with women affected twice as often as men. MS usually starts between the ages of 15 and 50 with the average age of onset at 30. The risk of MS varies for different geographic areas and tends to increase as one lives farther north or south of the equator. There are several types of MS, but most patients (85%) initially have relapsing remitting disease, with abrupt onset of neurological problems that later dissipate.

All forms of MS are associated with inflammation in the CNS that is accompanied by areas of demyelination. Multiple, randomly scattered lesions (referred to as plaques), representing sites of myelin destruction, accumulate in the brain and spinal cord and cause a variety of neurological problems. When the myelin is damaged, neurological transmission may be slowed or blocked completely, leading to diminished or lost function. During an attack, the neurological symptoms may last for days, weeks, or months. The initial symptom is often blurred or double vision. Some individuals can also experience blindness. Nearly all MS patients experience numbness and muscle weakness in the limbs and difficulty with coordination and balance. These symptoms can be severe enough to impair walking and standing. Speech difficulty, fatigue, and dizziness are commonly present. The symptoms may be mild or severe and may appear in various combinations depending on the affected area(s) of the CNS.

Although genetic and environmental factors are known to contribute to MS, the cause of MS is unknown. Although MS is not inherited, the condition is more likely to be present if there is a close relative with the disorder. There is strong evidence that MS is linked to the immune system and that the patient's own immune system attacks the CNS. In MS, the main targets of the misguided immune system appear to be myelin and oligodendrocytes. Astrocytes contribute to the scar tissue in the plaques throughout the brain and spinal cord. The mediator of the autoimmune attack is the patients' T lymphocytes—a type of white blood cell derived from the thymus gland that normally responds to infection and offers long-term immunity. The abnormal autoimmune response involves activation of helper T cells and cytotoxic T cells, with a corresponding decrease in suppressor T-cell activity (see Chapters 11 and 12 for immune cell functions). Experimental autoimmune encephalitis (EAE) is an inflammatory immune disease of the CNS that serves as a model for MS. EAE is produced in animals by immunization with myelin proteins. Animal studies are now guiding the evolution of experimental gene therapies to delay, control, or prevent MS, and a number of promising immunotherapies are currently being evaluated for future use in MS. Local delivery of interleukins (IL-4, IL-10) by retroviral transduction or transfection of T lymphocytes has been shown to delay the onset and reduce the severity of EAE in mice immunized with myelin basic protein.

TABLE 9.4 Clinical Trial Examples with Neurotrophic Factor Administration

Disorder	Neurotrophic Factor
Alzheimer's	NGF
ALS	BDNF
Parkinson's	GDNF
ALS	CNTF
Diabetic neuropathy	NGF

CLINICAL TRIALS TESTING GENETICALLY MODIFIED CELLS AND NEUROTROPHIC FACTORS FOR NEURODEGENERATION

Therapeutic options for human neurodegeneration that involve gene transfer procedures are at an early developmental stage. A number of limited clinical trials have been conducted to evaluate the effects of neurotrophic factors for central as well as peripheral neural disorders. Table 9.4 lists some major central and peripheral neurological disorders that have used neurotrophic factors in various preclinical, phase I, II, and III trials. It should be pointed out that although NGF was identified and isolated more than 40 years ago, the notion of using neurotrophic factors for clinical application has only surfaced in the last 10 years. Major strides in cellular and molecular neuroscience and collaborative efforts with biotechnology companies such as Amgen, Genentech, and Regeneron have provided the thrust for the reality of using neurotrophic factors in clinical trials.

At this time, neurotrophic factors are delivered when the disorder is signficiantly advanced. Unlike the laboratory models of disease, for the majority of situations, we cannot predict the onset of a particular disorder. The best we can do at this time is hope for a particular factor or combination of factors to stop or slow down the sequence of cell degeneration and thereby limit the clinical symptoms associated with the neurological disorder.

In 1991, the first attempt to treat Alzheimer's with infusions of NGF was carried out by Lars Olson and colleagues at the Karolinska Institute in Stockholm, Sweden. NGF was infused into the lateral ventricle of the patient's brain over a 3-month period. Unfortunately, no overall significant improvement in cognition or memory was reported during this brief preliminary study. There were transient improvements during the NGF treatment, but these improvements were not evident after the NGF infusion. The patient had advanced Alzheimer's with a number of additional clinical conditions not related to the NGF infusion that complicated the clinical evaluations of the procedure. There were also side effects of appetite loss and pain associated with movement in this patient.

Based on promising nonhuman data, clinical trials have been conducted to evaluate the efficacy of BDNF and CNTF in ALS patients. The first CNTF safety and efficacy trials in humans were marred by the side effect of weight loss. Unfortunately, the phase III trials for CNTF and BDNF have both failed to show statistically significant clinical efficacy. Although the BDNF trial confirmed safety and tolerability, it showed no significant or clinically relevant difference in breathing capacity or survival between the treated and control group of patients. Combinations of CNTF and BDNF at lower doses are also currently being evaluated in multicenter trials as a potential therapy for the treatment of ALS.

Phase I trials involving the implantation of polymer capsules containing baby hamster kidney cells genetically engineered to secrete CNTF have been tested in ALS patients. These CNTF releasing implants were surgically placed within the lumbar intrathecal space. The cells released significant doses of CNTF into the CNS without unwanted peripheral side effects (loss of appetite) that were observed with systemic administration in the initial CNTF trials. Trials of this nature demonstrate that neurotrophic factors can be continuously delivered within the cerebrospinal fluid (CSF) of humans by an ex vivo gene therapy approach and hence, open new avenues for the treatment of neurological diseases.

The first clinical trial with GDNF in Parkinson's patients was announced in August, 1996, by Amgen. This initial trial based on the potent survival effects of GDNF on dopamine neurons in the animal models will determine the safety and tolerability of GDNF in patients with moderate to severe Parkinson's.

A number of clinical trials are in progress that use neurotrophic factors to target peripheral nerve disorders, referred to as peripheral neuropathies (disorders of motor and sensory functions in the peripheral nerves). Despite the fact that there is no direct evidence linking abnormal neurotrophic expression to a neuropathy, there is evidence that certain factors may be useful in certain clinical situations. NGF is showing promise for patients with diabetic peripheral neuropathy, a condition that affects the sensory neurons for the extremities and produces spontaneous unremitting pain, numbness, and abnormal sensations such as burning or tingling. Patients are susceptible to injury and show impaired healing. Phase II trials administering NGF to diabetic patients with peripheral neuropathy have shown significant improvement in neurological function and in the sensations of cooling detection and of heat measured by neurological function tests. On the basis of accessibility to the PNS and the current results from the clinical trials, the peripheral neuropathies may be the first nervous system disorders to receive effective therapy from the systemic administration of neurotrophic factors.

From these clinical trials it is apparent that our current animal models do not tell the whole story. As described above the administration of a trophic factor to the CNS of an animal can produce dramatic results in terms of neuronal protection and restorative functional behaviors. When applying and testing our knowledge in clinical trials, a different picture emerges. The dramatic reversal of neurological symptoms seen in the laboratory is not apparent and the issues of serious adverse side effects are realized. Administration of these factors represents a completely new group of pharmacological agents that carry numerous unknown parameters in terms of the exact cellular and molecular actions. Quickly we appreciate the gap between the animal model and the clinical setting.

FUTURE CONSIDERATIONS AND ISSUES

The conceptual framework for gene therapy in the nervous system has been outlined from a variety of perspectives. It is clear that recent advances in molecular biology and medicine have established gene therapy in the CNS as a realistic goal. We have identified many conditions that promote neuron survival, limit degeneration and offset neural dysfunction. The genetic expression of selected trophic factors or antiapoptotic gene products significantly enhances the survival and growth of neurons.

Although we have developed numerous ex vivo and in vivo neuroprotective gene transfer strategies in animal models, the current animal models of neurodegenerative events are not ideal representations of similar human conditions. Animal models must be further developed and refined to unravel the complexity of human CNS dysfunction. As a result, a large gap currently exists between the laboratory and the application of protective gene therapy strategies for human neurological diseases. While single molecules or gene products can be extremely functional on subsets of CNS neurons in the laboratory animal, a completely different set of circumstances may be responsible for neuronal degeneration seen in the analogous neuronal groups affected in human neurological disease. We simply do not have enough knowledge at this time to make definitive statements regarding the cause(s) of neuronal degeneration or the specific formula of gene products that will cure or prevent diseases such as Alzheimer's, ALS, or MS. As our knowledge base of the neurological disease mechanisms expands, parallel experiments will evaluate the effectiveness of new gene products in the nervous system and increase the efficacy of CNS gene graft therapy.

At this time, the regulation of gene expression by many viral vectors is poorly understood. When transgenes are introduced into the nervous system, the expression is often down-regulated. We need to identify factors that influence and control the level of gene expression in vivo. Likewise, the characterization of cell-specific promoters and inducible promoters will further enhance the utility of viral vectors in the nervous system. There are also immunological responses to vectors (particularly the recombinant adenoviral vectors) and at times the transgene itself. The safety of the vectors used for clinical purposes will always remain an issue in gene therapy because there is the potential for harmful activation by complementation or recombination with latent wild-type viruses.

It is likely that initial gene therapy protocols will be used to slow down the rate of neurodegeneration in Parkinson's and Alzheimer's. Promising progress has surfaced for neurotrophic factor therapy in cases of the peripheral neuropathies. However, like gene therapy in general, our understanding of this therapeutic modality is just beginning. Gene therapy technology that can dampen the symptoms of neuronal degeneration will represent a significant step for those individuals who have a neurodegenerative disorder and are well aware of the limitations of current therapies.

KEY CONCEPTS

- The conceptual framework for gene therapy in the nervous system has been outlined and the interface between molecular biology and medicine has established gene therapy in the CNS as a realistic goal. Many conditions that promote neuron survival have been identified. The genetic expression of selected trophic factors or antiapoptotic gene products significantly enhances the survival and growth of neurons.

- Although numerous ex vivo and in vivo neuroprotective gene transfer strategies have been developed in animal models, the current animal models of neurodegenerative events are not ideal representations of similar human conditions. While single molecules or gene products can be extremely functional on subsets of CNS neurons in the laboratory animal, a completely different set of circumstances may be responsible for neuronal degeneration seen in the analogous neuronal groups affected in human neurological illness.

- The cause(s) of neuronal degeneration or the specific formula of gene products that will cure or prevent diseases such as Alzheimer's, ALS, or MS are unknown.

As our knowledge base of neurological disease mechanisms grows, parallel experiments will evaluate new gene products in the nervous system and increase the efficacy of CNS gene/cell therapy.

· At this time, the regulation of gene expression by many viral vectors is poorly understood. When transgenes are introduced into the nervous system, the expression is often down-regulated. The characterization of cell-specific promoters and inducible promoters will further enhance the utility of viral vectors in the nervous system. There are also immunological responses to vectors (particularly the recombinant adenoviral vectors) and at times the transgene itself. The safety of the vectors used for clinical purposes will always remain an issue in gene therapy because there is the potential for harmful activation by complementation or recombination with latent wild-type viruses.

· It is likely that initial gene therapy protocols will be used to slow down the rate of neurodegeneration in Parkinson's and Alzheimer's. Promising progress has surfaced for neurotrophic factor therapy in cases of the peripheral neuropathies.

· Neural stem cells exist in the adult nervous system of mammals. Future therapeutic directions will include activation of stem cells to induce self-repair or transplants of genetically modified stem cells that fully integrate in the brain.

ABBREVIATIONS

ALS	amyotrophic lateral sclerosis
APP	amyloid precursor protein
BDNF	brain-derived neurotrophic factor
CAG	cytosine adenine guanine
CNTF	ciliary neurotrophic factor
EAE	experimental allergic encephalitis
EGF	epidermal growth factor
FGF	fibroblast growth factor
GABA	γ-aminobutyric acid
GDNF	glial-cell-line-derived neurotrophic factor
IAP	inhibitors of apoptosis
IDPN	β,β'-iminodipropionitrile
IGF-2	insulinlike growth factor
LSD	lysosomal storage disease
MBP	myelin basic protein
MS	multiple sclerosis
NGF	nerve growth factor
NT4/5	neurotrophin 4/5
PCD	programmed cell death
SOD1	superoxide dismutase
TGF-β	transforming growth factor β
trk	tyrosine receptor kinase
6-OHDA	6-hydroxydopamine

SUGGESTED READINGS

Neurotrophic Growth Factors

Apfel SC (Ed.). Clinical Applications of Neurotrophic Factors. Lippincott-Raven, New York, 1997, p. 209.

Bock GR, Goode JA. Growth Factors as Drugs for Neurological and Sensory Disorders. Ciba Foundation, Chichester, 1996.

Lindsay RM, Wiegand SJ, Altar CA, DiStefano PS. Neurotrophic factors: From molecule to man. Trends Neurosci 17:182–190, 1994.

Oppenheim RW. The concept of uptake and retrograde transport of neurotrophic molecules during development: History and present status. Neurochem Res 21:769–777, 1996.

Snider WD, Wright DE. Neurotrophins cause a new sensation. Neuron 16:229–232, 1996.

Gene Therapy in the CNS

Blömer U, Naldini L, Verma IM, Trono D, Gage FH. Applications of gene therapy to the CNS. Hum Mol Genet 5(Rev):1397–1404, 1996.

Chiocca EA, Breakefield XO. Gene Therapy for Neurological Disorders and Brain Tumors. Humana, Totowa, NJ, 1998.

Doering LC. Gene therapy and neurodegeneration. Clin Neurosci 3:259–321, 1996.

Kaplitt MG, Loewy AD. Viral Vectors, Gene Therapy and Neuroscience Applications. Academic, San Diego, 1995.

Apoptosis and Grafting

Blömer U, Kafri T, Randolph-Moore L, Verma IM, Gage FH. *Bcl-xL* protects adult septal cholinergic neurons from axotomized cell death. Proc Natl Acad Sci 95:2603–2608, 1998.

Deveraux QL, Reed JC. IAP family proteins-suppressors of apoptosis. Genes Dev 13: 239–252, 1999.

Gage FH, Fisher LJ. Intracerebral grafting: A tool for the neurobiologist. Neuron 6:1–12, 1991.

Kostic V, Jackson-Lewis V, de Bilbao F, Dubois-Dauphin M, Przedborski S. Bcl-2: Prolonging life in a transgenic mouse model of familial amyotrophic lateral sclerosis. Science 277:559–562, 1997.

Alzheimer's Disease

Seiger Å, Nordberg A, von Holst H, et al. Intracranial infusion of purified nerve growth factor to an Alzheimer patient: The first attempt of a possible future treatment strategy. Behav Brain Res 57:255–261, 1993.

Winkler J, Thal LJ, Gage FH, Fisher LJ. Cholinergic strategies for Alzheimer's disease. J Mol Med 76: 555–567, 1998.

Huntington's Disease

Emerich DF, Winn SR, Hantraye PM, Peschanski M, Chen EY, Chu Y, McDermott P, Baetge EE, Kordower JH. Protective effect of encapsulated cells producing neurotrophic factor CNTF in a monkey model of Huntington's disease. Nature 386:395–399, 1997.

Huntington's Disease Collaborative Research Group. A novel gene containing a trinucleotide repeat that is expanded and unstable on Huntington's disease chromosomes. Cell 72:971–983, 1993.

Parkinson's Disease

Dunnett SB, Björklund A. Prospects for new restorative and neuroprotective treatments in Parkinson's disease. Nature 399(Suppl):A32–A39, 1999.

Lindvall O, Brundin P, Widner H, Rehncrona S, Gustavii B, Frackowiak R, Leenders KL, Sawle G, Rothwell JC, Marsden CD, Björklund A. Grafts of fetal dopamine neurons survive and improve motor function in Parkinson's disease. Science 247:574–577, 1990.

Polymeropoulos MH, Lavedan C, Leroy E, et al. Mutation in the α-synuclein gene identified in families with Parkinson's disease. Science 276:2045–2047, 1997.

Stem Cells

Clarke DL, Johansson CB, Wilbertz J, Veress B, Nilsson E, Karlström H, Lendahl U, Frisen J. Science 288:1660–1663, 2000.

Gage FH. Mammalian neural stem cells. Science 287:1433–1438, 2000.

McKay R. Stem cells in the nervous system. Science 276:66–70, 1997.

Snyder EY, Taylor RM, Wolfe JH. Neuronal progenitor cell engraftment corrects lysosomal storage throughout the MPSVII mouse brain. Nature 374:367–370, 1995.

Vescovi AL, Snyder EY. Establishment and properties of neural stem cell clones: Plasticity *in vitro* and *in vivo*. Brain Pathol 9:569–598, 1999.

CHAPTER 10

Gene Therapy in the Treatment of Cancer

SIMON J. HALL, M.D., THOMAS F. KRESINA, PH.D., RICHARD TRAUGER, PH.D., and BARBARA A. CONLEY, M.D.

BACKGROUND

Approximately 50% of the human gene therapy protocols approved by the National Institutes of Health (NIH) Recombinant DNA Committee and the Food and Drug Administration (FDA) have been in the field of cancer. This is due to the intense research effort into the elucidation of mechanism(s) of carcinogenesis and malignancy. With a fuller understanding of these processes, it now appears that the generation of cancer is a multistep process of genetic alterations. The genetic alterations vary according to the type and stage of cancer. But once determined, they provide targets for therapy. Currently, surgery, radiation, and chemotherapy (drug therapy) form the medical management of cancer. With the emphasis of human protocols in cancer gene therapy, successful treatment of cancer with gene therapy may be on the horizon.

INTRODUCTION

Cancer arises from a loss of the normal regulatory events that control cellular growth and proliferation. The loss of regulatory control is thought to arise from mutations in genes encoding the regulatory process. In general, a genetically recessive mutation correlates with a loss of function , such as in a tumor suppressor gene. A dominant mutation correlates with a gain in function, such as the overexpression of a normally silent oncogene. Either type of mutation may dysregulate cell growth. It is the manipulation of these genetic mutations and the enhancement of normal cellular events that is the goal of cancer gene therapy. Thus, gene therapy for the treatment of cancer has been directed at (1) replacing mutated tumor suppressor genes, (2) inactivating overexpressed oncogenes, (3) delivering the genetic component of targeted prodrug therapies, and (4) modifying the antitumor immune response.

An Introduction to Molecular Medicine and Gene Therapy, Edited by Thomas F. Kresina
ISBN 0-471-39188-3 © 2001 Wiley-Liss

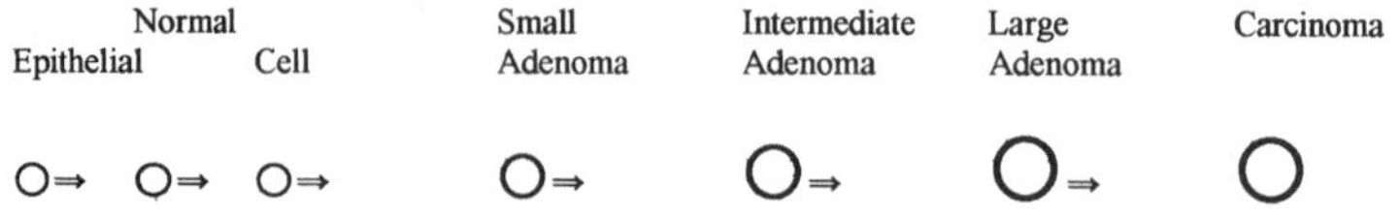

| Normal | | | Small | Intermediate | Large | Carcinoma |
| Epithelial | Cell | | Adenoma | Adenoma | Adenoma | |

O⇒ O⇒ O⇒   O⇒   O⇒   O⇒   O

Stage Specific Genome Alleles

APC/APC		APC/APC	APC/APC	APC/APC	APC/APC
MSH2/MSH2		MSH2/MSH2	MSH2/MSH2	MSH2/MSH2	MSH2/MSH2
RAS/RAS		RAS/RAS	RAS/RAS	RAS/RAS	RAS/RAS
DCC/DCC		DCC/DCC	DCC/DCC	DCC/DCC	DCC/DCC
p53/p53		p53/p53	p53/p53	p53/p53	p53/p53

FIGURE 10.1   Genetic basis of carcinogenesis. Diagrammatic representation of sequential mutations needed to develop colorectal carcinoma from normal epithelial cells. Abbreviations: APC, adenomatous polyposis coli gene; MSH2, mammalian DNA repair gene 2; *Ras*, oncogene; DCC, deleted in colorectal carcinoma gene; *p53* tumor suppressor gene. Mutations in DNA repair genes would occur initially in normal cells (bold) with subsequent mutations in the APC (italics) occurring as an early event developing the small adenoma. Mutation of the RAS oncogene (activation by point mutation) develops the intermediate adenoma with subsequent deletion of DCC gene in the large adenoma stage. The last mutation is in the *p53* tumor suppressor gene to form the carcinoma.

GENETIC BASIS OF CARCINOGENESIS

Alterations in the normal cellular proces\ses of proliferation, differentiation, and programmed cell death, apoptosis, contribute to the development of cancer. Tissue-specific and cellular-specific factors as well as other gene products mediate the processes of differentiation, growth, and apoptosis. Alterations in these gene products can lead to premalignant, benign tumors or malignancy. Thus, numerous genes can be implicated in oncogenesis, or the development of a malignant tumor. These include oncogenes, or the activation of growth-promoting genes, and tumor suppressor genes, or the inactivation of growth-suppressing genes. Two important characteristics in carcinogenesis are integral to the genetic alterations: (1) multistep oncogenesis and (2) clonal expansion. The mulitstep formation of tumor development requires that several genetic alterations or, "hits," occur in sequence for normal cells to progress through various stages to malignancy, as represented in Figure 10.1. Clonal expansion indicates that a growth advantage is conferred to a cell by virtue of a genetic alteration (mutation) that occurs as part of the multistep carcinogenesis.

Cell Cycle

The cell cycle is comprised of five phases based on cellular activity (Fig. 10.2). A period of deoxy-ribonucleic acid (DNA) replication occurs in the S phase and mitosis occurs in the M phase. Two intervening phases are designated G_1 and G_2. Cells commit to a cycle of replication in the G_1 phase at the R (restriction) point. Also, from the G_1 phase cells can enter a quiescent phase called G_0. Regulation of the cell cycle is critical at the G_1/S junction and at the G_2/M transition. Cyclins regulate progression through the cell cycle in conjunction with cyclin-dependent kinases (CDK). Cyclins act as structural regulators by determining the subcellular

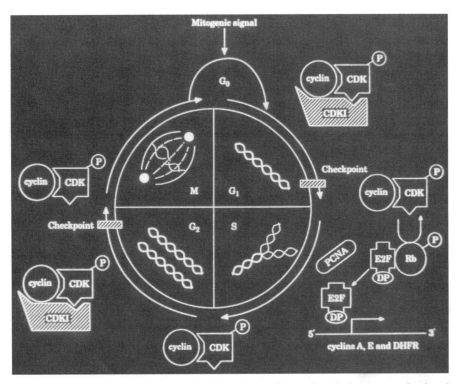

FIGURE 10.2   Cell cycle. Diagram of the five phases of the cell cycle, important check points for regulation and the interactions of cyclins and cyclin-dependant kinases (CDKs), CDKI (inhibitors), tumor suppressor genes such as *Rb* (retinoblastoma) and DHFR dihydrofolate reductase.

TABLE 10.1   Cyclins and the Cell Cycle

Cyclin	Cell Cycle Phase	Regulatory Action
C, D1-3, E	G_1/S	Determines when new cell cycle occurs
A	S, G_2M	Promotes mitosis
B1, B2	S, G_2M	Promotes mitosis

location, substrate specificity, interaction with upstream regulatory enzymes, and timing of activation of the CDK. Thus, each of the eight distinct cyclin genes (Table 10.1) regulate the cell cycle at its designated point by binding to CDKs and forming CDK/cyclin complexes. Cyclins are synthesized, bind, and activate the CDKs and then are destroyed. The CDKs phosphorylate subcellular substrates such as the retinoblastoma protein (pRb), which act to constrain the G_1/S transition in the cell cycle. pRb, therefore, is a tumor suppressor gene product. Phosphorylation of pRb, which occurs by the sequential action of cyclinD-CDK4/6 complex and cyclin E-CDK2 complex, inactivates the growth-inhibitory function of the molecule allowing for cell cycle progression. Thus, the synthesis of specific cyclins and complexing with CDKs could result in uncontrolled cell growth. For instance, cyclinD1 has been shown both in vitro and vivo to initiate oncogenic properties and is amplified and overexpressed in certain esophagus squamous cell carcinomas as well as other head, neck, bladder, and breast cancers. Other functions for the cyclins exist as well. The cyclin A gene is the site of integration of the hepatitis B virus (Chapter 6), thereby promoting hepatitis virus integration into the genome.

The inhibition of CDK phosphorylation is, therefore, an important goal for reducing cellular proliferation. Investigations have resolved other molecules that bind and inhibit CDKs. CDK-integrating protein (Cip1) binds multiple cyclin/CDK complexes and inhibits their activity. Cip1 is activated by the *p53* tumor suppressor gene product and by cell senescence. Thus, Cip1 is a candidate negative regulator of cell proliferation and division. Another inhibitor is p16 or multiple tumor suppressor (MTS-1), which specifically inhibits CDK4. It has a gene locus at chromosome 9p21. In esophageal and pancreas tumors, deletion or point mutations at this locus are observed. A naturally occurring CDK inhibitor is p27 or Kip1, which binds tightly to cycklinE/CDK2 and cyclinD/CDK4 complexes. Kip1 is also involved in the mediation of extracellular signals by transforming growth factor β1 (TGF-β1), thereby inferring a mechanism to the growth inhibitory properties of TGF-β. Since inhibitors of CDK phosphorylation modulate cell cycle activity, they represent target molecules for cancer gene therapy as molecules that can arrest cellular proliferative activity.

Apoptosis

Apoptosis, genetically programmed cell death, involves specific nuclear events. These include the compaction and segregation of chromatin into sharply delineated masses against the nuclear envelope, condensation of cytoplasm, nuclear fragmentation, convolution of the cellular surface, and formation of membrane-bound apoptotic bodies. The latter entities are phagocytosed by adjacent cells. In cell death there is cleavage of double-stranded DNA at linker regions between nucleosomes to produce fragments that are approximately 185 base pairs. These fragments produce a characteristic ladder on electrophoresis. The genetic basis for programmed cell death is being elucidated. An oncogene, *bcl-2*, protects lymphocytes and neurons from apoptosis. However, another protein, termed *bax*, forms a dimer with *bcl-2*, and *bax* contributes to programmed cell death. It is the cellular ratio of *bcl-2* to *bax* that determines whether a cells survives or dies. An additional protein, interleukin 1β-converting enzyme, ICE, promotes cell death on accumulation. Alternatively, *bak*, a proapoptotic member of the *bcl-2* gene family has been recently described. The use of *bax*, *bak*, *bcl-2*, or ICE or other apoptosis-related genes in targeted gene transfer techniques represent an approach to modify the viability of specific cellular populations. Cancer cells could be targeted for death by insertion of apoptosis genes. On the other hand, localized immune cells fighting malignant cells could provide added protection through the transfer of genes that protect from apoptosis.

Cellular Transformation

Cells are said to be "transformed" when they have changed from a normal phenotype to a malignant phenotype. Malignant cells exhibit cellular characteristics that are distinguished from normal cells. On a morphological basis, for example, normal epithelial cells are polar, nondividing, uniform in shape, and differentiated. In the transformation to a malignant phenotype, epithelial cells become nonpolar, pleomorphic, display variable levels of differentiation, contain mitotic figures, rapidly divide, and express tumor-associated antigens on the cell surface. The expression of tumor-associated antigens has been used to target tumor cells via monoclonal antibodies, liposomes, and the like for drug- or toxin-induced cell death. This targeting approach has also been used in gene therapy protocols (see below). Cells can also be transformed by chemical treatment, radiation, spontaneous mutations of endogenous genes, or viral infection. Transformed cells generated by these mechanisms display rounded morphology, escape density-dependent contact inhibition (clump), are anchorage independent, and are not inhibited in growth by restriction point regulation of the cell cycle (Fig. 10.3). In addition, transformed cells are tumorgenic when adoptively transferred to naïve animals. Viral transformation is a major

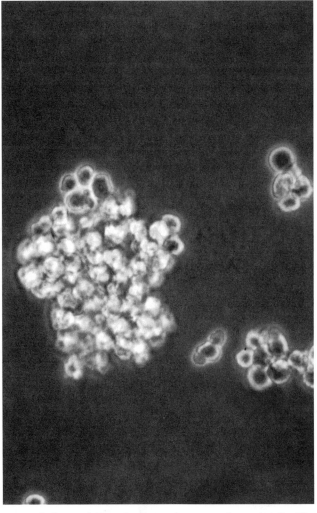

FIGURE 10.3 Morphology of Epstein–Barr virus transformed cells. Note the rounded morphology, aggregation, clumping, and satellite colonies of growth.

concern for gene therapy approaches that utilize viral vectors. Although replication-defective viral vectors are used in viral vector gene transfer (see Chapter 4), the remote possibility of viral recombination of vector with naturally occurring pathogenic virus to produce a competent transforming virus remains.

Oncogenes

Cellular oncogenes are normal cellular genes related to cell growth, proliferation, differentiation, and transcriptional activation. Cellular oncogenes can be aberrantly expressed by gene mutation or rearrangement/translocation, amplification of expression, or through the loss of regulatory factors controlling expression. Once defective, they are called oncogenes. The aberrant expression results in the development of cellular proliferation and malignancy. There have been over 60 oncogenes identified to date and are associated with various neoplasms. Salient oncogenes with related functions are listed in Table 10.2. Oncogenes can be classified in categories according to their subcellular location and mechanisms of action. An example of an oncogene is the normally quiescent *ras* oncogene which comprises a gene family of three members: Ki-*ras*, Ha-*ras*, and N-*ras*. Each gene encodes for a 21-kD polypeptide, the p21 protein, a membrane-associated GTPase (enzyme). In association with the plasma membrane, p21 directly interacts with the *raf* serine-theonine kinase. This complexing (*ras/raf*) starts a signal transduction cascade pathway. Along this pathway is the activation MAP kinase, which is translocated to the nucleous and posphorylates nuclear transcription factors. This pathway provides signaling for cell cycle progression, differentiation, protein transport, secretion, and cytoskeletal organization. *Ras* is particularly susceptible to point mutations at "hot spots" along the gene (codons 12, 13, 59, and 61). The result is constitutive activation of the gene and overproduction of the p21 protein. *Ras* mutations are common in at least 80% of pancreatic cancers, indicating that this genetic alteration is part of the multistep oncogenesis of pancreatic cells. A second oncogene is c-*myc*, which encodes a protein involved in DNA synthesis; c-*myc* in normal cells is critical for

TABLE 10.2 Categories and Function of Salient Oncogenes

Oncogene	Functional Category	Associated Neoplasia—Representative
sis, *int*-2, K53 FGF-5, *int-1*, Met	Growth factor related	Thyroid neoplasms
Ret, *erb*-B 1-2, *neu*, *fms*, *met*, *trk*, *kit*, *sea*	Receptor protein tyrosine kinases	Breast cancer
src, *yes*, *fgr fps/fes*, *abl*	Nonreceptor protein tyrosine kinases	Colon cancer
raf, *pim0-1*, *mos*, *cot*	Cytoplasmic protein-serine kinases	Small-cell lung cancer
Ki-*ras*, Ha-*ras*, N-*ras*, Gsp, *gip*, *rho* A-C	Membrane G protein kinases	Pancreatic ductal Adenocarcinoma
c-*myc*, N-*myc*, L-*myc*, *mby*, *fos*, *jun*, *maf*, *cis rel*, *ski*, *erb*-A	Nuclear	Squamous cell carcinoma

cell proliferation, differentiation, apoptosis through its activity as a transcription factor, and DNA binding protein. The c-*myc* cellular expression is associated with cellular proliferation and inversely related to cellular differentiation. It has been noted that constitutive expression of c-*myc* results in the inability of a cell to exit the cell cycle. In certain cancers, such as colon cancer, no genetic mutation in c-*myc* has been found. But messenger ribonucleic acid (mRNA) levels for the gene are highly elevated. Thus, loss of posttranscriptional regulation is, at least, partially responsible for cellular proliferation. In all cases, the genetic abnormalities of oncogene expression represent specific targets for gene therapy.

Oncogenes can also be found in RNA tumor viruses (retrovirus). Some retrovirus contain transforming genes called v-*onc*, for viral oncogene, in addition to the typically encoded genes such as *gag, pol,* and *env* (see Chapter 4). Viral oncogenes are derived from cellular oncogenes with differences arising from genetic alterations such as point mutations, deletion, insertions, and substitutions. Cellular oncogenes are presumed to have been captured by retroviruses in a process termed retroviral transduction. This occurs when a retrovirus inserts into the genome in proximity to a cellular oncogene. A new hybrid viral gene is created and, after transcription, the new v-*onc* is incorporated into the retroviral particles and introduced into neighboring cells by transfection. For example, the oncogenes HPV-16 E6/E7 are derived from human papilloma virus and their expression initiates neoplastic transformation as well as maintains the malignant phenotype of cervical carcinoma cells.

Tumor Suppressor Genes

Tumor suppressor genes encode for molecules that modify growth of cells through various mechanisms including regulation of the cell cycle. An abnormality in a tumor suppressor gene could result in a loss of functional gene product and susceptibility to malignant transformation. Thus, restoration of tumor suppressor gene function by gene therapy, particularly in a premalignant stage, could result in conversion to a normal cellular phenotype. Possibly, the restoration of tumor suppressor gene function in malignant cells could result in the "reverse transformation" of a malignant cells to a nonmalignant cell type.

There are numerous tumor suppressor genes (Table 10.3), but the most notable are retinoblastoma (*rb*, discussed in Chapter 3) and *p53*. The *p53* tumor suppressor is a 393–amino-acid nuclear phosphoprotein. It acts as a transcription factor by binding DNA promoters in a sequence-specific manner to control the expression of proteins involved in the cell cycle (G_1/S phase). *p53* functions as the "guardian of the genome" by inhibiting the cell cycle via interactions with specific cyclin/CDK complexes or inducing apoptosis via the *bax, Fas* pathways. These activities are in response to DNA damage. Thus, by the action of *p53*, malignant cells or premalignant cells can be inhibited or killed and phagocytosed. Alternatively, loss of the *p53* gene by mutation, deletion, or inhibition of the *p53* tumor suppressor molecule has been implicated in tumor progression. Inactivation of *p53* can occur by various mechanisms including genetic mutation, chromosomal deletion, binding to viral oncoproteins, binding to cellular oncoproteins such as *mdm2*, or alteration of the subcellular location of the protein. It has been estimated that *p53* is altered, in some form, in half of all human malignancies. The appearance of *p53* mutations have been

TABLE 10.3   Short Listing of Tumor Suppressor Genes

Tumor Suppressor Gene	Genetic Loci
p53	17p
retinoblastoma, *rb*	13q
BRCA-1	17q
NFI	17q
Deleted in colon cancer, *DCC*	18q
MEN-1	11p
WT1	11p
c-*ret*	10p
MTS-1	9q
Adenomatous polyposis coli, *APC*	5q

associated with poor prognosis, disease progression, and decreased sensitivity to chemotherapy. For all of these reasons, individuals with *p53* abnormalities represent potential candidates for gene therapy.

DNA Repair Genes

Genetic defects in double-stranded DNA can be repaired by the products of DNA repair genes. These gene products act to proofread and correct mismatched DNA base pair sequences. Mismatched base errors, if not corrected, are replicated in repeated cell divisions and promote genomic instability. Four mammalian genes are known to date. They are *hMHL1*, *hMSH2*, *hPMS1*, and *hPMS2*. Mutations in these genes, resulting in defective gene products, have been noted in the germline in hereditary nonpolyposis colorectal cancer (HNPCC) syndromes. Mutations in the *hMSH2* gene (loci at chromosome 2p) and the *hHLH1* gene (loci at chromosome 3p) have been well documented in HNPCC where a large number (estimated to the tens of thousands) of somatic errors (random changes in DNA sequence) are apparent. Thus, mutations in DNA repair enzymes may be a mechanism for carcinogenesis in inherited neoplasms or cancers appearing in ontogeny.

GENE THERAPY APPROACHES TO THE TREATMENT OF CANCER

One strategy in the gene therapy of cancer is the compensation of a mutated gene. If a gene is dysfunctional through a genetic alteration, compensation can occur by numerous mechanisms. For a loss of function scenario, such as for a tumor suppressor gene, compensation would be provided by the transfer of a dominant normal gene or by directly correcting the gene defect. If a gene incurs a gain in function, such as for an oncogene or growth factor, then approaches at gene deletion or regulation of gene expression could be employed.

Augmentation of Tumor Suppressor Genes

Tumor suppressor genes are a genetically distinct class of genes involved in suppressing abnormal growth. Loss of function of tumor suppressor proteins results in loss of growth suppression. Thus, tumor suppressor genes behave as recessive onco-genes. Study of "cancer families" predisposed to distinct cancer syndromes has led to the identification of mutated tumor suppressor genes transmitted through the germline. Individuals from these families are more susceptible to cancer because they carry only one normal allele of the gene. The loss of tumor suppression func-tion requires only one mutagenic event. The most targeted tumor suppressor gene for gene therapy has been *p53* (see Table 10.4). This is because *p53* is the most com-monly mutated tumor suppressor gene in human cancer. The transfer of *p53* gene to tumor cells in vitro results in a transduction that suppresses growth, decreases colony formation, reduces tumorgenicity of the cells, and induces apoptosis. In addition, normal cells have been shown to remain viable after transfection and over-expression of the *p53* gene. These findings laid the groundwork for further studies in initial clinical trials.

Clinical studies with the *p53* gene have begun, and many obstacles to successful therapy need to be overcome. Numerous gene therapy delivery systems will be needed to match the clinical application for optimal therapy. Differing delivery systems will be needed for local intratumor delivery of tumors versus systemic delivery to blood-borne or metastatic disease.

Retrovirus For retroviral vectors, a significant advantage is the preferential inte-gration of the *p53* transgene into rapidly dividing tumor cells as compared to normal cells. However, this integration is genomic and thus represents a permanent modi-fication of the cells. In addition, one cannot discount the possibility of insertional mutagenesis of normal cells with the *p53* transgene. Retroviruses are also still

TABLE 10.4 Tumor Suppressor Factor Gene Therapy Using *p53*

Cancer	Vector	Cell Line/Xenograft	Efficacy
Breast	Adenovirus	MDA-MB; SK-BR-3; BT-549; T47-D; HBL-100; MCF-7; SkBr3; 184B5; MCF10	Decreased proliferation and colony formation, apoptosis in cells
	Retrovirus	MDA-MB; BT549	Decreased colony formation
	Adenovirus	MDA-MB	71–95% growth inhibition
	Liposomes	MDA-MB; MCF-7	40–75% growth inhibition in xenografts
Ovarian	Adenovirus	SK-OV3; 2774; Caov3,4; PA-1	Decreased proliferation and colony formation in cells
	Adenovirus	SK-OV3	Sensitized to irradiation and increased survival in xenografts
Cervical	Adenovirus	HeLa; C33A; HT3; C4-I; SiHa; CaSki; ME180; MS751	Decreased proliferation and colony formation in cells
	Adenovirus	C33A; HT3; HeLa; SiHa; MS751	100% tumor suppression—xenograft
Prostate	Adenovirus	C4-2; DU-145; PC-3; LNCaP; DuPro-1; Tsu-Prt	Decreased proliferation and augmented apoptosis in cells
	Adenovirus	C4-2; DU-145; PC-3; Tsu-Prt	90–100% tumor suppression in xenografts

TABLE 10.4   (*Continued*)

Cancer	Vector	Cell Line/Xenograft	Efficacy
Lung	Adenovirus	H23, 69, 266Br, 322, 358, 460, 596; H661; Calu-6; MRC-9; A549; WI-38;	Decreased proliferation in cells
	Retrovirus	H226Br; 322, 358, 460; WT226	Decreased proliferation in cells
	Adenovirus	H1299, 69, 358, 226Br	Growth inhibition with increased survival in xenografts
Head and neck	Adenovirus	Tu-138, 177; MDA 686-LN; TR146; MDA 886; CNE-1, 2Z	Decreased proliferation and increased apoptosis in cell lines
	Adenovirus	Tu138, 177; MDA886, 686-LN	67–100% tumor suppression in xenografts; apoptosis in tumors
Nervous system	Adenovirus	G55, 59, 112, 122, 124; U87 MG; SK-N-MC; SN-N-SH; U-251; T-98; U-87, 373 MG, 138 MG; A-172; LG; EFC-2; D54 MG; T98G	Decreased proliferation and increased apoptosis in cell lines
	Retrovirus	A673	Decreased colony formation in cells
	Adenovirus	G122	100% tumor suppression— xenograft
	Retrovirus	A673	Tumor suppression
Bladder	Adenovirus	HT-1376; 5637; J82; FHs 738B1	Reduced proliferation in cells
Colorectal	Adenovirus	DLD-1; HCT116; SW480, 620; RKO; KM12L4; SW837; Colo 205, 320D; EB	Decreased proliferation and increased apoptosis in cell lines
	Adenovirus	DLD-1; SW620; KM12L4	Growth inhibition and increased apoptosis in xenografts
Liver	Adenovirus	Hep3B, G2; HLE; HLF; SK-HEP-1	Decreased proliferation in cells
	Adenovirus	McA-RH7777	Growth inhibition in xenografts
Skin	Adenovirus	SK-MEL-24 SK-MEL-24	Decreased proliferation in cells Growth inhibition in xenografts
Muscle	Adenovirus	A673, SK-UT-1	Decreased proliferation in cells
Bone	Adenovirus	Saos-2	Decreased proliferation and increased apoptosis in cells
	Retrovirus	Saos-2	Decreased proliferation and colony formation in cells
	Adenovirus	Saos-2	100% tumor suppression— xenograft
	Retrovirus	Saos-2	100% tumor suppression— xenograft
Lymphomas	Adenovirus	JB6; k-562	Decreased colony formation in cells
	Retrovirus	Be-13	Decreased proliferation and colony formation in cells
	Vaccinia virus	HL-60	Decreased proliferation and increased apoptosis and differentiation in cells

plagued by low titer production processes and poor stability. Thus, improvements in current generation retrovirus vectors are needed for effective in vitro or ex vivo therapy with *p53*.

Adenovirus and Adenoassociated Virus For adenovirus–based gene delivery systems, adenovirus, adenoassociated virus, herpes, and vaccinia virus have been explored for gene therapy (see Chapter 4). For gene therapy using the *p53* trans-gene, adenovirus and vaccinia virus have been used. The significant advantages of theses vectors include (1) the transduction of dividing or quiescent cells, (2) wide tissue tropism, and (3) the ability to generate clinical-grade material at high concentrations. The adenovirus remains extrachromosomal, and thus transient transgene occurs with replication-defective recombinant adenoviruses. Short-term expression of *p53* may be advantageous for treatment of neoplasia if the induction of growth inhibition, reduction in colony formation, or reduction in tumorgenicity is permanent in targeted cancer cells. Certainly, if apoptosis is induced by transient *p53* expression, individual tumor cells would be clonally deleted. A difficult complication of therapy would be the observation of these biological processes in normal cells. However, replication-deficient adenovirus has been used in clinical studies without significant adverse side effects to normal cells. Another significant issue in the use of adenovirus is the host's immune response to the vector. Both neutralizing antibody and cytotoxic T-cell cells have been shown to inhibit the efficacy of adenovirus-based gene therapy. Most recent generations of adenovirus vectors have specifically addressed this issue and significantly reduced the immunogenicity of the vector construct. Thus, it is likely that delivery of the *p53* transgene by an adenovirus vector will provide the initial demonstration of effective gene therapy for cancer.

Nonviral Gene Delivery Systems For *p53*, these include the use of liposomes or the direct injection of *p53* DNA. Although less efficient, both systems are likely to be less toxic and less immunogenic than viral systems. Liposomes provide the best opportunity for use in metastatic malignancies through the ability to specifically target neoplastic cells. This is most effectively done by the incorporation of cancer-targeting molecules, such as antibodies to tumor-specific antigens, into the concentric lipid bilayers of the liposome. Liposomes made of conventional phosphatidylcholine can deliver gene(s) to specific intracellular organelles because they are not fusion active and are acid resistant (see Chapter 5). Thus, liposomes can bypass intracellular processing to provide gene delivery to the nucleus. In the context of gene therapy, the delivery of therapeutic genes by liposomes can also result in the inhibition of angiogenesis and the observation of an enhanced efficacy through the "bystander" effect (see blow). Both events would be advantageous for the therapy of metastatic neoplasia.

Inactivating Overexpressed Oncogenes

The over expression of oncogenes can be abrogated by approaches limiting their expression. Specific gene inhibition can be accomplished by the use of antisense molecules or ribozymes. An antisense oligonucleotide, specifically generated based on the sense sequence of an oncogene, would bind the oncogene. The target of the

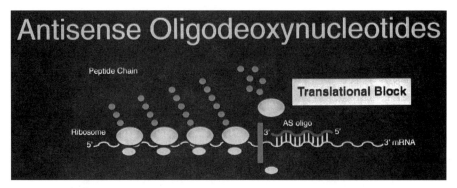

FIGURE 10.4 Diagram of an antisense oligonucleotide, specifically generated based on the sense sequence of an oncogene, binding the oncogene and inducing a translational block of RNA polymerase shown as a large oblong circle.

TABLE 10.5 Nucleic Acid-Based Gene Therapy Strategies for Cancer Treatment

Approach	Structure	Mechanism	Target
Antisense	DNA or RNA	Translation arrest RNase H activation	mRNA-oncogenes
Antigene	DNA or RNA	Triplex formation Transcription blockage	DNA-oncogenes Transcription factors
Aptamer	DNA or RNA	Binding and inhibition of function	Protein transcription factors
Ribozyme	RNA	mRNA cleavage	mRNA of growth factors; drug resistance gene oncogenes

oligonucleotide antisense molecule usually is a translational initiation site or a splicing site on the gene (Fig. 10.4). This binding represents an antigene approach that would inhibit genetic information flow (DNA–RNA–protein). The antigene approach is based on targeting genomic DNA, which comprises two copies of the oncogene. Inhibition of gene expression is achieved by forming a triplex (comprising the antisense molecule and the duplex double-stranded DNA). With formation of a stable triplex, translation to RNA of the oncogene would be inhibited. It can be noted that triplex formation is based on thermodynamic stable base pairings, and thus a function of complementarity and length of the antisense molecule. Antisense-based inhibition of gene expression can also occur at other transcriptional sites through the titration of regulatory proteins (sense and aptamer approaches, Table 10.5). Targets are the transcription factors and other nuclear regulatory proteins that promote gene expression. A final alternative approach for antisense-based inhibition of gene expression can target translational and posttranslational events. Translation of RNA to protein can be inhibited by targeting mRNA by an antisense oligonucleotide. This strategy is significantly more challenging in cancer therapy because of the large number of mRNA molecules for an oncogene in a malignant cell.

Regardless of the antisense approach taken, three steps need to be fulfilled prior to the use of antisense molecules. The first is the establishment of the relevance of the genetic basis of carcinogenesis. Second is the determination of the specific oncogene to target. Third is the determination of the specific sequence to target for antisense inhibition. Although the genetic basis of carcinogenesis is well established through the mulitstep formation of tumor development, the specific genetic pathway leading to the generation of an oncogene needs to be identified. It is important to identify the several genetic alterations that occur in sequence for progression to malignancy to ensure that mutation to oncogene is a relevant central event. Identification of genetic alterations comprises a molecular-based diagnostic strategy for an individual clinical management of cancer. Once the relevance of a genetic alteration is established in the carcinogenesis, identity of the oncogene is needed to provide information regarding gene regulation for determination of the best antisense approach. With the target identified, the specific nucleotide sequence is needed to provide a basis for antisense generation.

Specific examples of antisense gene therapy can be obtained from breast cancer, adenoma of the pancreas, and colon cancer studies. The clinical course of breast cancer is indicated by an early progression to distant metastasis. Prognostic factors for this event are important to distinguish metastatic cancer for adjuvant therapies. As noted in Table 10.2, the *erbB* oncogenes have an association with breast cancer. The amplification and overexpression of the *erbB* oncogenes have been suggested to play a fundamental role in the progression to metastasis in breast cancer. The fundamental role of *erb* oncogene activation is seen through the loss of cellular control of DNA replication, repair, and chromosomal segregation. The extent of these cellular changes have been shown to be determined by a gene dosage effect of the oncogene. Tumor cells observed to have a higher gene copy of oncogenes also show a propensity to metastasis and poor clinical outcome. A similar observation can be made for adenocarcinoma of the pancreas. Although cells derived from metastatic tumors are noted to have an inactivation of the *p53* tumor suppressor gene, chromosomal deletions at 18q, and point mutations at codon 12 of the K-*ras* oncogene, it is the overexpression of the *rhoC* oncogene that significantly correlates with a poor prognosis. Thus, the best targets of antisense gene therapy in cancer are overexpressed oncogenes that play a role in pathogenesis.

Numerous oncogenes have been targeted for antisense gene therapy. They include c-*fos* for brain cancer, c-*src* for colon cancer, c-*myb* for leukemia and tumors of the central nervous system, as well as c-*myc* for melanoma and ovarian cancer. Inhibition of targeted oncogene expression was noted in each case in cell lines and in xenografts grown in immune-deficient (*nude*, SCID) mice. Coupled with the reduced expression is a biological effect such as down-regulation of growth factor expression or increased sensitivity of the tumor cells to chemotherapy. Reducing expression of a growth factor such as vascular endothelial growth factor or transforming growth factor-α has significant effects on tumor angiogenesis and tumor growth, respectively. Use of antisense to c-*fos* in the brain has resulted in changes in neuronal function as well as behavior (see Chapter 9). For the case of increased sensitivity of the tumor cells to chemotherapy, the reduction in tumor cell proliferation and tumor colony formation has suggested that antisense gene therapy augments specific antineoplastic drugs as a "combination therapy."

Cellular uptake of the antisense oligodeoxyribonucleotide appears to be the limiting factor for effective therapy. In animal studies, enhanced uptake can be seen with the use of liposomes compared to intravenous administration. Thus, additional generations of antisense molecules are needed as well as new delivery techniques and methodologies. An expansion of the antisense technology is the use of ribozymes that are antisense RNA molecules that have catalytic activity (Fig. 10.5). Ribozymes function by binding to the target RNA moiety through antisense sequence-specific hybridization. Inactivation of the target molecule occurs by cleavage the phosphodiester backbone at a specific site (see Fig. 10.5 and Chapter 11). The two most thoroughly studied classes of ribozymes are the hammerhead and hairpin ribozymes, which are named from their theoretical secondary structures. Hammerhead ribozymes cleave RNA at the nucleotide sequence U-H (H = A, C, or U) by hydrolysis of a 3′–5′ phosphodiester bond. Hairpin ribozymes utilize the nucleotide sequence C-U-G as their cleavage site. A distinct advantage of ribozymes over traditional antisense RNA methodology is that the ribozyme is not consumed during the target cleavage reaction. Therefore, a single ribozyme can inactivate a large number of target molecules, even at low concentrations. Additionally, ribozymes can be generated from very small transcriptional units and, thus, multiple ribozymes targeting different genomic regions of an oncogene could be generated. Ribozymes also have greater sequence specificity than antisense RNA because the target must have the correct target sequence to allow binding. However, the cleavage site must be present in the right position within the antisense fragment.

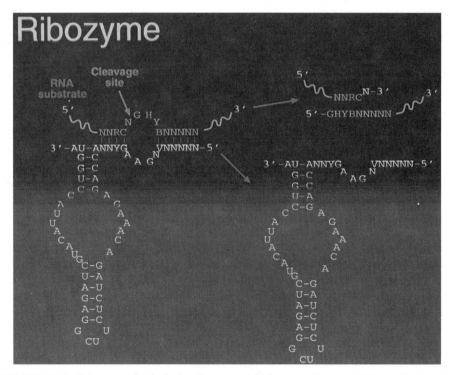

FIGURE 10.5 Diagram of a hairpin ribozyme, which are antisense RNA molecules that have catalytic activity. The cleavage site of RNA is C-N-G, where N = any nucleotide.

The functionality and the extent of catalytic activity of ribozymes, in vivo, for oncogenic RNA targets are presently unclear. This is because any alteration of the binding or cleavage sites within the target oncogene sequence required by the ribozyme for activity would render the ribozyme totally inactive. In the dynamic environment of carcinogenesis with numerous mutations and genetic alterations, genomic stability of the oncogene is a relevant issue.

Nevertheless, hammerhead ribozyme therapy in cancer cells has been investigated with *HER-2/neu* cellular oncogene in the context of ovarian cancer, *bcl-2* and induction of apoptosis in prostate cancer, *bcr-abl* oncogene in chronic myelogenous leukemia, c-*fms* in ovarian carcinoma, H-*ras*, c-*fos*, and c-*myc* in melanoma, N-*ras*, Ha-*ras*, and v-*myc* in transformed cell lines, as well as c-*fos* in colon cancer. In all cases, whether transfection of the cells with ribozyme occurred via polyamine beads, adenovirus, or retrovirus vector, the targeted oncogene expression was suppressed (Table 10.6). In addition, biological effects such as decreased proliferation, reversed cellular differentiation, augmented apoptosis in cancer cells and increased sensitivity to antineoplastic drugs were observed. Thus, ribozyme antisense gene therapy holds substantial promise for specific cancer treatment.

Another method of correcting an overexpressed oncogene effect is by interfering with the posttranslational modification of oncogene products necessary for function. For example, *ras* oncogenes, as mentioned above, are overexpressed in many tumors. However, in order to be active, *ras* must move from the cytoplasm to the plasma membrane. The addition of a farnesyl group, catalyzed by farnesyl transferase, to the *ras* protein is necessary in order to allow membrane localization of *ras*. Farnesly transferase can be inhibited by several tricyclic and other compounds

TABLE 10.6   Application of Ribozyme Therapy to Human Cancers

Vector	Promoter	Targeted Oncogene	Cancer Cells
Plasmid			
pHβApr-1 neo	β-actin	H-*ras*	Bladder and melanoma
		K-*ras*	Pancreatic
		c-*sis*	Mesothelioma
pMAMneo	MMTV-LTR	H-*ras*	Melanoma
		c-*myc*	Melanoma
		c-*fos*	Melanoma and ovarian
pLNCX	CMV	H-*ras*	Melanoma and pancreatic
pLNT	Tyrosinase	H-*ras*	Melanoma
pRc	CMW	Pleiotrophin	Melanoma
Adenovirus	CMV	H-*ras*	Melanoma
		K-*ras*	Pancreatic
Retrovirus	β-actin	*bcr/abl*	CML
	thymidine kinase	*bcr/abl*	CML
Liposome			
Lipofection		*bcr/abl*	CML
		AML1/MTG8	AML

now in development. Such inhibition results not only in growth inhibition in vitro but also results in growth inhibition of tumors in animal models of carcinogenesis. This inhibition occurs with little toxicity to normal cells. Like antisense therapy, it seems that farnesyl transferase inhibitors may augment the efficacy of cytotoxic chemotherpeutic drugs. In addition, such agents may be useful as chemopreventive agents in patients at high risk for tumors know to overexpress *ras*.

Targeted Prodrug Therapies

Targeted prodrug gene therapy against cancer is tumor-directed delivery of a gene that activates a nontoxic prodrug to a cytotoxic product by using tissue-specific promoters in viral vectors (Table 10.7). This approach should maximize toxicity at the site of vector delivery while minimizing toxicity to other, more distant cells. In animals, certain enzyme-activated prodrugs have been shown to be highly effective against tumors. However, human tumors containing similar prodrug-activating enzymes are rare. Gene-directed enzyme prodrug therapy (GDEPT) addresses this deficiency by attempting to kill tumor cells through the activation of a prodrug after the gene encoding for an activating enzyme has been targeted to a malignant cell (Fig. 10.6). Specific enzyme/prodrug systems have been investigated for cancer therapy using GDEPT. The requirements are nontoxic prodrugs that can be converted intracellularly to highly cytotoxic metabolites that are not cell cycle specific in their mechanism of action. The active drug should be readily diffusable to promote a bystander effect. Thus, adjacent nontransduced tumor cells would be killed by the newly formed toxic metabolite. The best compounds that meet these criteria are alkylating agents such as a bacterial nitroreductase.

The herpes simplex virus thymidine kinase (HSV*tk*) gene/ganciclovir system has

TABLE 10.7 Promoters Used for Targeted Gene Expression in Cancer Gene Therapy[a]

Cancer Cells	Promotors
Breast and Mammary carcinoma	*MMTV*-LTR; *WAP*-NRE; β-casein; SLPI; DF3(MUC1); c-*erbB2*
Neuroblastoma and glioblastoma	Calcineurin Aα; synapsin 1; HSV-*LAT*
Melanoma	Tyrosinase; *TRP-1*
B-cell leukemia	Ig heavy and κ light chain; Ig heavy-chain enhancer
Lung	*CEA*; *SLPI*; *Myc-Max* response element
Colon	*CEA*; *SLPI*
Liver	*AFP*
Prostate	*PSA*
Pancreas	c-*erbB2*
Bone and cartilage	c-*sis*

[a] Abbreviations: AFP, α-fetoprotein; CEA, carcinoembryonic antigen; DF3, high-molecular-weight mucinlike glycoprotein; HSV-LAT, herpes simplex virus latency-associated transcript; Ig, immonoglobulin; MMTV-LTR, mouse mammary tumor virus long terminal repeat; PSA, prostate specifc antigen; SLPI, secretory leukoprotease inhibitor; TRP-1, tryrosinase-related protein-1; WAP, whey acidic protein.

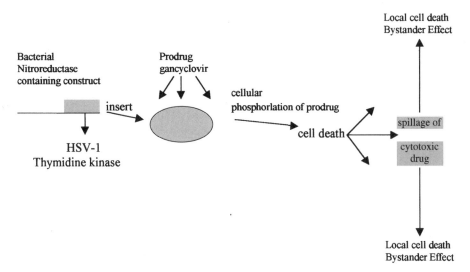

FIGURE 10.6   Gene-directed enzyme prodrug therapy (GBEPT).

been most commonly used for GDEPT. HSV*tk*, but not mammalian thymidine kinases, can phosphorylate ganciclovir to ganciclovir-triphosphate. Gancilovir triphosphate inhibits DNA synthesis by acting as a thymidine analog; incorporation into DNA is thought to block DNA synthesis. In addition to a direct cytotoxic effect upon HSV*tk*-transduced cells treated with ganciclovir, this approach produces the required bystander effect where nearby cells not expressing HSV*tk* also are killed. This may occur by the passage of phosphorylated ganciclovir from HSV*tk*-transduced cells to nonexpressing neighbors via gap junctions and/or through the generation of apoptotic vesicles taken up by neighboring cells. Vesicles could contain HSV*tk* enzyme, activated ganciclovir, cytokines, or signal transduction molecules such as *bax*, *bak*, or cyclins. In addition, the bystander effect may augment local immunity and promote killing of remaining tumor cells. Regardless of the mechanism, the bystander effect allows the efficient killing of tumor cells without treating every malignant cell.

Ganciclovir treatment of human leukemia cells transfected with HSV*tk* has been shown to inhibit cell growth. Both murine lung cancer cells and rat liver metastasis (an in vivo model of metastatic colon cancer) have been killed in vivo after transfection. Hepatoma cells have been successfully treated in vitro using varicella-zoster virus thimidine kinase (VZV*tk*), which converts nontoxic 6-methoxypurine arabinonucleoside (araM) to adenine arabinonucleoside triphosphate (araATP) a deadly toxin.

The success of these studies has lead to numerous clinical trials using HSV*tk*. Although growth suppression of the tumor has been well documented in these studies, cures remain elusive. It is likely that there is variability of the bystander effect in vivo compounded by limited tranduction efficiencies in vivo. However, the use of HSV*tk* has resulted in augmented sensitivity to chemotherapy, thus, suggesting a role of prodrug therapy in combination with antineoplastic drugs.

An additional prodrug system extensively investigated is the *Echerichia coli* cytosine deaminase (CD) gene plus 5-fluorocytosine (5-FC). The *CD* gene converts 5-

FC to the chemotherapeutic agent 5-flourouracil (5-FU). 5-FU has been a standard treatment for metastatic gastrointestinal (GI) tumors, and in the same manner this prodrug sytems has been tested. Systemic therapy with 5-FU results in the growth suppression of *CD*-transduced tumor cells with a significant bystander effect for 5-FU. Thus, strategies for metastatic GI tumors to the liver have focused on the regional delivery of *CD* to the tumor mass. For tissue-specific deliver to the liver, promoters for the carcinoembryonic antigen or α-fetoprotein genes are being explored for hepatic artery infusion of the *CD* vector. However, specific tumors are noted to develop resistance to repeated 5-FU treatment that will require additional methodological interventions.

Modifying the Antitumor Immune Response

Cell-Mediated Tumor Immunity The generation of cytotoxic T-cell-specific immunity is predicated on (1) the ability of the CD8 cells to recognize a pathogenic cell and (2) the activation and subsequent expansion of the antigen-specific CD8 cells. The selection and activation of the cell with the correct specificity for a particular antigen occurs in the lymph node. It is here that the T cells interact with antigen-presenting cells such as dendritic cells. Dendritic cells home to the lymph node after encountering pathogenic cells in the periphery. Dendritic cells are uniquely suited to this function since they express not only the MHC class I and II molecules but also specific co-stimulatory molecules such as B7.1, B7.2, CD40L, ICAM 1, 2, 3, VCAM-1, and LFA-3. With specific recognition and activation of the T cell, clone(s) migrate from the node and travel directly to the site of the pathogenic cells. As activated T cells, they now only require recognition, which occurs through the same signal delivered by the major histocompatibility (MHC) of the antigen-presenting cell (APC). Neoplastic cells themselves present a unique challenge to this system since these cells lack the co-stimulatory molecules needed for effective activation of the cytotoxic T cells. In addition, it has been shown that the delivery of the MHC signal without co-stimulation can anergize cells and may represent a separate mechanism by which tumor cells evade immune attack. One approach developed to overcome the lack of co-stimulatory molecules has been to transduce tumor cells with co-stimulatory molecules so they can function directly as APCs. These cells can either be directly administered to the host as vaccines, as discussed in the next section (usually through subcutaneous or intradermal injection), or modified in vivo via intratumoral injection of the gene for the co-stimulation molecule. Many reports of the success of this approach in animal models can be found in the literature, although there are also some reports of *B7.1*-modified cells failing to induce tumor-specific immunity. Nonetheless, clinical trials have been initiated with *B7.1*-transduced tumors in melanoma and colon cancer patients. The results to date would suggest that the use of *B7.1*-transduced irradiated tumor cells as vaccines can augment antitumor immune responses, although the clinical relevance of this effect remains to be proven. Another approach to boost the ability of tumor cells to function as APC has been to transduce a genetically mismatched histocompatablilty antigen into the cell. The net effect of this transfection would be to create a strong allo-response around the tumor, thus inducing the migration and activation of both APCs and T cells. In addition, local IL-2 production would be expected from the recruited T cells, further amplifying the local inflammatory response. Clinical trials using this approach are now in progress in melanoma patients.

Cytokines Cytokines are proteins secreted by immune cells that act as potent mediators of the immune response. Early clinical studies with these molecules demonstrated that significant toxicity could be expected at high doses when they were delivered systemically. It was therefore a natural extension of the early research on cytokines and cancer to use gene therapy to deliver cytokine gene(s) to tumor cells, thus creating an environment around the cell that would help to facilitate its destruction. To date, this has largely been accomplished via viral delivery through adenovirus and retrovirus constructs or through cationic lipids. Cytokine delivery has been both directly into the tumor (intratumoral) and into the tumor cells ex vivo. Virtually all of the cytokines studied have shown an effect on tumor growth and survival in some animal models. In most cases, the expression of the cytokine was only required in a small number of cells relative to the tumor challenge, suggesting that the cytokine was affecting an immune response against the tumor and not simply targeting or killing the transfected cells alone. This antitumor effect has mostly been attributed to the activation and expansion of existing antitumor immune cells in and around the tumor. However, it is also possible that some benefit was derived from the induction of an inflammatory response at the site of the tumor, resulting in an influx and activation of many types of cells at the tumor site. In addition, the delivery of cytokines to tumor cells ex vivo has provided a way to greatly enhance the immunogenicity of the tumor cells and opened the door for the use of these gene-modified tummor cells as vaccines.

Table 10.8 lists cytokines studied in clinical trials. As can be seen, the majority of trials employed IL-2. This 133-amino-acid polypeptide, originally described as the T-cell growth factor, is the primary cytokine produced by activated CD4 cells. IL-2 acts locally at the site of an immune response to expand the population of activated CD8 cells. Such T cells can be recovered directly from the tumor and have consequently been referred to as tumor-infiltrating lymphocytes (TILs). In addition, IL-

TABLE 10.8 Cytokines, Accessory Molecules, and Growth Factors Transfected to Augment Immunity

Cytokine	Biological Activity	Tumor System
IL-2	T-cell growth factor, expands CTLs	Brain, breast, colon, lung, small cell, melanoma, ovarian
IL-4	B-cell, T-cell growth factor	Advanced cancer, brain,
IL-7	CTL activation, down-regulates TGF-β	Colon, lymphoma, melanoma, Renal
IL-12	Actives Th1 response, CTL activation	Advanced cancer, melanoma
IFN-g	Activates CD8 cells, activates macrophages Up-regulates MHC class I, class II expression	Melanoma, prostate, brain
GM-CSF	Dendritic cell activation, macrophage activation	Renal, prostate, melanoma

2 can also expand natural killer (NK) cells, a subset of immune cells that are also potent killers of neoplastic cells. Other molecules in the interleukin family, which have similar effects and have also been studied, including interleukin 4 (IL-4), interleukin 7 (IL-7), and interleukin 12 (IL-12). IL-12 is a heterodimer consisting of 40,000 and 35,000 polypeptides. It has been most commonly associated with the Th1-type cell-mediated response and thus would be expected to synergize with other Th1-type cytokines such as IFN-g and IL-2. Another cytokine that has received considerable interest in recent years is granulocyte-monocyte stimulating factor (GM-CSF). This cytokine boosts APC activation and, thus, would be expected to indirectly expand CTLs through APC/CTL interations. Finally, although the direct modification of tumor cell vaccines to express cytokines has provided some encouraging preclinical and clinical results, it is apparent that the use of this approach on a large scale could be hampered by the variability of expresion of the cytokine of interest. To overcome this problem, cells such as fibroblasts can be engineered to express the cytokine of interest. These cells then can be co-injected with irradiated wild-type or modified tumor cells to boost the immune response at the site of injection. Likewise, the administration of cytokine secreting cells to the tumor bed through intratumor injection could also be accomplished. Phase I clinical trials with fibroblast secreting IL-2 have already been completed and would appear to suggest that the inclusion of these cells in a tumor cell vaccine preparation can augment anti-tumor-specific immune responses.

Immunosuppression   The success of a tumor development depends on its ability to escape the immune system. For example, immunosuppression is a common finding in patients with malignant brain tumors. Recent work has suggested that these impaired immune responses may be directly related to the intracranial tumor production of one or more distinct immunosuppressive cytokines. One such cytokine, which has been strongly implicated in this specific immunosuppression, is transforming growth factor β (TGF-β). There are at present three distinct isoforms of TGF-β, commonly referred to as TGF-β1, TGF-β2, and TGF-β3. In addition, a high-molecular-weight TGF-β has been reported that may represent a TGF-β1 molecule linked to larger cell protein. All TGF-β isoforms, except the high-molecular-weight species, typically are secreted as dimers and require cleavage, either through acidification or protease activity, to be active. Of the three isoforms of TGF-β reported, one isoform of this cytokine, TGF-β2 (previously called glioblastoma-derived T-cell suppressor factor), has been shown to be at high plasma levels in a bioactive form in immunosuppressed patients with anaplastic astrocytoma or glioblastoma multiforme. The source of this factor appears to be the glioma cells themselves, since high concentrations of the factor have been observed in glioma cell lines grown in vitro. In addition, it has also been demonstrated that some TGF-β levels fall and some degree of immunocompetence is restored upon tumor resection, a finding that further supports the tumor cells as the source of TGF-β2. Elevated levels of TGF-β1 have also been observed in plasma samples from colon cancer patients, and these increases are directly correlated to disease as measured by Duke's classification of tumor staging. Furthermore, these elevated TGF-β1 levels (11.9 ng/ml) approach normal levels (3.8 ng/ml) 4 weeks or more after surgical resection. One other potential immunosuppressive cytokine that has been found in patients with anaplastic astrocytoma or glioblastoma multiforme is interleukin 10

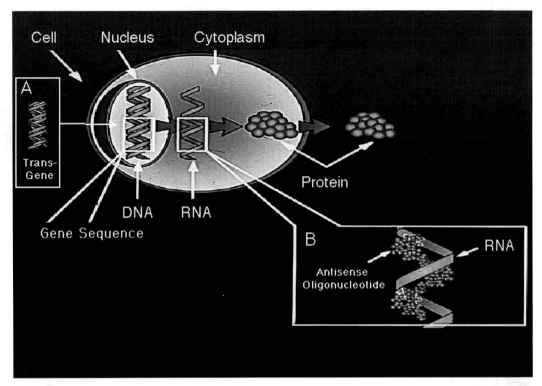

Figure 8.1 Gene therapy strategies. See page 184. (A) Gene transfer involves delivery of an entire gene, either by viral infection or by nonviral vectors, to the nucleus of a target cell. Expression of the gene via transcription into mRNA and translation into a protein gene product yields a functional protein that either achieves a therapeutic effect within a transduced cell or is secreted to act on other cells. (B) Gene blockade involves the introduction into the cell of short sequences of nucleic acids that block gene expression, such as antisense ODN that bind mRNA in a sequence-specific fashion and prevents translation into protein.

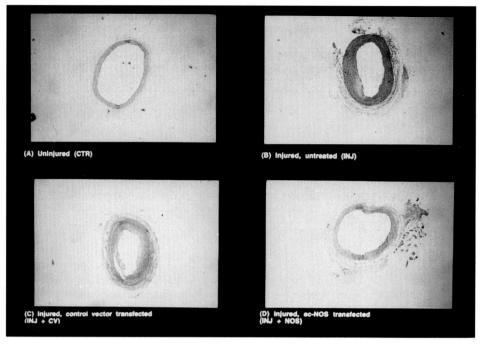

Figure 8.2 Inhibition of neointimal hyperplasia by in vivo gene transfer of endothelial cell-nitric oxide synthase (ecNOS) in balloon-injured rat carotid arteries. See page 189.

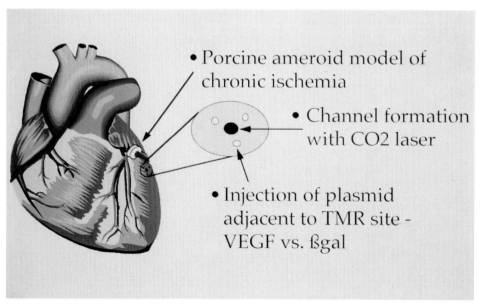

Figure 8.3 Combined gene transfer and transmyocardial laser revascularization (TMR), See page 191. Schematic representation of chronic ischemia induced by placement of Ameroid constrictor around the circumflex coronary artery in pigs. Ischemic hearts that underwent TN4R followed by injection of plasmid encoding VEGF demonstrated better normalization of myocardial function than either therapy alone.

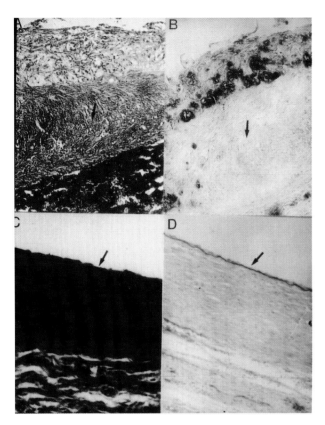

Figure 8.4 Control oligonucleotide-treated (A and B) and antisense oligonucleotide (against c and 2 kinase/PCNA)-treated vein grafts (C and D) in hypercholesterolernic rabbits, 6 weeks after surgery (370). See page 193. Sections were stained with hematoxylin/van Gieson (A and C) and a monoclonal antibody against rabbit macrophages (B and D). Arrows indicate the location of the internal elastic lamina.

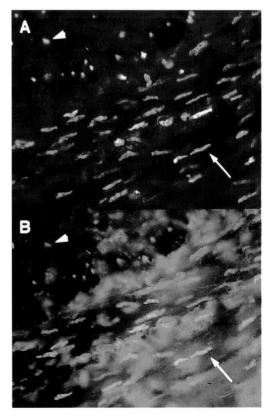

Figure 8.5 Interoperative pressure-mediated transfection of fluorescent-labeled ODN to saphenous vein graft cells. See page 194. (A) Hoechst 33,342 nuclear chromatin staining of vein graft in cross section, illustrating location of nuclei within graft wall (1003)(B) Same section of saphenous vein viewed under FITC-epifluoreseence at 1003. Note the pattern of enhanced green fluorescence in the nuclei of cells within the graft wall, indicating nuclear localization of labeled ODN.

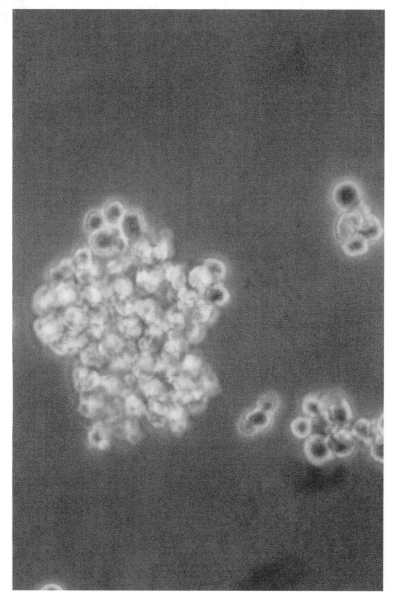

Figure 10.3 Morphology of Epstein-Barr virus transformed cells. See page 239. Note the rounded morphology, aggregation, clumping, and satellite colonies of growth.

(IL-10). The immunosuppressive activity of IL-10 is now well documented. It has recently been shown that IL-10 inhibits in vitro T-cell proliferation in response to soluble antigens and strongly reduces the proliferation of human alloreactive cells in mixed lymphocyte reactions (MLR). In addition, IL-10 induces a long-term antigen-specific anergic state in human CD4 + T cells. For these reasons, IL-10 might also hinder antitumor immune responses. There are now a number of reports that appear to suggest that down-regulation of TGF-β by antisense techniques can dramatically affect the immunogenicity of tumor cell vaccines. Such cells can be engineered ex vivo and applied alone or with cytokines, which have also been engineered into the tumor cells or into a carrier cell co-administered with the tumor cell vaccine. Current studies using this approach in patients with recurrent glioma multiforme should help to understand the clinical value of this strategy. In conclusion, the modification of antitumor immunity through gene therapy is being studied through a variety of strategies. Modification of tumors in vivo to express co-stimulatory molecules and/or cytokines has provided a way to increase immune reactivity directly at the site of the tumor. The use of either autologous or allogeneic tumor cells modified ex vivo as vaccines is also currently being studied. Such therapies would be applied postsurgery to kill any remaining transformed cells that could not be physically removed. It is also hoped that these vaccines may limit the development of metastatic tumors distal to the primary tumor. The next few years should provide a wealth of information regarding the clinical effects of gene modification of the antitumor response.

DNA CANCER VACCINES

The generation of a vaccine for cancer is a concept based on three principles: (1) a qualitative and/or quantitative difference exists between a normal cell and a malignant cell, (2) the immune system can identify the difference between cell types, and (3) the immune system can be programmed by immunization to recognize the differences between normal and malignant cells. A fundamental axiom of immunology is the active discrimination between self and nonself based on the presence of cell-mediated immunity and the expression of MHC antigens. Cancer vaccine efforts have focused in five areas related to augmenting host immunity through malignant cells recognition and memory: (1) immunization of irradiated malignant cells, with or without adjuvants, and potentially modified by transfection with cytokines or accessory molecules to further augment the immune response; (2) cellular immunization of tumor-associated proteins to allow phagocytosis by antigen presenting cells and presentation to killer cells via MHC alleles; (3) immunization or presentation of polypeptide tumor antigens or mutations as part of the antigen priming process; (4) the immunization with naked DNA or viral vectors containing cDNA, which encode tumor-associated antigens, accessory molecules, cytokines, or other molecules that could augment immunity; and (5) immunization of carbohydrate antigens associated with malignant cells. These vaccine strategies can be targeted directly to the cancer or to viral infections that are associated with the development of cancer. For instance, chronic infection with hepatitis C can result in the development of hepatocellular carcinoma. Thus, the generation of a vaccine to protect from hepatitis C infection would also reduce the incidence of liver cancer.

Vector-Based Vaccines

The immunological basis for the transfection of cells with cytokines or accessory molecules is the enhancement of the antitumor immune response. The target for enhancement of the immune response is the augmentation of antigen presentation. One such approach is the genetic engineering of tumor cells to present tumor antigens directly to cytotoxic T cells or helper T cells. Thus, a subpopulation of tumor cells would be turned into professional antigen presenting cells such as macrophages or dendritic cells. Many cytokines and growth factors (see Table 10.8) have been transfected into tumor cells based on the hypothesis that augmented cytokine expression at the site of the tumor will augment local antigen presentation and antitumor immunity, particularly CD8$^+$ cytotoxic T cells.

Primary factors implicated in the escape of tumor cells from the surveillance of cytotoxic T cells is the lack of expression of co-stimulatory molecules by tumor cells and an inappropriate cytokine milieu. For cytotoxic T cells to kill a tumor cell, two intercellular signals are required: (1) an antigen-specific signal mediated by the engagement of the T-cell receptor with the antigen MHC complex, and (2) an antigen nonspecific or costimulatory molecule provided by accessory receptors after engagement by ligands expressed on the antigen presenting cells. Thus, the presence of co-stimulatory molecules (T cell receptor CD28 and B7 family ligands on APCs) are crucial for T-cell expansion and immune responsiveness. Studies in animals have shown that transfection of melanoma cells with B7 co-stimulatory molecules promotes antitumor immunity as well as transfection with cytokines and growth factors such as IL-2, IL-4,IL-6, interferon-δ, and GM-CSF. With transfection, an immune response is observed comprising an eosinophilic infiltrate with CD4$^+$and CD8$^+$ T cells. In a specific system, acute myelogenous leukemia cells were transfected with a retrovirus containing a transgene for B7.1. and 10^4 to 10^5 cells administered to tumor bearing mice. All mice rejected their tumors and remained tumor free for 6 months. The rejection immune response comprised of IL-2 and interferon-δ as well as very active CD8$^+$ T cells. However, these studies also showed that DNA vaccines were not effective in animals with higher tumor burdens. In these animals, the vaccine efficacy could be enhanced by the addition of chemotherapy. These phase 1 successes have opened the door for clinical trials using recombinant cytokines and co-stimulatory molecules.

Cellular-Based Vaccination

Two cellular-based gene therapy approaches to the immunotherapy of cancer are gene-modified tumor vaccines and dendritic-cell-based vaccination. Both approaches require cellular discrimination (recognition) of the tumor and augmentation of the immune response. As presented earlier, vaccine strategies for tumor eradication span multiple gene therapy approaches when based on the augmentation of the immune response.

Gene-Modified Tumor Vaccines The original basis for this approach was to enhance tumor immunogenicity through the expression of additional specific cytokines. The cytokines would, the hypothesis goes, help in the process of antigen presentation and the generation of protective antitumor immunity. This hypothesis was put forward based on data showing that vaccination with regular nonmodified tumor cells did not augment antitumor immunity. The cytokine-induced protective immune response would comprise both T helper cells and cytotoxic T cells, based on the vaccination route. The T helper cells would be integral to the development of anti-tumor-specific antibodies, such as idiotypic or anti-idiotypic antibodies (see below), which could promote antibody-dependent cell-mediated cytotoxicity (ADCC). Mature cytotoxic T cells would be generated from naive cells through vaccination. Attempts at tumor cell vaccination to induce either established tumor regression or immunologic memory were unsuccessful with the suggestion that in situ cytokine levels could not reach "physiologic" levels by ex vivo transfection of autologous tumor cells.

Current studies suggest that the most efficient way to generate mature cytotoxic T cells is through tumor cell presentation. Tumor antigens can be presented through the release of tumor-cell-associated antigens upon cell death or apoptosis. Antigen is released from tumor cells through an inflammatory response resulting in tumor antigen degradation and cell death. This form of antigen priming is thought to be a major pathway for the induction of cytotoxic T cells. Thus, gene therapy approaches to augment the immune response via cytokine gene transfection is in effect an attempt to activate this antigen priming pathway for the induction of cytotoxic T cells. As noted earlier, efforts have been made to transfect the genes for IL-2, IL-4, IL-6, IL-7, γ-interferon, tumor necrosis factor, or granulocyte-macrophage colony stimulating factor. These efforts showed the induction of tumor-specific immunity in animals through the rejection of subsequent tumor challenge (lung or breast cancer). Additionally, efforts at transfecting the gene for B7-1 are targeted at enhancing tumor antigen presentation. For the case of common solid tumors that grow at particularly slow rates, virally induced transfection has not been optimal for the transfer of immune enhancing genes. For these tumors, a transfection rate between 10 and 15% has been achieved by using a plasmid DNA vector using the long terminal repeats of adenoassociated virus (AAV) incorporated into a liposome vehicle. Weekly vaccination with this construct in an animal model of metastatic lung cancer showed a reduction in lung metastases. Although these methods produce encouraging results, an alternative approach is to utilize the professional antigen presenting cell in vaccination strategies.

Dendritic Cell Vaccination   The use of dendritic cells in vaccination strategies to induce antitumor immunity is based on the hypothesis that cytotoxic T-cell priming is somehow defective or not efficient, thereby resulting in tumor proliferation. Thus, augmentation of tumor antigen expression by the dendritic cell would limit the need of antigen transfer from the tumor cell to the antigen presenting cell. In this case, tumor cell recognition by the innate immune system would not be necessary for the induction of antitumor T-cell immunity. The overall approach of dendritic cell vaccination is to utilize ex vivo gene transfer techniques to overexpress the tumor cell antigen(s) on the surface of the antigen presenting cell and to subsequently "vaccinate" the recipient to induce antitumor immunity. This approach requires optimization of numerous techniques and steps. These include the identification and characterization of tumor immunogens (antigens that induce immune responses), isolation, and in vitro growth of dendritic cells, gene or protein transfer techniques for dendritic cells, identification of vaccination methods, and screening for adverse effects related to vaccination including the induction of autoimmunity. The approach of dendritic cell vaccination has been utilized in animal models of human cancer. Most notable is the testing in the murine postsurgical metastasis model to prevent the growth of preexisting micrometastasis after excision of the primary tumor. In this model, treatment of the tumor bearing mice with dendritic cells expressing tumor-derived antigens either in the form of tumor cell protein extracts, specific tumor peptides, or RNA resulted in the induction of tumor-specific immunity.

The demonstrated efficacy of dendritic cell vaccination in an animal model of human cancer has resulted in translational research efforts to investigate this approach for cancer therapy in humans. Recent studies have investigated the localization of radiolabeld dendritic cells in humans based on the route of administration. Dendritic cells are administered intravenously, localized initially to the lungs and subsequently to the liver, spleen, and bone marrow. Cells administered intradermally were cleared from the injection site and migrate to regional lymph nodes. Thus, in humans the development of protective antitumor immunity by dendritic cell vaccination will depend on the type of tumor and the route of administration of vaccine.

Idiotype-Based Vaccines

The term *idiotype* denotes the array of antigenic determinants that can be serologically defined on a given antibody molecule . When these antigenic determinants are shared among antibodies, soluable factors, or cells, the term *cross-reactive idiotype* (CRI) is applicable. CRIs form the basis for regulatory networks for immunoregulation and communication among the network members. CRIs can define a major proportion of a given antibody population. The designation CRI_M, or dominant regulatory idiotype, is used. In the corollary, when a small fraction of antibodies expresses a CRI, a minor cross-reactive idiotype (CRI_m) is defined. The relative expression of idiotype infers a level of connectivity among members of the immune system (antibodies, factors, B cells, T cells). It is also the basis for the immunoregulatory aspects of the idiotypic immune network. The immunoregulatory aspect of idiotypy was originally proposed as a set of complementary interactions that form the basis for self-regulation of an autologous immune response (Table 10.9). Fundamental to the hypothesis was the dual nature of the antibody molecule. The primary antibody molecule recognizes and binds antigen through the antigen combining site. Also, at this location is the expression of idiotypy. Thus, acting as antigen, idiotypic molecules (Ab_1) induce a second population of antibody molecules (Ab_2). These Ab_2 molecules are serologically complimentary to the Ab_1 antibody molecules. The Ab_2 antibody populations are termed anti-idiotypic. A unique subpopulation of anti-idiotypic antibodies are those members that are serologically defined by the initial antigen. This subpopulation is complementary to the antigen binding site of the Ab_1 population and binding to idiotypic antibodies is inhibited by antigen. As such, these molecules represent an internal image of the antigenic epitope. As internal images of antigen, in this case tumor-specific antigens, it follows that these molecules could represent candidates for vaccine molecules in the immunotherapy of cancer.

Idiotypes expressed by tumor cells in B-cell malignancies can be regarded as

TABLE 10.9   Serological Aspects of Immunoglobulin, B and T Cells

Idiotypic	Anti-Idiotypic	Anti-Anti-Idiotypic
Ab1	Ab2	Ab3
Binds antigen	Binds idiotype	Binds antigen
Induced by antigen idiotype	Induced by idiotype	Induced by anti-idiotype
Expresses CRI other	Defines CRI	Express CRI and idiotypes (expanded repertoire)
Individual molecules	Subpopulations may be internal image of antigen	Individual molecules may neutralize cancer cell; on a population basis may be more effective than Ab_1 (expanded repitoire)

tumor-specific antigens and targets for vaccine imunotherapy. Haptens, adjuvants, and cytokines have been used to increase idiotype immunogenicity and established a protective anti-idiotypic immune response. These results have been extended by the use of DNA technology for the development of fusion proteins and naked DNA vaccines comprising components of idiotype–anti-idiotype networks. Thus, idiotype vaccination has been shown to be efficacious in individuals with B-cell lymphoma and multiple myeloma. In these patients a prolongation of disease-free period with increased survival and the generation of idiotype-specific immunity was noted.

Initial animal studies demonstrated the existence of the idiotype–anti-idiotype network. This network comprises antigen in the form of tumor-specific antigen, Ab_1 (idiotypic) antibody, Ab_2 (anti-idiotypic antibody), and Ab_3 (anti-anti-idiotypic) antibody. For idiotype vaccination, one uses the immunoglobulin heavy- and light-chain hypervariable regions that contain the idiotopes. These antigenic determinants can be immunized directly or small synthetic polypeptides can be made and conjugated to a carrier immunogen to produce an antitumor immune response. Both antitumor antibody and $CD4^+$ (helper) and $CD8^+$ (cytotoxic) T cells are generated that specifically recognize the idiotype of the original tumor-specific antigen (immunogen). Immunization with growth factors such as granulocyte-macrophage colony stimulating factor, augments the antitumor immune response, particularly with regard to tumor killing T cells ($CD8^+$). In addition, when animals are immunized with anti-idiotype antibodies (Ab_2) antibodies derived from a tumor-specific antigen, an anti-anti-idiotype (Ab_3) antibody response is generated. This antibody response is amplified with greater antigen binding diversity (expanded repertoire) compared to the Ab_1 antibodies and functionally decreases tumor growth and colonization in vivo. Immunization with DNA constructs encoding the lymphoma idiotype results in specific anti-idiotype antibody responses. These Ab_2 antibodies protect animals from tumor challenge. The immunization with DNA constructs can take the form of naked DNA encoding the human antibody variable region administered intradermally. In a long-term clinical trial, idiotype vaccination resulted in tumor regression in cancer patients and cancer immunity in patients in remission. Thus, idiotype vaccination, on an individual basis for multiple myeloma and lymphoma patients, represents a methodology to induce tumor immunity to prevent recurrent disease.

SUMMARY

Numerous gene-based therapies for cancer are in clinical trials and are based on the augmentation of the host's antitumor immunity or the augmentation of sensitivity to antineoplatic drugs. The protocols include both ex vivo and in vivo gene therapy techniques for cytokine or accessory molecule gene transfer, the gene transfer of prodrug-induced cytotoxicity, genetic vaccination, and the molecular correction of the genetic alterations of carcinogenesis. The latter include the inactivation of oncogene expression and the gene replacement for defective tumor suppressor genes. The data generated to date indicate that in patients with advanced cancers that are refractory to conventional therapies, cancer gene therapy techniques may mediate tumor regression with acceptable low toxicity and side effects. Important areas for development remain, however. Viral vectors need modification to reduce toxicty and immunogenicity and transduction efficiencies need to be increased for both viral and nonviral vectors. Tumor targeting and specificity need to be advanced and a further understanding of gene regulation, apoptosis, and the synergy between gene therapy and chemotherapy will augment the approaches for gene-based therapy of cancer.

KEY CONCEPTS

- Cancer arises from a loss of the normal regulatory events that control cellular growth and proliferation. The loss of regulatory control is thought to arise from mutations in genes encoding the regulatory process. In general, a genetically recessive mutation correlates with a loss of function, such as in a tumor suppressor gene. A dominant mutation correlates with a gain in function, such as the overexpression of a normally silent oncogene.

- Gene therapy for the treatment of cancer has been directed at (1) replacing mutated tumor suppressor genes, (2) inactivating overexpressed oncogenes, (3) delivering the genetic component of targeted prodrug therapies, and (4) modifying the antitumor immune response.

- Cell cyclins act as structural regulators of the cell cycle by determining the subcellular location, substrate specificity, interaction with upstream regulatory enzymes, and timing of activation of the cyclin-dependent kinases.

- Cancer cells could be targeted for death by insertion of apoptosis genes. On the other hand, localized immune cells fighting malignant cells could be provided added protection through the transfer of genes that protect from apoptosis.

- Cellular oncogenes are normal cellular genes related to cell growth, proliferation, differentiation, and transcriptional activation. Cellular oncogenes can be aberrantly expressed by gene mutation or rearrangement/translocation, amplification of expression, or through the loss of regulatory factors controlling expression. The aberrant expression results in the development of cellular proliferation and malignancy. There have been over 60 oncogenes identified to date and are associated with various neoplasms. The overexpression of oncogenes can be abrogated by approaches limiting their expression by the use of antisense molecules or ribozymes.

- Tumor suppressor genes encode for molecules that modify growth of cells through various mechanisms including regulation of the cell cycle. An abnormality in a tumor suppressor gene could result in a loss of functional gene product and susceptibility to malignant transformation. Thus, restoration of tumor suppressor gene function by gene therapy, particularly in a premalignant stage, could result in conversion to a normal cellular phenotype or "reverse transformation" of a malignant cells to a nonmalignant cell type.

- Targeted prodrug gene therapy against cancer is tumor-directed delivery of a gene that activates a nontoxic prodrug to a cytotoxic product. This approach should maximize toxicity at the site of vector delivery while minimizing toxicity to other, more distant cells. Specific enzyme/prodrug systems have been investigated for cancer therapy. The requirements are nontoxic prodrugs that can be converted intracellularly to highly cytotoxic metabolites that are not cell cycle specific in their mechanism of action. The active drug should be readily diffusable to promote a bystander effect. Thus, adjacent nontransduced tumor cells would be killed by the newly formed toxic metabolite. The best compounds that meet these criteria are alkylating agents such as a bacterial nitroreductase.

- The generation of a vaccine for cancer is a concept based on three principles: (1) a qualitative and/or quantitative difference exists between a normal cell and a malignant cell, (2) the immune system can identify the difference between cell types, and (3) the immune system can be programmed by immunization to recognize the differences between normal and malignant cells.

SUGGESTED READINGS

Cancer Gene Therapy

Cai Q, Rubin JT, Lotze MT. Genetically marking human cells—results of the first clinical gene transfer studies. Cancer Gene Ther 2:125–136, 1995.

Christian MC, Pluda JM, Ho PT, Arbuck SG, Murgo AJ, Sausville. Promising new agents under development by Division of Cancer Treatment, Diagnosis, and Centers of the National Cancer Institute. Semin Oncol 2:219–240, 1997.

DeCruz EE, Walker TL, Dass CR, Burton MA. The basis for somatic gene therapy of cancer. J Exp Ther Oncol 1:73–83, 1996.

Gough MJ, Vile RG. Different approaches in the gene therapy of cancer. Forum (Geneva) 9:225–236, 1999.

Hall, SJ, Chen S-H, Woo SLC. The promise and reality of cancer gene therapy. Am J Hum Genet 61:785–789, 1997.

HwU P. Current challenges in cancer gene therapy. J Intern Med Suppl 740:109–114, 1997.

McCabe RP, Curiel DT. Gene therapy. In Rustgi Ak (Ed.), Gastrointestinal Cancers: Biology, Diagnosis and Therapy. Lippincott-Raven, 1995, pp. 619–629.

Runnebaum IB. Basics of cancer gene therapy. Anticancer Res 17:2887–2890, 1997.

Genetic Basis of Carcinogenesis

Hauses M, Schackert HK. Gene therapy and gastrointestinal cancer: Concepts and clinical facts. Langenbecks Arch Surg 384:479–488, 1999.

Nielsen LL, Maneval DC. P53 tumor suppressor gene therapy for cancer. Cancer Gene Therapy 5:52–63, 1998.

Roth JA, Swisher SG, Meyn RE. p53 tumor suppressor gene therapy for cancer. Oncology (Huntingt) 13(Suppl):148–154, 1999.

Rustgi AK. Oncogenes and tumor suppressor genes. In Rustgi AK (Ed.), Gastrointestinal cancers: Biology, diagnosis and Therapy. Lippincott-Raven, 1995, pp. 65–76.

Weinstein IB. Relevance of cyclin D1 and other molecular markers to cancer chemoprevention. J Cell Biochem Suppl 25:23–28, 1996.

Cancer Gene Therapy and the Cell Cycle

Strauss BE, Costanzi-Strauss E. Efficient retrovirus mediated transfer of cell-cycle control genes to transformed cells. Braz J Med Biol Res 32:905–914, 1999.

Antisense Cancer Gene Therapy

Irie A, Kijima H, Ohkawa T, Bouffard DY, Suzuki T, Curcio LD, Holm PS, Sassani A, Scanlon KJ. Anti-oncogene ribozymes for cancer gene therapy. Adv Pharmacol 40:207–257, 1997.

Warzocha K, Wotowiec D. Anitsense strategy: Biological utility and prospects in the treatment of hematological malignancies. Leuk Lymphoma 24(3/4):267–281, 1997.

Farnesyl Transferase Inhibition

Beaupre DM, Kurzrock R. Ras and leukemia: From basic mechanisms to gene-directed therapy. J Clin Oncol 17:1071–1079, 1999.

Prodrug Cancer Therapy

Connors TA. The choice of prodrugs for gene directed enzyme prodrug therapy of cancer. Gene Therapy 10:702–709, 1995.

Vector-Based Vaccines

Dunussi-Joannopoulos K, Weinstein HJ, Arcesi RJ, Croop JM. Gene therapy with B7.1 and GM-CSF vaccines in a murine AML model. J Ped Hematol/Oncol 19:536–540, 1997.

Idiotype-Based Vaccines

Bianchi A, Massaia M. Idiotypic vaccination in B-cell malignancies. Mol Med Today 3:435–441, 1997.

Hsu FJ, Caspar CB, Czerwinski D, Kwak LW, Liles TM, Syrengelas A, Taida-Laskowski B, Levy R. Tumor-specific idiotype vaccines in the treatment of patients with B-cell lymphoma—long term results of a clinical trial. Blood 89:3129–3135, 1997.

Gene Therapy for HIV Infection

BRUCE BUNNELL, M.D.

BACKGROUND

In the previous chapters of this text, the technological aspects of gene therapy have been discussed. The application of these technologies to specific genetic disorders has also been presented. In this chapter, the application of this technology for the treatment of an infectious agent will be discussed. Specifically, gene therapy approaches to limit replication of the human immunodeficiency virus (HIV-1), the causative agent in acquired immunodeficiency syndrome, will be presented.

INTRODUCTION

Acquired immunodeficiency syndrome (AIDS) is a rapidly expanding global pandemic. Approximately 15 million people worldwide are infected with HIV-1. Despite more than a decade of intense research efforts aimed at understanding the HIV-1 virus and developing an effective therapy for AIDS, HIV-1 infection remains an incurable and fatal disease. However, significant progress has been made in the management of HIV-1 replication using traditional drug-based therapies. Most notable is the advent of the triple-drug regiment, which is composed of three drugs that inhibit the HIV-1 life cycle at two different stages. A protease inhibitor, which blocks the normal processing of proteins necessary to generate new HIV-1 particles, and AZT and 3TC, which are nucleoside analogs that inhibit replication of the viral genome, are typically the components of the triple-drug cocktail. The high rate of mutation in the viral genome and the generation of drug-resistant strains of HIV-1 are the major factors that prevent the development of effective drug-based therapies. The triple-drug regiment has not been sufficiently tested to assess the ability of the HIV-1 to form drug-resistant mutants. The inability of traditional drug-based therapies to effectively inhibit the HIV-1 replication has made it necessary to develop new and innovative therapies for this deadly disease.

As part of the normal virus life cycle, the HIV-1 virus integrates into the host

An Introduction to Molecular Medicine and Gene Therapy, Edited by Thomas F. Kresina
ISBN 0-471-39188-3 © 2001 Wiley-Liss

cell's genome and remains there permanently. Thus AIDS can be considered as an acquired genetic disorder. As previously discussed, gene therapy holds considerable potential for the treatment of hereditary and acquired genetic disorders. Human gene therapy can be defined as the introduction of new genetic material into the cells of an individual with the intention to produce a therapeutic benefit for the patient. Therefore, AIDS may be amenable to treatment by gene therapy approaches to inhibit the replication of HIV-1.

The ultimate goal of gene therapy is to inhibit HIV-1 viral replication and the resulting AIDS pathogenesis. For gene therapy of HIV infection to be successful, it will be necessary to introduce genes that are designed to specifically block or inhibit the gene expression or function of viral gene products such that the replication of HIV is blocked or limited. This concept was originally denoted as intracellular immunization and is currently being investigated as a therapeutic approach for a wide variety of infectious agents. In addition to intracellular interventions, gene therapy may be employed to intervene with the spread of HIV at the extracellular level. Inhibition of viral spread could be accomplished by sustained expression in vivo of a secreted inhibitory protein or by stimulation of an HIV-specific immune response.

GENETIC ORGANIZATION OF HIV

The HIV-1 virus is a member of the family of viruses denoted as retroviruses. The retrovirus classification encompasses a heterogeneous group of viruses containing a single-stranded, positive-sense ribonucleic acid (RNA) genome and the enzyme reverse transcriptase. Reverse transcriptase functions by copying the viral genomic RNA into double-stranded deoxyribonucleic and (DNA), which is a critical phase in the life cycle of retroviruses. Retroviruses have historically been subdivided into three groups primarily based on the pathologic outcome of infection. The oncovirus subgroup includes retroviruses that can cause tumor formation in the infected host; however, this group also includes several apparently benign viruses. Lentiviruses cause slowly progressing, chronic diseases that most often do not contain a tumor-forming component. The spumavirus subgroup, although causing marked foamy cytopathic effect in vitro, have not yet been clearly associated with any disease. Upon intense investigation into the pathology of HIV infection, it has become clear that the virus is a member of the lentivirus subgroup. Lentiviruses were initially isolated in the 1960s when it was found that certain slowly evolving, degenerative diseases in sheep were communicable. Interestingly, unlike the oncogenic retroviruses, the lentiviruses did not form tumors but were cytopathic (caused cells death). Several members of the lentivirus family have been isolated and described. Members of the lentivirus family include Visna virus, Simian immunodeficiency virus, human immunodeficiency virus 1 and 2, caprine arthritis-encephalitis virus, and equine infectious anemia virus.

As with all other retroviruses, HIV is an enveloped virus that contains two copies of single-stranded, positive-sense RNA (Fig. 11.1). The genomic organization of HIV is shown in Figure 11.2. At the ends of the genome are two identical genetic regions similar to those found in all retroviruses. The genetic elements are called long terminal repeats (*LTRs*). The *LTRs* contain elements that are responsible for the proper regulation of gene expression during virus replication such as promoters,

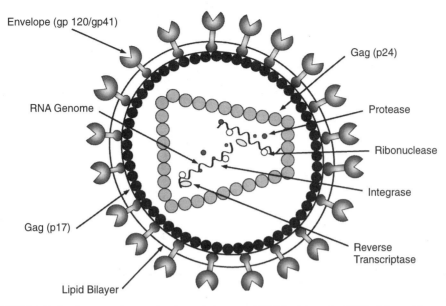

FIGURE 11.1   Structural organization of a mature HIV-1 virion. An HIV virion with structural and virion accessory proteins identified. HIV particles are approximately 110 nm in diameter. They are composed of a lipid bilayer membrane surrounding a conical nucleocapsid. Two copies of single-stranded positive sense RNA are contained with the nucleocapsid.

enhancers, and elements required for efficient messenger RNA (mRNA) polyadenylation. Between the LTRs are the genes that encode all of the viral proteins. The HIV genome encodes three sets of viral proteins; the structural proteins (Gag, Pol, and Env), the regulatory proteins (Tat, Rev, and Nef), and the maturation proteins (Vif, Vpu, and Vpr).

As shown in Figure 11.2, the structural proteins can be subdivided into three groups: core proteins, enzymes, and envelope proteins. These three groups of proteins are encoded by the *gag*, *pol*, and *env* genes, respectively. The *gag* gene refers to the group antigen and produces the viral core proteins that have antigens cross-reacting with other antigens within large retrovirus groups. The Gag proteins are all produced as a large single polyprotein that is then cleaved into individual proteins by a virus-encoded protease (p24, p18, and p15). The *pol* gene products are also encoded from a single open reading frame as a large polyprotein that is cleaved into the virus-associated enzymes—protease, reverse transcriptase (RT), ribonuclease, and integrase. The *env* gene products are surface glycoproteins that are produced as a polyprotein (gp160), however, they are cleaved by cellular enzymes to produce the two HIV surface glycoproteins (gp120 and gp41).

In addition to the structural elements necessary to assemble the virus particle, the virus genome codes for several nonstructural proteins that play vital roles in the regulation of the viral life cycle. The nonstructural proteins produced by the HIV can be divided into two classes, the regulatory proteins and the maturation proteins. The regulatory proteins include Tat, Rev, and Nef. The Tat protein was the first viral regulatory protein to be described. The Tat protein, which is encoded by the *tat* gene, is a strong transactivator of viral gene expression. In other words, the Tat protein

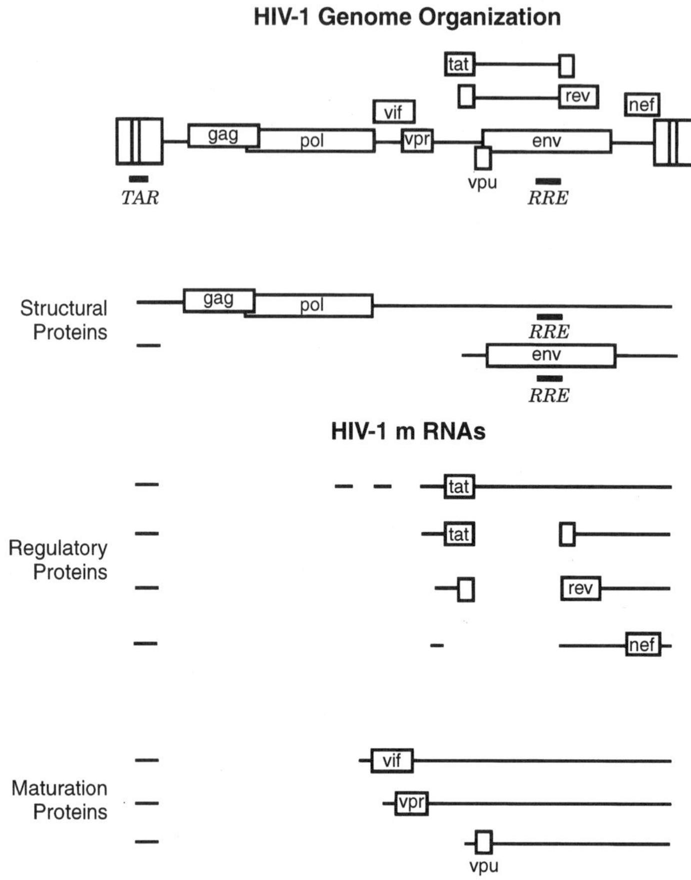

HIV-1 Genome Organization

FIGURE 11.2 Genomic organization and mRNA expression pattern of HIV-1. The diagram depicts the organization of the nine predominant genes of HIV-1. The diagram represents the major RNAs derived from the HIV-1 genome by alternative splicing of the HIV-1 genome. Three distinct classes of viral proteins are generated from these mRNAs: structural proteins, regulatory proteins, and maturation proteins. The structural proteins include the viral envelope protein (gp 120, gp 41) which is encoded by the *env* gene and the core proteins (p6, p9, p17, and p24) which are encoded by the *gag* gene. The *pol* gene generates the viral-associated reverse transcriptase, integrase, RNase H, and protease enzyme activities. The viral-associated regulatory proteins are encoded by the *tat*, *rev*, and *nef* genes, respectively. The Tat and Rev proteins are powerful regulatory proteins. The Tat protein interacts with the TAR (*tat*-responsive) element, which leads to a strong transactivation of viral gene expression, while the Rev protein interacts with the RRE (*rev* response element), which enhances the nuclear export of unspliced and single-spliced viral mRNA. The third class of viral proteins are the maturation proteins that are encoded by the *vif*, *vpr*, and *vpu* genes.

regulates the function of genes that are not immediately adjacent to its own gene. The Tat protein binds to the *trans*-activation response (TAR) element. The TAR element corresponds to an RNA stem-loop structure present within the untranslated leader sequence of all HIV-1 transcripts, including the RNA genome, and is required for HIV-1 Tat function. The interaction between Tat and TAR can lead to a potent transactivation (increasing expression of viral genes by 1000 times their level of expression in HIV-1 mutants lacking the *tat* gene) by inducing transcriptional initiation and/or elongation.

A second important regulatory protein is Rev, which produced by the *rev* gene. The Rev protein is produced early in the replication phase of HIV and interacts with a 234-nucleotide region of the *env* open reading frame in mRNA called the *Rev* response element (RRE). The interaction of the Rev protein with the RRE markedly enhances nuclear export of single-spliced and unspliced viral mRNAs from the nucleus; these RNAs encode the viral structural proteins. The production of Rev protein is an absolute requirement for the replication of the HIV virus, since mutants of the Rev protein are incapable of inducing synthesis of the viral structural proteins and are, thus, replication defective.

The last member of the regulatory protein family is the Nef protein. The role of the Nef protein in HIV-1 replication cycle remains unclear. However, the *nef* gene product is not required for HIV-1 replication in vitro or SIV in vivo. It is clear that the *nef* gene plays a role in the down-regulation of CD4 gene expression in infected cells. It is also hypothesized that *Nef* may be involved in the ability of HIV-1 to turn off its growth and reside dormant in the host cell genome.

In addition to the Gag, Pol, and Env, the late gene products encoded by HIV include the maturation proteins Vif, Vpu, and Vpr. Both the Vif (virion infectivity factor) and Vpu (viral protein U) proteins play roles in the maturation and production of infectious HIV virion particles. The Vpr (viral protein R) protein has recently been described as playing an integral role in causing the cell cycle arrest of HIV-infected cells. Expression of Vpr alone was sufficient to cause arrest of the cell cycle at the G_2/M transition phase of the cell cycle. Thus, HIV-infected cells are unable to progress normally from the G_2 phase of the cell cycle through mitosis to complete the cell cycle. The cell cycle arrest after infection by HIV causes the infected cell to remain in an activated state and, thus, may maximize virus production from the infected cell.

LIFE CYCLE AND PATHOGENESIS OF HIV-1 INFECTION

As shown in Figure 11.3, the initial stage of infection (the early phase) begins with the binding of the viral gp120 protein to its cell surface receptor, the CD4 protein. CD4 is present in high concentration on the surface of peripheral blood lymphocytes (PBL) and at lower concentrations on other cells that can be infected by HIV, including monocytes, macrophages, and dendritic cells. However, CD4 is not the sole mediator of HIV infection. Previous work in murine cell lines expressing human CD4 are not infected by HIV, which suggested the existence of a human specific cofactor. The HIV infection co-factor has recently been identified. This co-factor, termed fusin (CXCR4), is absolutely required, in addition to CD4, for the entry of HIV in to human cells. Fusin is an integral membrane glycoprotein and a member of the chemokine receptor family. Several of these co-factor proteins (CXCR4, CCR5, and CCR3) have now been identified on various cell types. The binding of the HIV gp120/gp41 envelope protein induces conformational changes that allow interaction with the co-receptor and subsequent fusion of the virus with the host cell plasma membrane. The HIV-1 nucleocapsid is internalized into the cytoplasm where the viral-genome is uncoated. The RNA genome is reverse transcribed into

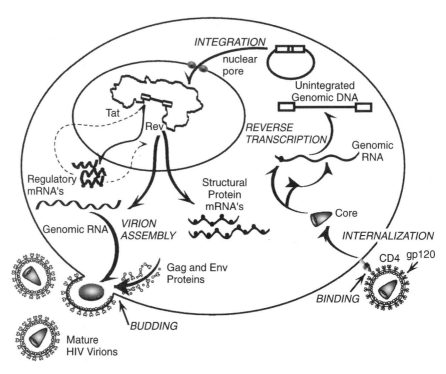

FIGURE 11.3 Life cycle and replication of HIV-1.

a single, negative strand of DNA, by the RT protein encoded by the *pol* sequences. The viral-encoded ribonuclease then degrades the viral genomic RNA. The RT enzyme then encodes the second (positive strand) of DNA, and this double-stranded viral genome is circularized and transported through the nuclear pore and into the nucleus of the infected cell. The newly synthesized viral DNA genome then randomly integrates into the host cell genome by the virally encoded integrase protein; this integrated form of the virus is denoted as the provirus. The provirus can replicate immediately or remain latent for extended periods of time and in so doing is passed along to all progeny cells derived from the original infected cell.

Although the mechanism of proviral activation is unclear, once the provirus is activated the intermediate stage of viral infection begins. Activation induces transcription of multiply spliced viral RNAs, which are utilized to produce the Tat and Rev proteins that act as powerful regulatory proteins during virus replication. As discussed previously, the Tat protein enhances the transcription elongation of viral RNA within the nucleus of the infected cell. Whereas, the Rev protein enhances nuclear export of single-spliced and unspliced viral mRNAs from the nucleus; these RNAs encode the viral structural proteins.

The late phase of HIV-1 infection begins upon the accumulation of significant amounts of structural proteins. The late phase consists of assembly of virus particles containing two copies of the viral RNA genome. The assembled particles are transported to the cell membrane where the mature virus particles bud off from the plasma membrane. In theory, the life cycle of the HIV-1 virus can be interrupted by blocking or inhibiting the function of one or more or the key viral proteins or their cis-acting regulatory elements.

HIV can kill an infected CD4$^+$ T lymphocyte in one of two ways. As progeny virus particles are budded off from the cell membrane, the external envelope protein gp120 reacts with CD4 molecules found on the surface of the infected cell to disrupt the integrity of the cell membrane in the areas with high concentrations of CD4. Disruption of the cell membrane kills the infected cell. Alternatively, an infected cell may interact with an uninfected cell through the HIV envelope proteins embedded in their cell surface membranes. The interaction is again through the CD4 molecules found on the surface of the uninfected cell. As the cell fusion occurs, hundreds of CD4 cells may eventually be involved in the formation of a large syncytium. All of the cells that fused into the syncytium die, and thus the cytopathic effects of HIV can extend beyond cells directly infected with the virus. It is predominantly through these two mechanisms that loss of CD4$^+$ lymphocytes occurs in HIV-infected patients. The outcome of HIV infection in monocyte–macrophage lineage cells is unclear. It appears as though the virus is capable of replication, but it does not appear to have any obvious cytopathic effects as in T lymphocytes. Similar to infected T cells, the formation of multinucleated syncytium of macrophage-like cells is observed in HIV-infected tissues. Macrophages that contain replicating virus may not be destroyed, but evidence suggests that they become dysfunctional.

GENETIC APPROACHES TO INHIBIT HIV REPLICATION

Approaches to gene therapy for HIV can be divided into three broad categories: (i) protein approaches such as transdominant negative proteins and single-chain antibodies, (ii) gene therapies based on nucleic acid moieties, including antisense DNA/RNA, RNA decoys, and catalytic RNA moieties (ribozymes), and (iii) immunotherapeutic approaches using genetic vaccines or pathogen-specific lymphocytes (Table 11.1). It is further possible that combinations of the aforementioned approaches may be used simultaneously to inhibit multiple stages of the viral life cycle or in combination with other approaches, such as hematopoietic stem cell transplantation or vaccination. The extent to which gene therapy approaches will be effective against HIV-1 is the direct result of several key factors: (i) selection of the appropriate target cell in which to deliver the gene therapy, (ii) the efficiency of the gene delivery system on the target cell, (iii) appropriate expression, regulation, and stability of the anti-HIV gene product(s), and (iv) the strength of the inhibition of viral replication by the therapeutic entity.

TRANSDOMINANT NEGATIVE PROTEINS

Transdominant negative proteins (TNPs) are mutant versions of regulatory or structural proteins that display a dominant negative phenotype that can inhibit replication of HIV. By definition, such mutants not only lack intrinsic wild-type activity but also inhibit the function of their cognate wild-type protein in trans. Inhibition may occur because the mutant competes for an essential substrate or co-factor that is available in limiting amounts, or, for proteins that form multimeric complexes, the

TABLE 11.1   Gene Therapy Strategies to Inhibit HIV Replication

Anti-HIV Strategy	Potential Mode of Action
Protein-Based Approaches	
Transdominant Negative Proteins	
Rev	Nuclear export of viral mRNA
Tat	Viral genome transcription/processing
Gag	Viral assembly
Env	Viral assembly
Endogenous Proteins	
Soluble CD4	Receptor binding/viral assembly
CD4-KDEL	Trapping of Env and Rev in ER
E1F-5A	
Intrabodies	
anti-gp120	Maturation and function of Env protein
Nucleic Acid Approaches	
Antisense RNA	
Antisense Tat/Rev	Translation of Tat and Rev proteins
Antisense Gag	Translation of Gag protein
Ribozymes	
5′ leader sequence	Translation of viral RNA
Multitarget	Translation of viral RNA
Antisnese oligonucleotides	
RNA decoys	Translation of viral RNA
TAR decoy	Viral genome transcription/processing
RRE decoy	Nuclear export of viral mRNA
Immunity Augmentation	
DNA Vaccines	Induction of cellular and humoral response
Env	Augments cytotoxic activity to HIV
Virus Specific CTL	

mutant may associate with wild-type monomers to form an inactive mixed multi-mer. A potential drawback in the use of transdominant viral proteins is their possible immunogenicity when expressed by the transduced cells. The protected cells may consequently induce an immune response that might result in their own destruction. This may diminish the efficacy of antiviral gene therapy using transdominant proteins. HIV-1 regulatory (Tat and Rev) and structural proteins (Env and Gag) are potential targets for the development of TNPs.

The most thoroughly investigated TNP is a mutant Rev protein denoted RevM10. The Rev protein is rendered a TNP through a series of mutations introduced into the *rev* gene (Fig. 11.4). The RevM10 still retains the ability to multimerize and bind to the RRE; but as a result of these mutations, the RevM10 protein can no longer efficiently interact with a cellular co-factor that activates the Rev function. Cell lines stably expressing RevM10 are protected from HIV-1 infection in long-term cell culture assays. Transduction of RevM10 into T-cell lines or primary PBL delays virus replication without any detectable negative effects on the cells. Recently, it has been demonstrated that RevM10 inhibits HIV-1 replication in chronically infected T cells. A different TNP Rev protein developed by Morgan et al. (1994) inhibited HIV-1

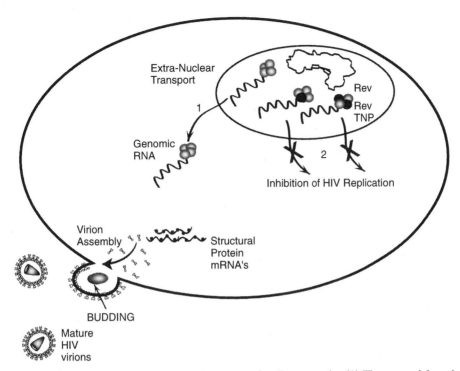

FIGURE 11.4 Activity of a transdominant negative Rev protein. (1) The normal function of the Rev protein is to form multimeric complexes (gray circles) which increase the efficiency of extranuclear transport of genomic viral RNA(s) and (2) the transdominant negative Rev protein (black circles) forms inactive mixed multimeric complexes with the wild-type Rev protein (gray circles). These inactive Rev complexes interfere with the normal functioning of the wild-type Rev complexes and inhibit the extra-nuclear transport of unspliced and singly spliced HIV RNA(s).

replication in T-cell lines and PBL when challenged with both laboratory and clinical HIV-1 isolates. A third type of Rev TNP was generated by deletion of the nucleolar localization signal sequence. This sequence functions as a signal to direct the Rev protein to the nucleolar region of the nucleus of an infected cell. This TNP Rev is retained in the cytoplasm and prevented the localization of wild-type Rev to the nucleus by forming inactive oligomers.

The HIV-1 regulatory protein Tat was also utilized to generate TNPs. A TNP Tat was mutated in its protein binding domain. Upon transduction into T-cell lines, the TNP Tat inhibited HIV-1 replication for up to 30 days. The mechanism through which this Tat TNP may function is by sequestration of a cellular factor involved in Tat-mediated transactivation. Interestingly, in this study a retroviral vector was developed that was capable of expressing both a Tat and Rev TNP. The multi-TNP vector was more effective at blocking HIV-1 replication than retroviral vectors expressing either TNP Tat or Rev alone. This study suggests that the inhibition of Tat and Rev simultaneously may be a more effective HIV-1 gene therapy. Recently, a double transdominant Tat/Rev fusion protein (Trev) was designed in an attempt to inhibit two essential HIV-1 activities simultaneously. Upon transfection or transduction of the *Trev* gene into T cells, they were protected from the cytopathic effects of HIV-1. Simultaneous inhibition of two HIV-1 functions may have potential advantages over single-function TNPs.

TNP moieties based on structural proteins have also been investigated for their anti-HIV-1 functions. The HIV-1 structural proteins (Gag and Env) oligomerize into multimeric complexes during viral assembly. Multimerization makes them ideal candidates for the generation of TNPs. Several Gag TNPs have been investigated and all are capable of inhibiting HIV-1 replication. The Gag TNPs function by disrupting distinct stages of the viral life cycle, such as viral assembly, viral budding, uncoating of the viral genome, or initiation of reverse transcription. Due to inherently low levels of transcription of *gag* genes in the absence of the HIV-1 Rev protein, the application of Gag TNPs has been limited. The low levels of mutant Gag expressed are insufficient to effectively block HIV-1 replication. Env TNPs have been generated as well but in initial testing showed only low levels of antiviral activity.

Single-Chain Antibodies (Intrabodies)

One of the more novel classes of antimicrobial gene therapies involves the development of intracellularly expressed single-chain antibodies (also called intrabodies). The single-chain variable fragment of an antibody is the smallest structural domain that retains the complete antigen specificity and binding site capabilities of the parental antibody. Single-chain antibodies are generated by cloning of the heavy- and light-chain genes from a hybridoma that expresses antibody to a specific protein target. These genes are used for the intracellular expression of the intrabody, which consists of an immunoglobulin heavy-chain leader sequence that targets the intrabody to the endoplasmic reticulum (ER), and rearranged heavy- and light-chain variable regions that are connected by a flexible interchain linker. Since the single-chain antibody cannot be secreted, it is efficiently retained within the ER, probably through its interaction with the ER-specific BiP protein. The BiP protein binds incompletely folded immunoglobulins and may facilitate the folding and/or oligomerization of these proteins. Intrabodies can directly bind to and prevent gene function or may sequester proteins in inappropriate cellular compartments so that the life cycle of HIV is disrupted.

Expression of an intrabody specific for the CD4 binding region of the HIV-1 gp120 (Env) markedly reduced the HIV-1 replication by trapping the gp160 in the ER and preventing its maturation by cleavage into the gp120/gp41 proteins (Fig. 11.5). Intrabodies developed to the Rev protein trapped Rev in a cytoplasmic compartment and blocked HIV-1 expression by inhibiting the export of HIV-1 RNAs from the nucleus. Additionally, intrabodies containing an SV40 nuclear localization signal sequence were developed to Tat. The anti-Tat single-chain antibody blocked Tat-mediated transactivation of the HIV-1 LTR and rendered T-cell lines resistant to HIV-1 infection.

Endogenous Cellular Proteins as Anti-HIV Agents

Proteins derived from cellular genes have been identified that exhibit specific gene inhibitory activity (Fig. 11.5). These activities may act by preventing the binding of HIV to cells, by binding directly to the regulatory/structural proteins, or indirectly

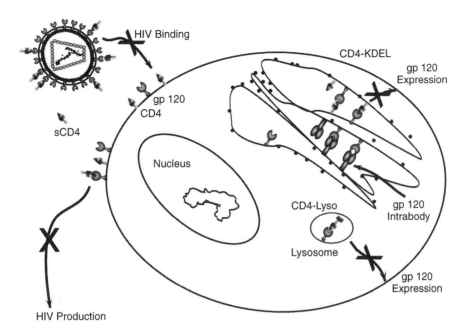

FIGURE 11.5 Cellular protein-based approaches for the inhibition of HIV-1 replication. The intracellular production of a soluble CD4 protein (sCD4) can prevent both the binding of infectious HIV particles and the production of new virus particles from an infected cells by saturating all of the available envelope protein. The attachment of a endoplasmic reticulum (ER) retention signal (KDEL) to the CD4 protein brings about the inhibition of virus replication by retaining the gp160 envelope complexes with the endoplasmic reticulum. The incorporation of a lysosomal targeting sequence into the CD4 protein leads to the inhibition of gp160 expression through targeted degradation in the lysosomes. The expression of single-chain antibodies (intrabodies) can also lead to the retention of viral proteins in the ER by specific interaction with the BiP protein.

by inducing or repressing cellular factors that in turn influence viral gene expression. One of the most successful in vitro uses of endogenous cellular proteins to inhibit an infectious agent is the use of a soluble version of the HIV receptor CD4 (sCD4). The T helper cell antigen CD4 functions as the receptor for the HIV through the physical interaction of the HIV envelope glycoprotein gp120 and the CD4 protein. Based on these results, investigators have demonstrated that sCD4 protein can effectively bind to and inhibit HIV infection in CD4+ cells. The effect of this strategy is to compete for binding of HIV to cellular CD4 with high concentrations of sCD4. In order for this strategy to be efficacious, a high level of continuous expression of sCD4 will be required. Retroviral vectors expressing sCD4 have been shown to protect T-cell lines from HIV infection in vitro. A significant limitation to this strategy is the ability to achieve sufficiently high levels of sCD4 to neutralize HIV effectively. The use of sCD4 for the gene therapy of HIV infection in a clinical setting has been disappointing. The intravenous infusion of recombinant sCD4 protein in HIV-infected patients failed to show efficacy in phase I clinical trials. In contrast to the laboratory strains of the HIV virus, clinical isolates have shown a significant increase resistance to the neutralizing characteristics of soluble CD4 protein.

A variation on the CD4-based anti-HIV gene therapy approach is the development of a mutated soluble CD4 molecule that contains a specific ER retention signal (Lys-Asp-Glu-Leu or KDEL). This hybrid molecule blocked secretion of gp120 and cell surface expression of gp120/41, when expressed intracellularly (Fig. 11.5). The CD4-KDEL/gp120 complex was retained within the endoplasmic reticulum, thus preventing maturation of infectious HIV-1 particles. It has also been demonstrated that the mutations that decrease the affinity of CD4 for gp120 discussed above have little effect on the ability of CD4-KDEL to retain gp120 in the ER. A similar approach is to specifically target soluble CD4/gp160 complexes to the lysosomes through the incorporation of lysosome targeting domains onto the soluble CD4. The CD4-lysosomal domain/gp160 complexes are degraded in the lysosomes of the cells and production of mature HIV particles is diminished (Fig. 11.5).

It has recently been demonstrated that the Rev protein interacts specifically with cellular factors in order to perform its normal function in the infected cell. The eukaryotic initiation factor 5A (eIF-5A) is a cellular transcription factor that interacts with Rev by binding to the Rev activation domain. The interaction between mutants of the eIF-5A and Rev can effectively inhibit HIV-1 replication in vitro. Utilization of the interactions between cellular factors and HIV could provide an additional approach for the development of HIV genetic therapies. Since these are endogenous cellular proteins, they are nonantigenic. Therefore, cells engineered with these cellular inhibitory genes may not be eliminated by the patient's immune system. This is an advantage as compared to the use of genes encoding potentially immunogenic, trans-dominant viral proteins for gene therapy.

A novel strategy exploiting the interaction of CD4/fusin with the gp120/gp41 protein of the HIV virus has been developed. As a consequence of HIV replication, infected cells express the gp120/gp41 envelope protein on their surface in order for the assembly of new virus particles to occur. The expression of gp120/gp41 on the cell surface lead investigators to hypothesize that a virus could be engineered to contain the HIV receptor and co-receptor in its envelope in place of endogenous viral envelope proteins. Generation of these hybrid virus particles would specifically target these hybrid virus particles to infected cells where replication of the virus would kill the cell. The vesicular stomatitis virus (VSV) has been engineered in this manner. In the VSV studies, deletion of the *VSV* gene and substitution with the genes for CD4 and CXCR4 lead to the formation of recombinant VSV particles that specifically infected HIV-infected cells. Upon infection, the replication of VSV was cytopathic in HIV-infected cells.

NUCLEIC-ACID-BASED GENE THERAPY APPROACHES

RNA Decoys

This approach disrupts the normal interaction of the HIV regulatory proteins with their cis-acting regulatory elements through the overexpression of short RNA molecules (decoys) that compete with viral RNA elements for binding of proteins that are required for virus replication (Fig. 11.6). The TAR (transactivation response)

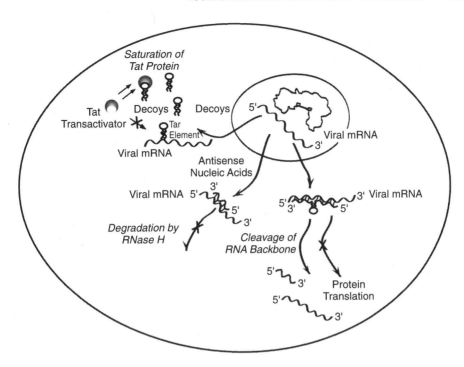

FIGURE 11.6 Nucleic-acid-based gene therapies for HIV-1. Anti-HIV genes can be expressed in the context of antisense nucleic acids, as antisense oligonucleotides, antisense RNA, or ribozymes. All of these antisense approaches promote the destruction of the target sequence by RNAses or block the translation of the mRNA. Ribozymes function by specific interaction with a target sequence within the RNA and functionally inactivate it by cleavage of the phosphodiester backbone. Overexpression of short RNA molecules (decoys) that correspond to the viral cis-acting elements Transactivation response element (TAR) or the Rev response element (RRE) inhibit the binding of the viral protein to the cognate sequences found on the viral mRNAs.

and RRE (Rev response element) are two such viral regulatory elements found in HIV and are the binding sites for the transactivating proteins Tat and Rev, respectively. The antiviral activity of the TAR element decoys was examined by retroviral-mediated gene transfer into T-cell lines in vitro. Overexpression of the TAR decoys inhibited the transcriptional activation mediated by the Tat protein that, in turn, markedly reduced HIV replication of laboratory HIV isolates. The TAR decoy inhibition of virus replication results from the decreased binding of the Tat protein to the endogenous TAR elements, which in turn inhibits the transcriptional activation necessary for efficient virus replication. Expression of a tandem repeat TAR decoy composed of as many as 50 TAR repeats has also been demonstrated to effectively inhibit HIV-1 replication in both T-cell lines and primary lymphocytes. Furthermore, the multimerized TAR decoy was shown to efficiently inhibit virus replication in lymphocytes from late-stage AIDS patients.

The enhanced expression of RRE decoys by retroviral vectors resulted in long-term inhibition of HIV replication in human T-cell lines. RRE decoys may function by preventing the binding of REV to the normal RRE sequences, which decreases the levels of singly spliced and unspliced HIV-1 mRNAs that are exported from the nucleus of an infected cell. It is clear that the overexpression of RRE decoys has strong antiviral activity, but there is some concern as to the long-term effects that the expression of the RNA decoys will have on the normal function of the cell. In addition to viral proteins, both TAR and RRE bind cellular co-factors. The overexpression of the decoys may have negative effects on cell viability or activity through the sequestration of proteins required for normal cell function. To limit the interaction between the RRE decoy and cellular proteins, a minimal RRE decoy composed of only 13 nucleotides that retained the rev binding domain was tested for antiviral activity. This minimal RRE decoy was shown to effectively suppress HIV-1 replication in vitro.

Antisense DNA and RNA

Antisense nucleic acid technology encompasses a broad spectrum of methods all directed toward the specific silencing of gene expression. The silencing of gene expression is achieved through the introduction into the cell or tissue of an antisense RNA or single-stranded DNA moiety (oligodeoxynucleotide), which is complementary to a target mRNA (Fig. 11.6). In theory, the antisense nucleic acids utilize Watson–Crick nucleic acid base pairing to block gene expression in a sequence-specific fashion. One of the most intensely investigated approaches for application of antisense RNA is the introduction of DNA oligonucleotides that have been chemically modified in an attempt to increase their stability (half-life) within a cell. A variety of synthetic antisense oligonucleotides have been designed that inhibit the replication of HIV-1. However, their use for the inhibition of HIV-1 has been extremely limited because uptake of free oligonucleotides from the extracellular environment in vivo is extremely inefficient, and effective oligonucleotide delivery systems have not yet been devised. Also, the oligonucleotide moieties that are internalized into the target cells are ultimately degraded by cellular enzymes such that any inhibitory activity on gene expression is only transient. An additional problem with the use of DNA oligonucleotides is that the gene inhibition that is observed is often nonspecific. In other words, the inhibition of expression is most often not the direct result of the interaction between the oligonucleotide and the target sequence, but the interaction with RNA in a broad nonspecific manner.

The other approach for antisense nucleic-acid-mediated inhibition of gene expression is the direct introduction or intracellular production of antisense RNA in cells or tissues of the organism. The direct introduction of RNA transcripts into cells can be accomplished through microinjection of an in vitro transcription product or as a chemically modified oligonucleotide. The direct administration of antisense RNA transcripts in vivo is not plausible for gene therapy due to the vast number of cells that need to receive the therapeutic RNA.

An alternative approach to the delivery of antisense RNA for gene therapy is the use of vector-based systems, which produce the antisense RNA within the cell or tissue of the organism. Most often recombinant viral vector systems, such as retroviruses, are used because they efficiently target large numbers of cells. The use of retrovirus vector-based systems for the intracellular production of antisense RNA has an additional advantage. That is, the vector integrates into the host cell genome and, thus, the antisense effects are more prolonged in comparison to oligonucleotides. Also, the use of regulatable or inducible promoters would permit the levels of inhibition to be tightly controlled.

Although the mechanism of antisense-mediated inhibition of gene expression is not completely understood, it is hypothesized that RNA duplexes (antisense RNA and target RNA) are degraded by RNase H or by blocking subsequent translation of the mRNA. The limitations of antisense RNA transcripts are similar to those observed with oligonucleotides. With higher levels of expression of antisense transcripts the gene inhibition observed is often nonspecific. There is another major limitation to the use of stable expression of antisense sequences as a therapy for HIV-1 infection. Long-term high levels of antisense expression are required in order to effectively inhibit viral replication. The mechanism through which antisense moieties inhibit gene expression requires that one antisense molecule efficiently bind to one target molecule. The stoichiometry of antisense sequences to target sequences must be at a minimum: 1 to 1 (antisense to target), but ratios of 5 to 1 or 10 to 1 lead to more effective inhibition of gene expression. Thus, the antisense gene expression must be quantatively higher than the levels of HIV-1 gene expression for an antisense gene therapy strategy to be effective. Standard gene therapy vectors containing pol II promoters often do not produce sufficient levels of antisense sequence to inhibit viral replication. To subvert this problem, vectors containing alternative promoter systems have been developed. A retroviral vector containing a pol III promoter has been demonstrated to significantly increase the levels of expression. The transcription of transfer RNAs (tRNAs) and small nuclear RNAs (snRNAs) found in eukaryotic cells are controlled by pol III promoters. Pol III promoters are multipartite (composed of two distinct parts). Interestingly, they are found internal to the transcriptional start site. This means RNA polymerase III reaches backwards to initiate transcription of a pol III gene. A pol-III-based retroviral vector expressing an antisense to the TAR sequence has successfully inhibited HIV-1 replication in vitro. Alternatively, coordinating the expression of antisense RNA with HIV-1 infection would permit the efficient expression of antisense sequences within cells following infection. This could be accomplished by the development of a retroviral vector in which the HIV-1 LTR is used as a promoter. This vector permits the efficient expression of antisense sequences within cells following infection of a lymphocyte by HIV-1.

A number of antisense transcripts have been designed to target various regions of the HIV-1 genome. Stable intracellular expression of antisense HIV-1 transcripts is currently the most efficient method by which antisense technology can be utilized to inhibit HIV-1 gene expression. A number of studies have shown only limited antiviral activity using antisense transcripts to the viral genes *tat*, *rev*, *vpu*, and *gag*. An in-depth analysis of potential HIV-1 antisense gene sequences was performed in which various antisense RNAs targeted to 10 different regions of the HIV-1 genome were compared for their antiviral effects. The antisense gene sequences with the greatest antiviral activity were those that targeted a 1-kb region within the *gag* gene and a sequence specific for a 562-bp genomic fragment encompassing the *tat* and *rev* genes. Further analysis of the antiviral effects of the antisense *tat/rev* gene fragment has demonstrated a strong inhibition of HIV-1 replication in a T-cell line and primary CD4+ PBL, but loss of the protective effects was observed as the number of infectious HIV particles used to infect the protected cell population was increased.

Ribozymes (Catalytic Antisense RNA)

Ribozymes are antisense RNA molecules that have catalytic activity. They function by binding to the target RNA moiety through antisense sequence-specific hybridization. Inactivation occurs by cleavage of the phosphodiester backbone at a specific site (Fig. 11.6). The two most thoroughly studied classes of ribozymes are the hammerhead and hairpin ribozymes (the names are derived from their theoretical secondary structures). Hammerhead ribozymes cleave RNA at the nucleotide sequence U-H (H = A, C, or U) by hydrolysis of a 3′–5′ phosphodiester bond. Hairpin ribozymes utilize the nucleotide sequence C-U-G as their cleavage site. A distinct advantage of ribozymes over traditional antisense RNA is that they are not consumed during the target cleavage reaction. Therefore, a single ribozyme can inactivate a large number of target molecules. Additionally, ribozymes can be generated from very small transcriptional units. Thus, multiple ribozymes targeting different genomic regions could be incorporated into the same vector. Due to their unique catalytic properties, ribozymes have the potential to be highly efficient inhibitors of gene expression, even at low concentrations. Ribozymes also have greater sequence specificity than antisense RNA because the target must have the correct target sequence to allow binding. In addition, the cleavage site must be present in the right position within the antisense fragment. However, the functionality and the extent of catalytic activity that ribozymes actually have for their RNA targets in vivo is presently unclear. However, a potent limitation to the use of ribozymes for HIV-1 gene therapy is that they are inherently limited in effectiveness due to the high rate of mutation associated with HIV-1 replication. Any alteration of the binding or cleavage sites within the target sequence required by the ribozyme for activity could render the ribozyme totally inactive.

The first investigation into ribozymes designed to inhibit HIV-1 was performed by transfecting a hammerhead ribozyme targeted to the viral *gag* sequence into human fibroblasts that express CD4 antigen. Upon challenge with HIV-1, the cells were demonstrated to express reduced levels of full-length *gag* RNA molecules and markedly reduced levels of the *gag*-derived protein p24. Ribozymes developed to target the 5′ leader sequence of HIV-1 were shown to significantly inhibit HIV-1 replication in T-cell lines and PBL. These ribozymes inactivate incoming viral RNAs prior to integration into the genome. Thus, ribozymes targeted to the 5′ leader sequence of HIV prevent the establishment of infection. The ability to prevent infection by HIV-1 in the long term may allow uninfected cells to become permanently resistant to HIV-1 infection. Interestingly, this ribozyme may have the potential to globally inhibit viral gene expression because the leader sequence is contained within all of the HIV-1-derived RNAs. Multitarget ribozymes have also been developed in which a single ribozyme cleaves at multiple highly conserved targets within the HIV-1 genome. A multitarget ribozyme to conserved regions of the *env* sequences has been shown to effectively inhibit replication of several HIV-1 isolates.

Ribozyme transcription units are small enough that several ribozymes could be incorporated into a single vector, thus ribozymes targeted to several regions of the HIV-1 genome can be delivered within the same cell. Improved ribozyme-mediated inhibition of HIV replication may be achievable by the development of ribozymes that co-localize with their HIV-1 target RNA to the same subcellular compartment.

As a test of this strategy, a ribozyme transcript that contained the retroviral packaging signal was demonstrated to efficiently inactivate newly synthesized MoMLV genomic RNA prior to particle assembly resulting in a marked decrease in the release of mature particles.

GENETIC APPROACHES TO ENHANCE IMMUNITY IN AIDS STIMULATION OF AN HIV-SPECIFIC IMMUNE RESPONSE

DNA Vaccines

A novel nucleic-acid-based approach for gene therapy is to attempt to elicit an immune response to native proteins of the HIV synthesized by the transfer of plasmid DNA into cells; in other words, DNA-based vaccinations. The preliminary observations leading to the development of genetic vaccination were made based on the determination that plasmid DNA encoding marker genes could be expressed following intramuscular injection in mice. Although the levels of gene transfer were low, it was determined that the internalized plasmid persisted and was expressed for the life span of the animal. The generation of an immune response to marker proteins encoded by plasmids was demonstrated by two groups using plasmid DNA introduced into the skin of mice by a biolistic gene delivery system. The development of a protective immune response by immunization with a genetic vaccine was initially demonstrated in mice that underwent intramuscular injection of naked plasmid DNA encoding the internal nucleoprotein of the influenza virus. The potential efficacy of DNA vaccination into postmitotic muscle cells has since been demonstrated in a variety of murine and animal models infected with bacterial, viral, or parasitic pathogens. The rationale behind these gene vaccines is to generate both a specific, cytotoxic T-cell response as well as a humoral response.

There are a number of theoretical advantages of the DNA-based vaccination technology over traditional vaccine strategies. These include (i) the ease of production and preparation of plasmid DNA, (ii) the expression of antigens in their native form, which leads to the efficient generation of both cytotoxic and helper T cells, (iii) the long-term immunity elicited suggests the potential to reduce the number of doses of vaccine required to generate a protective immune response, and (iv) the cells need not be the target cells that are normally infected by the infectious agent.

The potential disadvantages of DNA vaccination include (i) accidental introduction of the plasmid DNA into other than intended cell types, (ii) generation of anti-DNA antibodies to the plasmid used for the vaccination, and (iii) random integration of the injected DNA into the target cells, which may activate an oncogene or inactivate a tumor suppressor gene by insertional mutagenesis.

This approach is being actively investigated as a technique to optimize HIV-1 vaccination strategies. Introduction of the HIV-1 *env* gene via injection of naked plasmid DNA into cells led to the formation of highly specific humoral and cellular immune responses in mice. In addition, expression plasmids encoding either the HIV-1 envelope glycoprotein or a defective noninfectious HIV particle were shown to produce transient antibody to Env and expression of the defective HIV genome raised persistent cytotoxic T-cell activity to the HIV Gag p24 protein.

Recently, results from a study using a DNA-based anti-HIV vaccine approach in chimpanzees indicated that although the DNA injections yielded a variable immune response, the vaccinated animals remained infection-free following HIV challenge. Taken together, these studies elucidate the potential for the formation of strong HIV-directed immune responses. However, the ability of such an immune response to persist and protect against polymorphic HIV-1 strains remains to be demonstrated.

HIV-Specific Cytotoxic T-Lymphocytes

The use of the infected individual's own cells (CD4$^+$ and CD8$^+$ lymphocytes, CD34$^+$ hematopoietic stem cells, or antigen presenting cells such as macrophages) for the passive restoration of the immune system function is another technique in the development of genetic therapies for HIV infection. Direct immunotherapy approaches involve the ex vivo expansion of selected T-cell populations, either CD4 or CD8 lymphocytes, followed by reinfusion of the expanded lymphocyte population into the HIV-1-infected individual. The major area of focus for adoptive cell therapy for HIV-1 infection has been the use of CD8 cells. Although the importance of MHC class 1 restricted CD8 CTL in controlling HIV-1 infection is not understood, it is clear that early in the infection the increase in HIV-specific CD8 cells is correlative with the resolution of viremia. This data strongly suggests that MHC class 1 restricted CD8 cells play a role in limiting infection during the acute phase (early stages) of infection. The development of MHC class 1 restricted CD8 specific for several HIV proteins, including Env, Gag, Pol, Vif, and Nef, has been demonstrated in HIV-1-infected individuals. HIV-specific cytotoxic T lymphocytes (Tc) are generated by the ex vivo expansion of pools of CD8$^+$ T cells in the presence of HIV-1 antigens (gag peptides, env peptides, etc.). Individual clones of antigen-specific CD8 cells are isolated, expanded, and used for autologous reinfusion into the HIV-1-infected individual. The data to support the use of HIV-1 specific individual Tc clones to limit HIV-1 infection is based on observations that CD8 T cells can inhibit replication of HIV-1 in human PBL *in vitro*.

PRACTICAL ASPECTS OF GENE THERAPY FOR HIV

Cellular Targets for Gene Therapy

The dominant reservoir of HIV-1 infection and replication are the cells of lymphoid (lymphopcytes) and myeloid (macrophages, monocytes) origin (see also Chapter 6). In order for HIV-1 gene therapy to be effective, it is vital that cells derived from these lineages be utilized as recipients for anti-HIV-1 gene therapies.

The pluripotent hematopoietic stem cells (HSCs) generate all cells of lymphoid and myeloid origin. Therefore, these cells are the ultimate candidates for use in gene therapy. In theory, permanent protection from HIV-1 infection could be achieved through the introduction of anti-HIV-1 genes into HSCs because these cells are self-regenerating and therefore, will indefinitely produce a population of HIV-resistant cells. However, it is not currently possible to isolate pure HSC populations, but several enrichment techniques based on selection for CD34$^+$ cells have been developed. CD34$^+$-enriched cells can be isolated directly from the bone marrow, from mobilized peripheral blood cells, or from umbilical cord blood. CD34$^+$ cells isolated from all of these various sources have successfully been used for in vitro analysis of HIV-1 gene therapies. Unfortunately, the levels of gene transfer obtained after these cells are reintroduced in vivo is very low (1 to 5%). It has been determined that CD34$^+$ cells that initially express the introduced gene may silence gene expression over time. It is believed that the silencing of gene expression may be the result of methylation of the gene therapy construct. Thus, improvements in the gene transfer technology and enhanced gene expression may eventually make HSCs viable candidates for use in gene therapy.

Due to the inefficiency of gene transfer into HSCs, investigators have turned to the predominant host cell of HIV-1, the mature CD4$^+$ T cell, as an alternative target for gene therapy. The CD4$^+$ lymphocytes are desirable due to their ease of isolation from the peripheral blood and ease of enrichment by depletion of CD8$^+$ cells. Using CD4$^+$ cells, high levels of transduction can be achieved, and the cells can be selected for during expansion in tissue culture prior to reinfusion into a patient. The questions surrounding the use of CD4$^+$ lymphocytes for gene therapy deal with the in vivo growth potential and length of life span of the cells. Preliminary investigations using nonhuman primates revealed a small number of transduced autologous T cells that could be recovered from the peripheral blood of rhesus monkeys 2 years after a single injection of gene-marked cells. Human studies using gene-marked tumor infiltrating lymphocytes (TIL) demonstrated that reinfused transduced TIL cells survived several weeks in vivo. Recent primate studies in which autologous CD4$^+$ lymphocytes transduced with an anti-HIV-1 retroviral vector indicate that a low level of transduced lymphocytes survive for several months in the peripheral blood and lymph nodes of these animals. A gene-marking clinical protocol involving identical twins suggests that gene-marked lymphocytes survive at low levels for up to 14 weeks in the HIV-1-infected individuals. Results from a adenosine deaminase (ADA-SCID) clinical human gene therapy trial indicate that the infused lymphocytes survive and divide for up to 4 years postinfusion. The transduced lymphocytes used in the ADA clinical trial may have an increased life span because it is hypothesized that the transduced cells have a survival advantage over nontransduced lymphocytes due the presence of the ADA gene product. Taken together, there is mounting evidence to indicate that long-term persistence of mature lymphocytes can be achieved in some experimental settings.

Gene Transfer Systems

In order for gene therapy to be effective, it is necessary to efficiently deliver genes into the target cells. The most efficient gene delivery systems are based on recombinant viral vector systems. A variety of gene delivery systems, including viral vectors based on retrovirus, herpes virus, adenovirus, and adenoassociated virus, have been developed (see Chapter 4). Several nonviral-based gene transfer techniques are currently being evaluated for their effectiveness for gene delivery. The viral-derived gene transfer systems have been extensively tested on both HSCs and PBL. Only the retroviral vectors effectively deliver genes into these target cells, albeit at low levels in HSCs. Thus, retroviral-mediated gene transfer is currently the optimal gene transfer system available for use in HIV-1 gene therapy. The limitations of retroviral vectors include the inability to infect nondividing cells due to the requirement for DNA replication in order to efficiently integrate into the host cell genome.

Nonviral-mediated gene transfer systems include liposomes, molecular conjugates, receptor ligands, direct injection of naked DNA, and particle-mediated gene transfer (see Chapter 5). These gene transfer methods are effective in situations where transient expression of the gene product is desired. In general, most of these systems are not effective on lympho-hematopoietic cells. To this point, investigators have used the particle bombardment technology to deliver the RevM10 transdominant negative protein to human CD4$^+$ lymphocytes. The initial levels of gene transfer ranged from between 0.1 and 10%, which are below the levels of gene transfer that are achievable with retroviral vectors. Also, the gene expression resulting from particle bombardment was transient and any therapeutic benefit was only transient. This experiment demonstrates the potential for the use of particle-mediated gene transfer in HIV-1 gene therapy, but the aforementioned technological problems must be overcome before the technology is widely used as a gene transfer system.

ANTI-HIV GENE THERAPY CLINICAL TRIALS

Eighteen different anti-HIV gene therapy protocols have been reviewed and approved by the Recombinant DNA Advisory Committee (RAC) (as of November 1997). The clinical protocols can be divided into three categories: (i) gene marking studies; (ii) immunotherapy; these gene therapy strategies are aimed at stimulating an anti-HIV-1 immune response; and (iii) inhibition of virus replication; these anti-HIV-1 strategies are aimed at the intracellular inhibition of virus replication (Table 11.2). All of the proposed clinical trials use the technologies discussed earlier, such as transdominant negative proteins, ribozymes, virus-specific cytotoxic T cells, antisense nucleic acids, or single-chain antibodies. Examples of a few of the current HIV-1 gene therapy trials are discussed below.

1. *Marking of Sygeneic T cells* A gene marker study on the safety and survival of the adoptive transfer of genetically marked syngeneic lymphocytes in HIV-discordant identical twins (one HIV-infected twin and an uninfected twin) has been initiated. The objective of this phase I/II pilot project is to evaluate the distribution and survival, tolerance, safety, and efficacy of infusions of activated, gene-marked syngeneic T lymphocytes obtained from HIV-seronegative identical twins on the functional immune status of HIV-infected twin recipients. This protocol represents the initial step in a sequence of studies designed to evaluate the potential value of genetically modified T lymphocytes (CD4$^+$ and CD8$^+$) in an attempt to prevent or control HIV infection. This study will provide the initial baseline data needed to prospectively evaluate the fate of activated CD4$^+$ and CD8$^+$ cells after reinfusion in HIV-infected individuals.

2. *Marking of Cytotoxic T cells* A second gene marking study involves adoptive immunotherapy using genetically modified HIV-specific CD8$^+$ T cells for an HIV-seropositive patient. The objectives of this trial are (i) to evaluate the safety and toxicity of administering increasing doses of autologous CD8$^+$ class I

TABLE 11.2   Clinical Trials for HIV-1

Protocol Description	Status	Investigator	Institute
Gene Marking Protocols			
Safety of adoptive transfer of syngeneic gene-modified lymphocytes in HIV-1-infected identical twins	Open	Walker	National Institutes of Health
Safety of adoptive transfer of syngeneic gene-modified cytotoxic T cells in HIV-1 infected identical twins (phase I/II)	Open	Walker	National Institutes of Health
Transduction of CD34$^+$ cells from the bone marrow of HIV-1-infected children: comparative marking by an RRE decoy (phase I)	Open	Kohn	Children's Hospital, Los Angeles
Transduction of CD34$^+$ autologous peripheral blood progentior cells from HIV-1-infected persons: a study of comparative marking using a ribozyme gene and neutral gene (phase I)	Open	Kohn	Childrens Hospital, Los Angeles
Immunotherapy Protocols			
Safety of cellular adoptive immunotherapy using gentically modified CD8$^+$ HIV-specific T cells	Open	Greenberg	Fred Hutchinson Cancer Center
Safety and biologic effects of murine retroviral vectors encoding HIV-1 IT(V) in asymptomatic individuals (phase I)	Closed	Viagene, Inc.	
Safety and biologic activity of HIV-1 IT(V) in HIV-1-infected individuals (phase I/II)	Closed	University of California, San Diego	
Repeat dose safety and efficacy study of HIV-IT(V) in HIV-1-infected individuals with ≥100 CD4$^+$ cells (phase II)	Open	Haubrich	Multiinstitute
Double-blinded study to evaluate the safety and optimal CTL inducing dose of HIV-IT(V) in HIV-infected subjects (phase I/II)	Closed	Merritt	VIRx, Inc. Viagene, Inc.

TABLE 11.2   (*Continued*)

Protocol Description	Status	Investigator	Institute
Safety of cellular adoptive immunotherapy using autologous unmodified and genetically modified CD8$^+$ HIV-specific T cells in seropositive individuals (phase I)	Open	Riddell	Fred Hutchinson Cancer Center
Inhibition of Replication Protocols			
Effects of a transdominant negative from of Rev (phase I)	Open	Nabel	University of Michigan
Safety and effects of a ribozyme that cleaves HIV-1 in HIV-1 RNA in infected humans (phase I)	Open	Wong-Staal	University of California, San Diego
Retroviral mediated gene transfer to deliver HIV-1 antisense TAR and transdominant Rev protein gene to syngeneic lymphocytes in HIV-1-infected identical twins (phase I)	Open	Morgan	National Institutes of Health
Intracellular antibodies against HIV-1 envelope protein for AIDS (phase I)	Open	Marasco	Dana Farber Cancer Institute
Autologous CD3$^+$ hematopoietic progenitor cells transduced with an anti-HIV ribozyme (phase I)	Open	Rosenblatt	University of California, Los Angeles
Randomized, controlled, study of the activity and safety of autologous CD4-Zeta gene-modified T cells in HIV-infected patients (phase II)	Open	Connick	Multiinsitute
Safety and in vivo persistence of adoptively transferred autologous CD4$^+$ T cells genetically modified to resist HIV replication (phase I)	Open	Gilbert	Fred Huthchinson Cancer Center
Intracellular immunization against HIV-1 infection using an anti-Rev single-chain variable fragment (Sfv) (phase I)	Open	Pomerantz	Thomas Jefferson University

MHC-restricted HIV-specific cytotoxic T-cell clones transduced by retroviral-mediated gene transfer to express a marker/suicide gene, (ii) to determine the survival of adoptively transfered HIV-specific T-cell clones, and (iii) to evaluate markers of HIV disease activity in these recipients.

The importance of MHC class-I-restricted CD8$^+$ CTL in controlling infection has not been as well documented for HIV as for other viruses for which small animal models exist but is supported by correlative data from HIV-1-infected patients. Prior to developing AIDS, HIV-seropositive patients commonly have MHC class-I-restricted CD8$^+$ CTL detectable in high frequency in peripheral blood, specific for numerous HIV proteins, including Env, Gag, Pol, Vif, and Nef. The rationale for using HIV-specific T-cell clones to limit HIV-1 infection is based on the observations that CD8$^+$ T cells inhibit replication of HIV in human lymphocytes in vitro. In addition, adoptive immunotherapy using in vitro expanded CMV-specific clones has proven effective for reconstituting CMV-specific T-cell responses following BMT. Hence, adoptive immunotherapy with in vitro expanded HIV-specific CD8$^+$ CTL may have a beneficial antiviral effect.

3. *Trans-Dominant Rev*   Based on the encouraging preclinical data (discussed previously in this chapter) obtained with the Rev M10 transdominant mutant, a clinical protocol was proposed, whereby CD4$^+$ T lymphocytes from an HIV-1-infected individual will be engineered with Rev M10 expression vectors. In this study, the efficacy of intracellular inhibition of HIV-1 infection by the M10 trans-dominant mutant Rev protein will be evaluated. The aim of this proposal is to determine whether expression of M10 can prolong the survival of PBL in AIDS patients, thus conferring protection against HIV-1 infection. CD4$^+$ T lymphocytes will be genetically modified in patients using either particle-mediated gene transfer or retroviral-mediated gene transfer. In each case, a control vector identical to the Rev M10 but with a frameshift that inactivates gene expression will be used to transduce a parallel population of CD4$^+$ cells. Retroviral transductions and particle-mediated transfections will be performed after stimulation of CD4$^+$-enriched cells with IL-2 and either anti-CD3 or anti-CD28 antibodies. Activation of endogenous HIV-1 is inhibited by addition of reverse transcriptase inhibitors plus an HIV-specific toxin gene (CD4-PE40). The engineered and expanded cells will be returned to the patient, and the survival of the cells in each group compared by limiting dilution PCR. The effect of Rev M10 on HIV-1 status and immunological parameters will also be evaluated.

4. *Trans-dominant Rev in Combination with Antisense TAR*   To specifically inhibit the function of Rev, a novel trans-dominant Rev mutant (RevTD) based on the previously described Rev M10 mutant was genereated. It was shown that the presence of just one point mutation in the activator domain was sufficient to confer a dominant negative phenotype. To inhibit Tat function, an antisense strategy was developed and targeted at the HIV-1 transactivation response (TAR) element. A retroviral vector was constructed that expresses a chimeric tRNA$_i$-Met-antisense TAR (pol III promoter-antisense TAR) fusion transcript complementary to the HIV-1 TAR region. Using transient and stable transfection assays, it was shown that antisense TAR inhibited Tat-mediated transactivation of HIV-1 LTR-containing expression vectors. The exact mechanism involved in this inhibition is not fully understood but may involve inhibition of Tat binding on the TAR element or RNase degradation of the RNA duplex between the antisense TAR and its complementary target sequence. This RNA duplex may also inhibit ribosome binding and consequently inhibit translation.

An additional clinical protocol for AIDS gene therapy uses retroviral-mediated gene transfer to deliver antisense TAR and RevTD genes to syngeneic lymphocytes in identical twins discordant for HIV-1 infection. This phase I pilot study is based on the preclinical data obtained with the antisense TAR and RevTD retroviral vectors and on the adoptive transfer of neomycin-marked syngeneic CD4$^+$ T cell in HIV-1 discordant identical twins described above. In this clinical trial, the safety, survival, and potential efficacy of the adoptive transfer of genetically engineered syngeneic lymphocytes obtained from HIV-seronegative identical twins on the functional immune status of HIV-infected twin recipients will be evaluated.

5. *Anti-HIV Ribozyme*   A clinical protocol for AIDS gene therapy using the HIV-1 leader-specific hairpin ribozyme has been proposed. In this phase I clinical trial, the safety and efficacy of ribozyme gene therapy will be evaluated in HIV-1-infected patients by reinfusing autologous CD4$^+$ T cells that have been transduced ex vivo with a retroviral vector that expresses a ribozyme to the HIV-1 leader sequence. Transduction of HIV-1-infected cells in vitro will require culture conditions that inhibit the spread of endogenous HIV-1, which can be accomplished through the addition of the anti-HIV agents nevirapine and CD4-PE40. The in vivo kinetics and survival of ribozyme-transduced cells will be compared by limiting dilution PCR with those of a separate aliquot of cells transduced with a control vector that is identical except for the ribozyme cassette. The level and persistence of ribozyme expression will also be assessed. The results will determine whether this ribozyme can protect CD4$^+$ T cells in patients with HIV-1 infection and will aid the design future trials of hematopoietic stem cell gene therapy for AIDS.

6. *Gene Vaccines*   Two related clinical protocols have been approved by the RAC/Food and Drug Administration (FDA) to test the safety and potential efficacy of genetic vaccination in HIV-1-infected individuals. In one protocol, HIV-1-infected patients will have their fibroblasts removed for ex vivo transduction with a potentially immunotherapeutic MoMLV-based retroviral vector encoding the HIV-1 Env/Rev proteins (designated as HIV-IT). In the other protocol, HIV-IT will be injected intramuscularly into the HIV-1-infected patient to achieve in situ transductions. The ex vivo genetic vaccination phase I clinical protocol involves three successive doses (and a booster set of three successive doses) of HIV-IT-transduced autologous fibroblasts. These fibroblasts will be obtained from a skin biopsy and subsequently transduced with the HIV-IT vector, selected, irradiated, quality control tested, and returned to the donor. The direct in vivo injection protocol is a phase I placebo-controlled clinical trial involving the administration of the HIV-IT vector or a diluent control to HIV-infected, seropositive, asymptomatic individuals not currently receiving antiretroviral treatment. Direct vector treatment consists of a series of three monthly intramuscular injections (using a two-tier dosing schedule). Treated individuals will be evaluated for acute toxicity and for normal clinical parameters, CD4 levels, HIV-specific T-cell responses and viral load prior to, during, and following treatment. Preliminary clinical data suggest that HIV-infected patients treated with vector-transduced autologous fibroblasts show augmented HIV-1 IIIB Env specific CD8$^+$ CTL responses. It is hoped that the retroviral vector-mediated immunization will result in a balanced in vivo immune attack by HIV-specific CTL and antibody responses that may eliminate HIV-infected cells and clear cell-free virus from an HIV-1-infected individual.

7. *Intracellular Antibodies*   As described previously, intracellular antibodies can be constructed that target a variety of HIV proteins. A protocol proposes to use the antienvelope (gp120) intracellular antibody sFv105 in an anti-HIV gene therapy trial. The choice of the HIV envelope as a target for attack is supported by the potential detrimental role of gp160 in syncytium formation, single-cell killing, and potential virus-independent cytopathology. This study plan is to enroll six patients who will undergo lymphopheresis from which CD4$^+$ enriched PBMC will be obtained. Again, as in the other protocol using HIV-infected cells, the anti-HIV drugs nevarapine and CD4-PE40 will be used to inhibit in vitro HIV expansion.

Two identical aliquots of lymphocytes will then be transduced with either the sFv105 expressing retroviral vector or with a control *Neo* gene-containing vector (LN). Following transduction, it is proposed to enrich for gene-engineered cells by selection for the *Neo* gene by growth in G418-containing medium. Large numbers of transduced and culture expanded cells are proposed to be returned to the patient. Patients will subsequently be monitored by limiting dilution PCR to quantitative transduced cells in the circulation, to evaluate in vivo expression of the sFv105 transgene in transduced lymphocytes, and to make preliminary observations on the effects of gene therapy on HIV viral burden and CD4$^+$ lymphocyte levels.

CONCLUSIONS

A large variety of anti-HIV-1 gene therapy strategies have been developed that effectively inhibit HIV-1 in vitro. Significant progress has recently been made in demonstrating that primary CD4$^+$ T lymphocytes can be protected from infection with HIV-1, including primary patient isolates, using gene therapy approaches based on transdominant mutant HIV-1 proteins, antisense, and ribozymes. The recent data obtained by vector-mediated immunization are also encouraging since long-term persistence of CTL and cross-protection against heterologous polymorphic HIV-1 strains has been demonstrated in animal models. Based on these preclinical findings, several anti-HIV gene therapy strategies have received RAC/FDA approval for testing in HIV-1-infected individuals. It is hoped that these clinical trials will be able to address the question of whether rendering a cell resistant to HIV-1 infection by gene therapy will have a therapeutic benefit to the patient.

KEY CONCEPTS

- Intracellular immunization refers to the introduction of genes that are designed to specifically block or inhibit the gene expression or function of viral gene products such that the replication of HIV is blocked or limited.
- A number of gene therapy strategies have been developed for the inhibition of HIV replication. These approaches include (i) protein approaches such as transdominant negative proteins and single-chain antibodies and (ii) gene therapies based on nucleic acid moieties, including antisense DNA/RNA, RNA decoys, and ribozymes.
- Several immunotherapeutic gene therapy strategies directed at restoration or supplementation of the immune response to HIV-1 have also been developed and include (i) the development of DNA vaccines and (ii) passive restoration of immune function using pathogen-specific cytotoxic T lymphocytes.

- Many of these anti-HIV gene therapy strategies have been translated into human gene therapy clinical trials. The clinical protocols can be divided into three categories: (i) gene marking studies, (ii) immunotherapy, and (iii) inhibition of virus replication.
- Due to the complexity of the pathogenesis of HIV, the extent to which gene therapy approaches will be effective against HIV-1 is the direct result of several key factors: (i) selection of the appropriate target cell, (ii) the efficiency of the gene delivery system, (iii) appropriate expression, regulation, and stability of the anti-HIV gene product(s), and (iv) the strength of the inhibition of viral replication by the therapeutic entity.

SUGGESTED READINGS

HIV Infection and Gene Therapy

Bridges SH, Sarver N. Gene therapy and immune restoration for HIV disease. Lancet 345:427–432, 1995.

Levy JA. Pathogenesis of human immunodeficiency virus infection. Microbiol Rev 57:183–289, 1993.

Morgan RA. Genetic strategies to inhibit HIV. Mol Med Today 5:454–458, 1999.

Schnell MJ, Johnson JE, Buonocore L, Rose JK. Construction of a novel virus that targets HIV-1-infected cells and controls HIV-1 infection. Cell 90:849–857, 1997.

Singwi S, Ramezani A, Ding SF, Joshi S. Targeted RNases: A feasibility study for use in HIV gene therapy. Gene Therapy 6:913–921, 1999.

Smith C, Sullenger BA. AIDS and HIV infection. In Dickson G (Ed.), Molecular and Cell Biology of Human Gene Therapeutics. Chapman and Hall, London, 1995, pp. 195–236.

VandenDriessche T, Chuah MKL, Morgan RA. Gene therapy for acquired immune deficiency syndrome. AIDS Updates 7:1–14, 1994.

Yu M, Poeschla E, Wong-Staal F. Progress towards gene therapy for HIV infection. Gene Therapy 1:13–26, 1994.

Enhanced Immunity

Hadida F, DeMaeyer E, Cremer I, Autran B, Baggiolini M, Debre P, Viellard V. Acquired constitutive expression of interferon beta after gene transduction enhances immunodeficiency virus type 1-specific cytotoxic T lymphocyte activity by a RANTES-dependent mechanism. Hum Gene Therapy 10:1803–1810, 1999.

Kim JJ, Nottingham LK, Tsai A, Lee DJ, et al. Antigen-specific humoral and cellular immune responses can be modulated in rhesus macaques through the use of INF-gamma, IL-12 or Il-18 gene adjuvants. J Med Primatol 28:214–223, 1999.

Riddell SR, Greenberg PD. Therapeutic reconstitution of human viral immunity by adoptive transfer of cytotoxic T lymphocyte clones. Curr Top Microbiol Immunol 189:9–34, 1994.

Trans Dominant Molecules

Ragheb JA, Bressler P, Daucher M, Chiang L, Chuah MKL, VandenDriessche T, Morgan RA. Analysis of trans-dominant mutants of the HIV type 1 Rev protein for their ability to inhibit Rev function, HIV type 1 replication, and their use as anti-HIV gene therapeutics. AIDS Res Hum Retro 11:1343–1353, 1995.

Sawaya BE, Khalili K, Rappaport J, Serio D, Chen W, Srinivasan A, Amini S. Suppression of HIV-1 transcription and replication by a Vpr mutant. Gene Therapy 6:947–950, 1999.

Shimano R, Inubushi R, Oshima Y, Adachi A. Inhibition of HIV/SIV replication by dominant negative Gag mutants. Virus Genes 18:197–201, 1999.

Intracellular Antibodies and Intracellular Immunization

Baltimore D. Intracellular immunization. Nature 335:395, 1988.

BouHamdan M, Duan LX, Pomerantz RJ, Strayer DS. Inhibition of HIV-1 by an anti-integrase single chain variable fragment (SFv): Delivery by SV40 provides durable protection against HIV-1 and does not require selection. Gene Therapy 6:660–666, 1999.

Rondon I, Marasco WA. Intracellular antibodies (intrabodies) for gene therapy of infectious diseases. Annu Rev Microbiol 51:257–283, 1997.

Steinberger P, Andris-Widhopf J, Buhler B, Torbett BE, Barbas CF. Functional depletion of the CCR5 receptor by intracellular immunization produces cells that are refactory to CCR5-dependnet HIV-1 infection and cell fusion. Proc Natl Acad Sci USA 97:805–810, 2000.

Yamada O, Yu M, Yee J-K, Kraus G, Looney D, Wong-Staal F. Intracellular immunization of human T cells with a hairpin ribozyme against human immunodeficiency virus type 1. Gene Therapy 1:38–45, 1994.

Gene Therapy Decoys

Browning CM, Cagnon L, Good PD, Rossi J, Engelke DR, Markovitz DM. Potent inhibition of human immunodeficiency virus type 1 (HIV-1) gene expression and virus production by an HIV-2 tat activation-response RNA decoy. J Virol 73:5191–5195, 1999.

Fraisier C, Irvine A, Wrighton C, Craig R, Dzierzak E. High level inhibition of HIV replication with combination RNA decoys expressed form an HIV-Tat inducible vector. Gene Therapy 5:1665–1675, 1998.

Morgan RA, Baler-Bitterlich G, Ragheb JA, Wong-Staal F, Gallo RC, Anderson WF. Further evaluation of soluble CD4 as an anti-HIV type1 gene therapy: Demonstration of protection of primary human peripheral blood lymphocytes from infection by HIV type 1. AIDS Res Hum Retro 10:1507–1515, 1994.

Sullenger BA, Gallardo HF, Ungers GE, Gilboa E. Overexpression of TAR sequences renders cells resistant to human immunodeficiency virus replication. Cell 63:601–608, 1993.

Antisense

Bunnell BA, Morgan RA. Development of retroviral vectors expressing antisense RNA to inhibit replication of the human immunodeficiency virus. In Weiss B (Ed.), Antisense Oligodeoxynucleotides and Antisense RNA. CRC Press, Boca Raton, FL, 1997, pp. 197–212.

Galderisi U, Casino A, Giordano A. Antisense oliogonucleotides as therapeutic agents. J Cell Physiol 181:251–257, 1999.

DNA Vaccines

Lewis PJ, Babiuk LA. DNA vaccines: A review. Adv Virus Res 54:129–188, 1999.

Montgomery DL, Ulmer JB, Donnelly JJ, Liu MA. DNA vaccines. Pharmacol Ther 74:195–205, 1997.

Mossman SP, Pierce CC, Robertson MN, Watson, et al. Immunization against SIVmne in macaques using multigenic DNA vaccines. J Med Primatol 28:206–213, 1999.

Wyeth-Lederle Vaccines, Malvern. HIV gp160 vaccine gene therapy. Drugs 1:451–452, 1999a.

Wyeth-Lederle Vaccines, Malvern. HIV gp120 vaccine gene therapy. Drugs 1:448, 1999b.

CHAPTER 12

Gene Therapy Approaches for Rheumatoid Arthritis

THOMAS PAP, M.D., ULF MÜLLER-LADNER, M.D., RENATE E. GAY, M.D., and STEFFEN GAY, M.D.

INTRODUCTION

Rheumatoid arthritis (RA) is a chronic, systemically spreading disorder that predominantly affects the joints and as the most prominent feature leads to their progressive destruction. A major problem in the treatment of RA is the lack of agents that interfere specifically with key pathological processes of disease. The complex pathology of systemic joint destruction, which is still not completely understood, limits the generation of effective therapies. To date, most pharmacological approaches focus on interfering with pain and inflammation. Drug discovery efforts resulting in useful agents have come into rheumatology empirically rather than through specific design. However, none of the drugs used today to ameliorate the pain and suffering of arthritis clearly prevents the progressive destruction of joints.

Recently, there has been enormous progress in elucidating the molecular and cellular basis of rheumatoid joint destruction. Advances in molecular biology, the utilization of novel animal models, and the diagnosis of early disease have provided insights into key biological mechanisms elucidating the destruction of extracellular matrix in RA. Based on these data, novel strategies to inhibit rheumatoid joint destruction have been proposed and developed. Among them, gene therapy is of particular interest. There is great potential in the technology of gene therapy for specifically modifying disease mechanisms in the context of the aggressive behavior of cells resulting in the rheumatoid joint destruction.

GENERAL CONSIDERATIONS ABOUT GENE THERAPY IN RA

Gene therapy can be defined as the delivery of genes to cells for the purpose of treating diseases. Originally, the application of gene therapy was for the treatment

An Introduction to Molecular Medicine and Gene Therapy, Edited by Thomas F. Kresina
ISBN 0-471-39188-3   © 2001 Wiley-Liss

of inherited diseases, such as cystic fibrosis, through the correction of the underlying genetic abnormalities. However, gene transfer methods can be used as a general vehicle for the delivery of a variety of gene products, thereby increasing the scope of diseases for which gene therapy can be used, such as for acquired diseases.

Genetic factors have been frequently implicated in the pathogenesis of RA, and recent reports suggest mutations of tumor suppressor genes to play a role in the pathogenesis of RA. However, RA is not caused by a specific genetic mutation. RA is an acquired disorder to which genetic susceptibility contributes to an unspecified level. Thus, gene therapy approaches, in diseases such as RA, differ from those in defined genetic disorders. Apart from the problem of how to correct a specific genetic abnormality or deliver a certain gene construct, the question of which pathogenic pathway to modulate becomes crucial. Therefore, a general overview of recent concepts on the pathogenesis of RA is essential for understanding gene therapy approaches for this disease. However, it needs to be emphasized that the use of gene transfer methods in RA has two important aspects: First, it is a tool that is used to study the role of different molecules in the pathogenesis of the disease. The use of gene transfer methods has helped elucidate key mechanisms of the disease and provide significant progress in developing novel strategies for the treatment of RA. Second, it is clear that although being the ultimate aim of all current research efforts, the utilization of gene transfer methods to treat human RA represents a perspective rather than an achievement in the near future. Thus, the term *gene transfer* is preferred to *gene therapy* in the context of therapy for RA.

Pathogenesis of RA

The pathogenesis of RA comprises the mutually interacting phenomena of chronic inflammation, altered immune responses, and synovial hyperplasia. Although recent data provide evidence that the pathological mechanisms of inflammation and articular damage may differ, there are no definitive answers as to what is the initiating step and what the perpetuating force in the pathogenesis of RA. However, the progressive destruction of joint cartilage and bone represents a unique and most prominent feature of this disease. It distinguishes RA from other arthritides as well as determines its outcome in most patients.

T cells have been assigned a key role in initiating and driving the destructive events of RA for decades. Recent data provide growing evidence that fibroblast-like cells within the RA synovial membrane play an important role.

Intriguingly, rheumatoid arthritis synovial fibroblasts (RA-SF) differ substantially from normal fibroblasts both in their morphological appearance and their behavior. They are characterized by a large, more round shape than normal synovial fibroblasts, and have large pale nuclei with prominent nucleoli. These changes, together with some alterations in their behavior, suggest these cells to be "transformed-appearing" fibroblasts. By investigating the specific properties of RA-SF on a molecular level, it has been understood that cellular activation and escape from normal regulation is a major event in the "transformation" of normal synovium into RA synovium. Although these cells are activated, they do not reveal an increased rate of proliferation. The elevated expression of proto-oncogenes and transcriptional factors suggest a further key role for cellular activation in synovial fibroblasts. Such events mirror the transformation events that occur in carcinogenesis (see Chapter 11). For example, some immediate early response genes, such as c-*fos* and *egr-1*, are constitutively expressed at elevated levels in RA synovial fibroblasts, and their expression levels correlate with collagenase expression in these cells. Other studies have revealed high transcription of proto-oncogenes, such as c-*jun* and c-*myc* in RA-SF. The c-*fos* and c-*jun* involve the formation of the AP-1 transcription factor which, in turn, is responsible for the up-regulation of several matrix-degrading enzymes. High deoxyribonucleic acid (DNA) binding activity of AP-1 has been demonstrated in RA-SF. Several lines of evidence suggest that the increased expression of AP-1 related molecules Jun and Fos is mediated through the action of upstream oncogenes such as ras, src, and raf. The latter oncogenes induce a prolonged activation of molecules involved in the mitogen-activated protein kinase (MAPK) pathway. Consequently, interest has focused on the investigation of intracellular signaling through the MAPK pathway. However, the exact mechanisms and pathways that mediate the cellular activation of RA-SF are only partially understood. Interestingly, recent reports indicate that jun D counteracts c-jun by having opposing effects on cell proliferation of fibroblasts as well as on cytokine and matrix metalloproteinase (MMP) production. Thus, it has been suggested that defective or unbalanced expression of *jun D* may contribute to the activation of RA-SF.

There is also evidence that apoptotic pathways are altered in the RA synovium. Apoptosis may alter the synovial lining layer that mediated the progressive destruction of cartilage and bone. Less than 1% of lining cells exhibit morphological features of apoptosis as determined by ultrastructural methods, and several studies have reported the expression of antiapoptotic molecules such as bcl-2 and sentrin in synovial cells. Although some unanswered questions concerning the regulation of apoptosis in RA remain, it appears that apoptosis suppressing signals outweigh proapoptotic signaling in activated RA-SF causing a dysbalance of pro- and antiapoptotic pathways. This dysbalance may lead to an extended life span of synovial lining cells as well as result in a prolonged expression of matrix-degrading enzymes at sites of joint destruction.

Activation of synovial cells also appears to be a major reason for the attachment of RA-SF to cartilage and bone. This process is pivotal for RA as compared with other nondestructive arthritides and is mediated by various surface proteins such as integrins, VCAM-1, and CD44. Activation of synovial cells results in the up-regulation of these adhesion molecules. But, conversely, the expression of early cell cycle genes such as c-*fos* and c-*myc* is stimulated further by cell adhesion molecules. In addition, the expression of adhesion molecules such as VCAM-1 by RA-SF contributes to T-cell anergy and the induction of angiogenesis. Therefore, the role of adhesion molecules in RA is not restricted to the attachment of synovium to cartilage and bone but involves the recruitment of inflammatory cells as well as the induction of MMPs.

The destruction of articular cartilage and bone by transformed-appearing RA-SF is mediated through the concerted action of different matrix-degrading enzymes. Among these, MMPs have been shown to play a major role. MMP-1 and MMP-3 have been found to be elevated in synovial fluid of patients with RA as compared to OA and are released in large amounts by synovial fibroblast-like cells in culture.

In situ studies revealed strong expression of MMP-1, MMP-3, and MMP-13 within rheumatoid synovium, both at the messenger ribonucleic acid (mRNA) and protein level. Recent data show that membrane-type (MT) MMPs are also expressed abundantly in cells aggressively destroying cartilage and bone. This is of particular importance because members of this MMP family such as MT1-MMP not only degrade extracellular matrix components but also activate other disease-relevant MMPs such as MMP-2 and MMP-13. It has also been shown that certain proto-oncogenes are directly involved in the up-regulation of different MMPs during the course of the disease.

To study the molecular properties of RA-SF in the absence of other human cells as well as their contribution to cartilage destruction, the severe combined immunodeficient (SCID) mouse co-implanted model was developed. In this model, isolated RA-SF are co-implanted together with fresh normal human cartilage under the renal capsule of SCID mice and kept there for 60 days. After sacrificing the mice, the implants are removed and the invasion of RA-SF into the cartilage is analyzed using conventional H&E staining as well as in situ hybridization and immunohistochemistry. Utilizing this model, it was shown that RA-SF maintain their phenotypic appearance and invasive behavior. In contrast to OA or normal synovial fibroblasts, RA-SF degrade cartilage progressively even in the absence of human inflammatory cells. These and other studies have demonstrated that RA-SF maintain their invasive behavior in the absence of human inflammatory cells.

Although recent studies indicate a partial "uncoupling" of inflammation and joint destruction, it must be emphasized that T cells and macrophages may modulate the behavior of RA-SF through the production of various cytokines. In this context, interleukin-1 (IL-1) and tumor necrosis factor-α (TNF-α) are of particular interest since they enhance inflammation and joint destruction in RA. IL-1 and TNF-α are produced predominantly by macrophage-like synovial lining cells and are capable of inducing a variety of other cytokines, chemokines, and prostaglandins. In addition, both cytokines may stimulate directly the production of matrix-degrading enzymes such as MMPs. Thus, the cellular interactions of neighboring macrophage-like cells, fibroblasts, and also chondrocytes appear to contribute to the perpetuation of chronic synovitis. Animal studies data as well as results from clinical studies using IL-1 and TNF-α inhibitors highlight the importance of these two cytokines during the course of the disease. However, there is evidence for a difference in the importance of TNF-α and IL-1 with respect to inflammation and joint destruction. While TNF-α appears responsible primarily for the extent of the synovitis, IL-1 seems to have a greater impact on the destruction of cartilage.

DELIVERY OF GENE TO SYNOVIAL CELLS

The transfer of genes into target cells such as RA-SF can be achieved by various methods. Viruses are naturally capable of delivering their genes into host cells, and therefore viral vectors have been used intensively for gene transfer approaches also in RA. However, the utilization of viral vectors for gene transfer requires substantial changes to the original viral genome. Apart from introducing the desired gene, these changes include modifications that disable replication of viral particles in infected cells. Optimal vectors should provide maximal safety together with high transduction efficacy and long-term expression of the transgene. Generally, retroviruses, adenoviruses, and adenoassociated viruses (AAV) are used for this purpose (see Chapter 4). To date, retroviral vectors have been extensively used for gene transfer of RA-SF because they ensure long-term expression of the transgene. For RA, the Moloney murine mouse leukemia virus (Mo-MuLV) is reconstructed into a replication-defective retrovirus and used as a vector to transduce fibroblast cells. In this vector, called MFG vector, the *env* gene, which is needed by the virus to synthesize proteins for its envelope, is replaced by the gene of choice. Transfection of packaging cells, which produce the virus envelope, results in the production of replication-deficient virus particles. These can be used to transduce RA-SF. The retroviral LXSN vector has also been used. Apart from the LTR promoter driving the transgene expression, a SV40 promoter is used for the expression of a neomycin(r) resistance gene. Such constructs allow for the selection of successfully transduced RA-SF using the antibiotic G418. The utilization of retroviral vectors such as the MFG or related vectors, however, has important limitations. Retroviral particles are unable to infect nondividing cells. Therefore, ex vivo approaches are used in which incubation of cultured RA-SF with the replication-deficient virus particles and subsequent selection of successfully transduced cells result in the long-lasting expression of the transgene. Currently, efforts are made to develop retroviral vectors for in vivo approaches. This would avoid multiple surgical interventions, which is a major disadvantage of ex vivo approaches. Another limiting factor, however, is the unpredictable site of insertion into the host genome resulting in, at least, a potential risk of insertional mutagenesis.

ANIMAL MODELS TO TEST GENE THERAPY APPROACHES

A suitable animal model for gene therapy approaches in RA should not only reflect relevant features of human disease but also permit to analyze the alteration of key disease processes as close as possible to the conditions found in humans. Since injection of various antigens can lead to the induction of a relapsing, erosive arthritis, models such as collagen-induced arthritis (CIA) and the streptococcal-wall antigen-induced arthritis (SCW-A) have been proposed and intensively studied as animal models for human RA. These models have provided important insights into molecular mechanisms of joint inflammation and helped elucidate key aspects of joint destruction. Therefore, these models have been used to study the effect of gene transfer approaches. However, as arthritis in all of these animal models is clearly antigen driven, they have some important limitations with respect to their use to study a potential effect of gene therapy for human RA.

As a consequence of this, the SCID mouse co-implantation model of RA has been used most recently to study the effect of gene transfer approaches to RA-SF. In this model, the behavior of genetically modified RA-SF cells can be investigated using the mouse as a "living culture flask." Cultured RA-SF are implanted together with fresh, normal human cartilage and thus can be used for retroviral as well as for adenoviral gene transfer approaches. Moreover, keeping some of the transduced RA-SF in culture during the entire implantation period allows analysis of these cells for transgene expression as well as for changes in the intracellular signaling.

CURRENT TARGETS FOR GENE THERAPY IN RA

Based on the aforementioned data, several approaches for the gene transfer into RA-SF have been developed. They comprise mainly three different strategies (Fig. 12.1):

1. Interfering with the stimulation of RA-SF by cytokines and growth factors
2. Modulating signaling and apoptosis regulating molecules
3. Direct inhibition of matrix-degrading enzymes such as MMPs and cathepsins

As shown, inhibition of inflammatory cytokines such as IL-1 and TNF-α constitutes an interesting approach to reduce cartilage degradation and inflammation. Fortunately, there is a naturally occurring inhibitor of IL-1, the IL-1 receptor antagonist (IL-1Ra). In normal synovium, IL-1Ra is mainly produced by fibroblasts and macrophages. By competitively binding to the IL-1 receptor without intrinsic activity, IL-1Ra counteracts the effects of IL-1 and contributes to cytokine homeostasis in the synovium. Since IL-1 is overexpressed in RA synovium, increasing the IL-1Ra may be a feasible approach to reduce the proinflammatory effects of IL-1. However, to achieve this goal, a large molar excess of IL-1Ra is required. This is

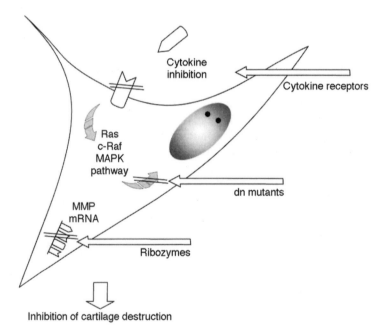

FIGURE 12.1   Gene transfer strategies to inhibit the cartilage destruction mediated by rheumatoid arthritis synovial fibroblasts (RA-SF). Current approaches involve (a) the inhibition of external stimulation of RA-SF by blocking the action of proinflammatory cytokines through the overexpression of cytokine receptors/antagonists, (b) the blocking of internal signaling cascades such as the Ras-c-Raf-MAPK pathway using dominant negative (dn) mutants of signaling molecules, and (c) the cleavage of mRNA for matrix degrading enzymes through the delivery of ribozymes into RA-SF.

because the affinity of IL-1Ra does not exceed that of IL-1, and IL-1 has a pronounced spare receptor effect. Several studies have investigated the effect of IL-1Ra overexpression in synovial cells. A MFG-IRAP construct was used to interfere with antigen-induced arthritis in rabbits in an ex vivo approach. In another study, the effect of gene transfer with IL-1Ra was determined using the development of bacterial cell wall arthritis in Lewis rats and a LXSN-based construct. These studies showed a marked anti-inflammatory effect of gene transfer with IL-1Ra.

Moreover, the former also revealed an increase in the proteoglycan synthesis. However, studies in the SCID mouse model using human RA-SF and fresh normal human cartilage revealed no alteration of the invasive behavior of RA-SF into the cartilage. In the SCID studies, transduction of RA-SF with MFG-IRAP did not lead to significant differences in the grade of invasion as compared to *lacZ*-transduced and mock-transduced cells. However, with respect to the perichondrocytic cartilage degradation, there was a clear effect. While mock- and *lacZ*-transfected RA fibroblasts showed considerable perichondrocytic cartilage destruction, this pericellular degradation was markedly reduced in IL-1Ra-transduced RA-SF. Therefore, the maintenance of the chondrocyte-matrix integrity was the main result of IL-1Ra gene transfer into human RA-SF. The data indicate that chondrocytic chondrolysis is an IL-1-driven process. In addition, it can be concluded that IL-1 independent pathways contribute most significantly to the adhesion of RA fibroblasts to cartilage and to its subsequent destruction.

TNF-α is another proinflammatory cytokine targetable for gene therapy in RA. Similar to IL-1, there exists a natural inhibitor for TNF-α, which is its shedded soluble receptor. There appears to be an imbalance between the expression of this naturally occurring inhibitor of TNF-α and TNF-α itself in rheumatoid synovium. This imbalance has been proposed to be the major cause of TNF-α-mediated upregulation of inflammation pathways. Thus, increasing the amount of TNF-α receptor proteins, for example, its p55 form (TNFRp55) is a promising way to reduce inflammation and possibly joint destruction. This approach has been supported by clinical studies. They have proven TNF-α blockage to be safe and effective. Among these are recent clinical trials that used fused proteins of the TNFRp55 as well as TNFRp75. However, transduction of RA synovial fibroblasts with TNFRp55 in the SCID mouse model resulted in neither a substantial reduction of invasiveness nor in a decrease of perichondrocytic cartilage degradation. Only few implants showed a slight reduction of invasiveness by synovial fibroblasts, which failed to reach statistical significance. These data support the hypothesis that inhibition of TNF-α, though significantly reducing the synovial inflammation, has only limited effects on rheumatoid joint destruction.

Interleukin 10 (IL-10) is an anti-inflammatory cytokine, which by downregulation of IL-1, TNF-α, IL-2, IL-6, and γ-interferon, exerts an inhibitory effect on inflammation and matrix degradation. In rheumatoid synovium, IL-10 is largely produced by macrophages and T cells. Despite its anti-inflammatory effects, IL-10 is elevated in RA synovial fluids and sera. Therefore, it has been speculated that this increase reflects an insufficient inhibitory response of the activated immune system in the synovium. As IL-10 has also been shown to inhibit tyrosine kinase activity and the *Ras* signaling pathway, which up-regulates cathepsin L, IL-10 appears to be a promising candidate for gene transfer in RA. IL-10-based gene therapy in RA has been suggested to be beneficial to both inflammation and joint destruction. Studies using systemic adenoviral administration of the IL-10 gene in mice with CIA demonstrated a positive effect on the onset of arthritis and suppressed the disease severity as well as joint destruction. Gene transfer studies using IL-10 in the SCID mouse model demonstrated a significant reduction of invasion with the grade of invasiveness being the lowest among all experiments performed. However, IL-10 transduction did not reduce perichondrocytic cartilage destruction. Therefore, it has been concluded that co-transduction of IL-1Ra and IL-10 may be an interesting approach combining the advantages of the gene transfer with IL-1Ra and IL-10. Currently, studies are being performed investigating the feasibility of this approach as well as the delivery of further cytokine genes.

Although the exact signaling pathways and cascades by which RA-SF are activated just begin to emerge, some pathways are already well documented. It has been shown that extracellular signals are forwarded by the *Ras-Raf*-MAPK cascade to the nucleus and certain proto-oncogenes therein. Therefore, blocking the *Ras-Raf*-MAPK pathway may constitute an interesting approach to reduce the production of MMPs and cathepsins (Fig. 12.1). One mechanism to block signaling pathways is the utilization of dominant negative mutants of signaling molecules such as c-*Raf*. Dominant negative (dn) mutants represent mutated variants of these molecules, which lack function. They are constructed by deleting parts of the gene that are important for their activity and can be used to study the effects of signaling molecules on the invasive behavior of RA-SF. Dominant negative mutants of c-*raf* were used to investigate the effect of this molecule on the aggressive behavior of RA-SF. In these experiments a retroviral LXSN-based construct was utilized to deliver the gene for dn c-*raf* mutant to RA-SF.

Transduction of RA-SF with the gene for dn c-*raf* mutant revealed no significant alterations in growth or apoptosis in the target cells. Using the transduced RA-SF in the SCID mouse model, a decrease in the invasiveness could be observed. However, dn c-*raf*-transduced cells still showed considerable destruction of the co-implanted human cartilage indicating that *Raf*-independent signaling pathways contribute to the activation of synovial fibroblasts in RA. Based on the aforementioned data on *jun D* expression in RA, gene transfer with *jun D* may have a potentially beneficial effect in RA.

Other current approaches are aimed at interfering directly with transcriptional activators such as NFκB. NFκB is expressed ubiquitously and involved in the regulation of a large variety of genes. Among them are those responsible for the inflammatory response in RA. It is regulated by its cytoplasmatic inhibitor IκBα. Activation of NFκB by molecules such as TNF-α involves the rapid degradation of IκBα. NFκB binding motifs not only have been described in the promoters of several MMP genes, but recent data suggest that inhibition of NFκB results in a decreased synthesis of urokinase-type plasmin activator (uPA), which has been associated with the activation of several MMPs. Based on these data, several strategies to interfere with the activation of NFκB in RA-SF have been developed. Using a retroviral construct overexpressing IκBα in several cell types, the feasibility of this approach was tested in preventing the IL-1-inducible degradation of endogenous IκBα as well as the activation of NFκB. The effect of transducing RA-SF with such a construct on the aggressive behavior of these cells is currently being investigated, but several lines of evidence point to a central role of NFκB in cellular activation. Moreover, recent data indicate that the antiapoptotic molecule *Bcl-2* may be involved in the activation of NFκB through degrading IκBα. These data point to a link between *Bcl-2* and NFκB signaling pathways in the regulation of apoptosis. As alterations in pathways leading to apoptosis appear to contribute to the aggressive phenotype of RA-SF, efforts are being made to interfere with apoptosis regulating molecules in RA. The recently described antiapoptotic molecule sentrin constitutes a candidate gene for RA gene transfer. It has been demonstrated that sentrin is up-regulated in RA-SF as compared to normal and OA synovial. Therefore, the use of mRNA antisense constructs against sentrin to inhibit the expression of sentrin is currently being investigated.

Lastly, modulating the terminal phase of MMP and cathepsin up-regulation by cleaving or blocking the mRNA for these enzymes has been a major target for gene transfer in RA (Fig. 12.1). This can be achieved by the delivery of MMP-specific mRNA antisense constructs or so-called hammerhead ribozymes. Hammerhead ribozymes are RNA molecules that share a special structural motive known as the "hammerhead" and are able to site specifically cleave RNA. Such ribozymes can be used to destroy specific messages inside cells. RA synovial fibroblasts expressing ribozymes capable of cleaving collagenase are likely to inhibit enzyme production or limit enzymatic levels in situ.

CLINICAL STATUS AND FUTURE PERSPECTIVE

The intital clinical trial addressing the safety and feasibility of gene transfer to human synovium occurred 3 years ago. In this phase I clinical trial, a retroviral MFG vector carrying the cDNA for the human IL-1Ra gene (MFG-IRAP construct) was used. A total of nine patients who needed total joint replacement of their MCP joints for severe RA were included in the study. In a first session, RA-SF were removed from MCP joints and transduced with the MFG-IRAP construct ex vivo. Transduced cells were then reinjected into the joints. One week later, the MCP joints were removed and the total joint replacement surgery was performed as scheduled. All steps were performed under strict safety condition including the screening for replication-competent retroviruses. The preliminary results of this study indicate that genes can be delivered to human joints safely and effectively. The final evaluation of the removed joints will be performed and include conventional histological evaluation as well as in situ hybridization and immunohisto-chemistry techniques.

Future perspective of gene therapy approaches in RA will mainly include the evaluation of disease-specific pathways as well as the design and utilization of novel viral vectors that have site-specific integration into the host genome with maximum safety and long-term duration of gene expression.

KEY CONCEPTS

- Rheumatoid arthritis (RA) is a chronic, systemically spreading disorder that predominantly affects the joints and as the most prominent feature leads to their progressive destruction. A major problem in the treatment of RA is the lack of agents that interfere specifically with key pathological processes of disease. Novel strategies to inhibit rheumatoid joint destruction have been proposed and developed. Among them, gene therapy is of particular interest. There is great potential in the technology of gene therapy for specifically modifying disease mechanisms in the context of the aggressive behavior of cells resulting in rheumatoid joint destruction.

- Several strategies for the gene transfer into rheumatoid arthritis synovial fluid have been developed. They comprise mainly three different strategies:

  1. Interfering with the stimulation of synovial cells by cytokines and growth factors

  2. Modulating signaling and apoptosis regulating molecules

  3. Direct inhibition of matrix-degrading enzymes such as matrix metalloproteinases and cathepsins.

- The initial phase I clinical trial addressing the safety and feasibility of gene transfer to human synovium indicates that genes can be delivered to human joints safely and effectively.

SUGGESTED READINGS

Rheumatoid Arthritis and Osteoarthritis

Blomberg P, Islam KB. Genes tame pro-inflammatory cytokines and promote bone healing. Acta Orthop Scand 70:415–418, 1999.

Malemud CJ, Goldberg VM. Future directions for research and treatment of osteoarthritis. Front Biosci 15:D762–D771, 1999.

Müller-Ladner U, Gay RE, Gay S. Cellular pathways of joint destruction. Curr Opin Rheumatol 9:213–220, 1997.

Müller-Ladner U, Gay RE, Gay S. Molecular biology of cartilage and bone destruction. Curr Opin Rheumatol 10:212–219, 1998.

Weyand CM, Goronzy JJ. Pathogenesis of rheumatoid arthritis. Med Clin North Am 81:29–55, 1997.

Gene Therapy and Arthritis

Evans CH, Ghivizzani SC, Kang R, Muzzonigro T, Wasko MC, Herndon JH, Robbins PD. Gene therapy for rheumatic diseases. Arthritis Rheum 42:1–16, 1999.

Evans CH, Robbins PD, Ghivizzani SC, Herndon JH, Kang R, Bahnson AB, Barranger JA, Elders EM, Gay S, Tomaino MM, Wasko MC, Watkins SC, Whiteside TL, Glorioso JC, Lotze MT, Wright TM. Clinical trial to assess the safety, feasibility, and efficacy of transferring a potentially anti-arthritic cytokine gene to human joints with rheumatoid arthritis. Hum Gene Therapy 7:1261–1280, 1996.

Firestein GS, Echeverri F, Yeo M, Zvaifler NJ, Green DR. Somatic mutations in the p53 tumor suppressor gene in rheumatoid arthritis synovium. Proc Natl Acad Sci USA 94:10895–10900, 1997.

Hummel KM, Gay RE, Gay S. Novel strategies for the therapy of rheumatoid arthritis. Br J Rheumatol 36:365–267, 1997.

Jorgensen C, Gay S. Gene therapy in osteoarticular diseases: Where are we? Immunol Today 19:387–391, 1998.

Pap T, Gay RE, Gay S. Gene transfer: from concept to therapy. Curr Opin Rheumatol 12: 205–210, 2000.

Tomita T, Takeuchi E, Tomita N, Morishita R, Kaneko M, Yamamoto K, Nakase T, Seki H, Kato K, Kaneda Y, Ochi T. Suppressed severity of collagen-induced arthritis by in vivo transfection of nuclear factor kappaB decoy oligonucleotides as agene therapy. Arthritis Rheum 42:2532–42, 1999.

CHAPTER 13

Federal Oversight of Gene Therapy Research

THOMAS F. KRESINA, PH.D.

BACKGROUND

Federal oversight of biomedical research is an evolving regulatory activity of the federal government. Federally supported research is regulated by the federal government in the context of animal care and their humane use, as well as for the safe and ethical use of humans in clinical trials. For experimental protocols that utilize animals, approval is required by the Institutional Animal Care and Use Committee (IACUC). The IACUC is a committee formed at the local institutional level comprised of scientists, clinical researchers, institutional officials, and lay and community representatives/leaders. It is the mission of the IACUC to review and approve all experimental protocols involving animal use. In a similar fashion, research protocols involving human use are required to receive review and approval by the Institutional Review Board (IRB). The IRB composition and approval procedure is directed by federal law (Protection of Human Subjects, Title 45 Code of Federal Regulations Part 46) and enforced by the Office for the Protection from Research Risks (OPRR). In order to highlight emphatically the role and authority of OPRR, as well ensure the protection of individuals in clinical trials, the OPRR has recently been elevated from an office at the National Institutes of Health (NIH) to a standing office of the Department of Health and Human Services.

Research protocols involving recombinant deoxyribonucleic acid (DNA) molecules provide additional risks and thus have additional review and approval requirements. A brief historical account of federal regulation is presented along with current regulatory requirements as well as potential future changes in review and approval procedures.

An Introduction to Molecular Medicine and Gene Therapy, Edited by Thomas F. Kresina
ISBN 0-471-39188-3 © 2001 Wiley-Liss

INTRODUCTION

As research advances provide new and exciting experimental techniques and protocols, the increasing risks in application of these technological advances need to be addressed. Already, sheep, cows, and primates have been cloned using nuclear transfer techniques (see Chapter 2). However, cloning using nuclear transfer techniques is not "new." The first experiments in nuclear transfer occurred in frogs in 1952. At that time, using the nuclei of tadpoles transferred into frog eggs, scientists raised cloned tadpoles and even adult frogs. The cloning was taken in stride because of the use of embryonic nuclei. Indeed such studies have continued and advanced for 45 years without controversy. Recent embryonic cloning work was published in 1996 when lambs were reported cloned from embryos. In the case of Dolly, modifications in the previously successful protocols resulted in the ability to clone using an adult cell, a mammary cell reprogrammed to "dedifferentiate," and thus permitting the development of an adult animal. Thus, the field entered the era of nonembryonic cloning. In March, 1997, scientist in the United States announced the cloning of primates from embryonic cells using nuclear transfer. These techniques have an obvious extension of cloning humans, and that has startled the research and lay communities alike. Quickly, 10 days after the adult cell cloning study was announced, President Clinton announced a ban on federal funds to support research on cloning of humans. He also requested the private sector to refrain from such research. The president went on to urge Congress to prohibit by law the cloning of humans. Three months later in June, 1997, the National Bioethics Advisory Commission concluded that, at this time, it is "morally unacceptable for anyone in the public or private sector, whether in research or clinical setting, to attempt to create a child using somatic cell nuclear transfer cloning" (see Suggested Readings). However, this has not stopped mavericks from announcing the attempt to open "Cloning Clinics" in Chicago or elsewhere. These clinics would be a for-profit venture with the noble cause of providing an option of parental cloning for infertile couples. Such announcements have created a public outcry and sent elected officials at both the state and federal levels scrambling to establish laws prohibiting the use of cloning technology. Thus, in the future, federal or local law may supercede or modify the Food and Drug Administration (FDA) jurisdiction over all recombinant DNA research (which included cloning) in the United States and its territories (see below). This is likely because the frontier continues to rapidly move forward in high profile.

Although nuclear transfer techniques, to date, do not utilize recombinant DNA technologies, they do involve genetic manipulation and thus are reviewed and approved through the mechanisms established for gene therapy protocols. This procedure continues to evolve over time and comprises submission to the Office of Recombinant DNA Activities (ORDA)—now a component of the Office of Biotechnology Activities—as well as some form of federal review and/or approval by the NIH and the FDA. In recognition of the expanding complexity of genetic manipulation in research, the NIH established in November, 1999, the Office of Biotechnology Activities, which monitors and coordinates research in recombinant DNA, gene transfer, genetic testing, and xenotransplantation.

OFFICE OF BIOTECHNOLOGY ACTIVITIES

As part of the Office of Science Policy at the National Institutes of Health, the Office of Biotechnology Activities is comprised of the Recombinant DNA and Gene Transfer Committee and two advisory committees of the Department of Heath and Human Services (DHHS)—the Secretary's Advisory Committee on Genetic Testing and the Secretary's Advisory Committee on Xenotransplantation. The Secretary's Advisory Committee on Genetic Testing (SACGT) was chartered in June, 1998, in response to recommendations of two working groups commissioned jointly by the NIH and the Department of Energy (DOE) for the Human Genome Project. The SACGT grew out of the identified need for broad-based public policy development to help address the benefits and challenges of genetic knowledge and genetic testing. The SACGT advises DHHS and the NIH on all aspects of the development and use of genetic tests, including the complex medical, ethical, legal, and social issues raised by genetic testing. The committee wrestles with issues such as the development of genetic testing guidelines. These include criteria regarding the risks and benefits of genetic testing, assisting institutional review boards (see below) in reviewing genetic testing protocols in both academic and commercial settings, the adequacy of regulatory oversight of genetic tests, provisions for assuring the quality of genetic testing laboratories, the need for mechanisms to track the introduction of genetic tests to enable accuracy and clinical effectiveness over time to be evaluated, and safeguarding the privacy and confidentiality of genetic information and preventing discrimination as well as stigmatization based on genetic information.

The Secretary's Advisory Committee on Xenotransplantation (SACX) was chartered in July, 1999, and is being formulated. The committee will consist of 15 voting members, including the chair. Members are currently being recruited from authorities knowledgeable in such fields as xenotransplantation, epidemiology, virology, microbiology, infectious diseases, molecular biology, veterinary medicine, immunology, transplantation surgery, public health, applicable law, bioethics, social sciences, psychology, patient advocacy, and animal welfare. Of the appointed members, at least one shall be a current member of the Xenotransplantation Subcommittee of the Food and Drug Administration Biologic Response Modifiers Advisory Committee and at least one shall be a current member of the Centers for Disease Control and Prevention (CDC) Hospital Infection Control Practices Advisory Committee. This is a newly formed advisory committee of the DHHS/NIH because DHHS has a vital role in safeguarding public health while fostering the development of promising strategies to treat tissue destruction, organ failure and other public health needs. The SACX will consider the full range of complex scientific, medical, social, and ethical issues and the public health concerns raised by xenotransplantation. These include ongoing and proposed protocols, and making recommendations to the secretary of DHHS on policy and procedures. The recommendations of the committee will facilitate efforts to develop an integrated approach to addressing emerging public health issues in xenotransplantation.

The Recombinant DNA and Gene Transfer Division of OBA monitors scientific progress in basic and clinical research involving DNA and human gene transfer. A component of this monitoring is performed by the Recombinant DNA Advisory Committee.

RECOMBINANT DNA ADVISORY COMMITTEE

In response to a report by the Committee on Recombinant DNA Molecules, established by the National Academy of Sciences in 1974, the DHHS chartered a committee to establish biological and physical containment practices and procedures for recombinant DNA research. The document generated by this committee became the basis for a set of guidelines to be used for the safe conduct of recombinant DNA research. The guidelines became known as "The NIH Guidelines for Research Involving Recombinant DNA Molecules" or the NIH Guidelines. The original DHHS committee developed into the NIH Recombinant DNA Advisory Committee or RAC. This action was based on the recommendation of the 1982 President's Commission for the Study of Ethical Problems in Medicine and Biomedical and Behavioral Research. The commission's report entitled "The Social and Ethical Issues of Genetic Engineering with Human Beings (Splicing Life)" explicitly stated that "the NIH should extend its purview over recombinant DNA research beyond environmental" (containment) "issues to human gene therapy." Thus, a Human Gene Therapy Committee was formed comprised of federal and nonfederal scientists and clinicians, lay individuals, and ethicists. This committee merged with the DHHS chartered committee to form the NIH RAC (Recombinant DNA Advisory Committee).

The RAC, in its original form, constituted a 25-member committee charged with the responsibility of reviewing and approving all research protocols involving human use and recombinant DNA molecules. Investigators, developing protocols utilizing human gene transfer for marking, nontherapeutic, and therapeutic studies, were required to utilize the NIH Guidelines. Thus, investigators were required to specify the research practices to be utilized for the constructing and handling of recombinant DNA molecules as well as organisms and viruses containing recombinant DNA molecules. Compliance with the NIH Guidelines was mandatory for all recombinant DNA research within the United States and its territories. Human gene transfer protocols were to be submitted in the format described in Appendix M of the NIH Guidelines entitled "Points to Consider in the Design and Submission of Protocols for the Transfer of Recombinant DNA Molecules into the Genome of One or More Human Subjects." The Points to Consider apply to any protocol conducted at or sponsored by an institution that receives any support for recombinant DNA research from NIH.

Based on the 1984 background study entitled "Human Gene Therapy" by the Office of Technology Assessment, the RAC was directed to consider for approval only somatic gene therapy protocols. The directive was based on the civic, religious, scientific, and medical community acceptance in principle of the appropriateness of gene therapy for somatic cells in humans for specific diseases. Somatic gene therapy was and is seen as an extension of current experimental therapeutic methods and potentially preferable to other elaborate technologies. Factors considered in the RAC review included the use of new vectors or new gene delivery systems, application to new diseases, unique application of gene transfer, and other issues that would require public attention and debate, such as unique ethical situations. The RAC recommendation, approval or disapproval, of a specific protocol is transmitted to the director of the NIH. The director provided a concurrence or nonconcurrence on the recommendation and forwarded the decision to the commissioner of the FDA. In 1993, protocols involving human gene transfer were also required to be simultaneously submitted to the FDA for Investigational New Drug (IND)

TABLE 13.1 Human Gene Transfer Protocols Received by NIH as of 5/99

Protocol	Number
Infectious diseases	
HIV	27
Monogenetic diseases	
α$_1$-Antitrypsin deficiency	1
Chronic granulomatous disease	3
Cystic fibrosis	18
Familial hypercholesterolemia	1
Fanconi anemia	2
Gaucher disease	3
Hunter syndrome	1
Ornithine transcarbamylase deficiency	1
SCID-ADA	1
Other diseases	
Peripheral artery disease	8
Rheumatoid arthritis	2
Coronary artery disease	5
Cancer therapy	
Antisense	5
Chemoprotection	9
In vitro transduction	60
In vivo transduction	59
Prodrug/HSV-TK and ganciclovir	30
Tumor suppressor gene	23
Oncogene down-regulation	3
Cell marking studies	34
Therapeutic protocols	277
Nontherapeutic	2

approval using the identical format. The first gene therapy protocol was a cancer gene marking study entitled "The Treatment of Patients with Advanced Cancer Using Cyclophosphamide, Interleukin-2 and Tumor Infiltrating Lymphocytes." The protocol received RAC approval on October 3, 1988, and NIH approval on March 2, 1989. Roughly 10 years later, as of May 1999, ORDA lists 313 protocols having been received and under review or having been reviewed (Table 13.1).

Amended Federal Oversight

On July 8, 1996, the director of the NIH proposed an amendment to the NIH Guidelines. This amendment modified the federal oversight of gene therapy research. It was proposed that in order to enhance NIH mechanisms for scientific and ethical oversight of DNA recombinant activities, the RAC would be discontinued and all approval responsibilities for recombinant DNA activities involving human gene transfer would be relinquished to the FDA. Thus, the FDA would retain statutory authority for authorizing IND. The enhancement of NIH oversight of human gene therapy research was proposed via various mechanisms. Most notably, a series of

Gene Therapy Policy Conferences would be initiated and intended to augment the quality and efficiency of public discussion of the scientific merit and ethical issues relevant to gene therapy clinical trials. These conferences would assemble individuals with scientific, ethical, and legal expertise to discuss and formulate policy on single topic issues. The conference policy statements would be submitted to the director of the NIH and, furthermore, made available to the DHHS agencies such as the FDA and the Office for Protection from Research Risks for implementation. The initial Gene Therapy Policy Conference occurred on September 11, 1997, and was entitled "Human Gene Transfer: Beyond Life Threatening Disease." The agenda included the scientific prospects for trait enhancement through gene therapy, assessing long-term safety and efficacy, the conceptual, ethical, and social issues of treatment versus enhancement, as well as to delineate the distinction between treatment and gene enhancement. Summaries of past Gene Therapy Policy Conferences and future agendas can be found at the Office of Biotechnology Activities home page www.nih.gov/od/oba.

On November 22, 1996, the NIH director published a document entitled "Notice of Proposed Action under the NIH Guidelines for Research Involving Recombinant DNA Molecules." This document detailed the retention of the RAC but modifying its roles and responsibilities. The RAC retains the function of a charted advisory committee to the director of NIH, but no longer has approval/disapproval protocol. It now meets quarterly to discuss novel human gene transfer experiments, identify novel scientific and ethical issues, and advise the NIH director on modifications to the NIH Guidelines. The RAC is now comprised of 15 members, with at least 8 members knowledgeable in molecular genetics, biology, or recombinant DNA research and at least 4 members knowledgeable in applicable legal aspects, public attitudes, and environmental safety issues and public health. In addition, RAC members co-chair the Gene Policy Conferences.

CURRENT REVIEW AND APPROVAL PROCESS

Based on the scientific scope of the developed protocol, a review process has been established and is required for all protocols involving DNA/RNA molecules, gene therapy, nucleic acids, or nuclear transfer. As shown in Table 13.2, the original review process required public review, RAC review, approval by the NIH, and/or FDA and institutional approval. As shown in Figures 13.1 and 13.2, an initial linear process is now being streamlined to allow for concurrent review to provide investigators minimal time between submission of the protocol for review and initiation of the experimental protocol. This streamlined approval process may now take a backseat to a rigorous approval process due to the recent disclosure of adverse events in human gene therapy clinical studies (see below). However, regardless of whether a linear process or concurrent process of review, a non-novel protocol requires both FDA and institutional approval prior to initiation.

FDA Approval

The FDA has defined gene therapy as "a medical intervention based on modification of the genetic material of living cells." Cells may be modified ex vivo for sub-

TABLE 13.2 Summary of Review Procedures Required for Recombinant DNA Protocols Prior to Initiation of Study

Timeline	Public Review	RAC Review	Approval	IRB/IBC
Original procedure	Every protocol	Yes	Yes	Yes
Consolidated simultaneous review	Some protocols	Triage	Yes	Yes
As of 12/97	Novel protocols	Yes	No	Yes
Proposed 9/99	Novel protocols	Yes	No	No

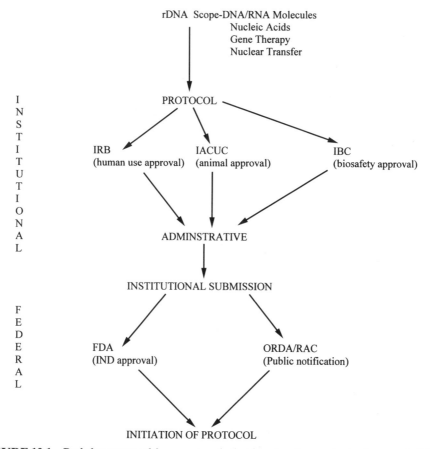

FIGURE 13.1 Path for approval for a protocols that involves human/animal use and rDNA.

sequent administration or may be altered in vivo by gene therapy products given directly to the subject. When the genetic manipulation is performed ex vivo on cells and subsequently administered to the patient, this is considered a form of somatic cell therapy. The genetic manipulation may be intended to "prevent, treat, cure, diagnose, or mitigate disease or injury in humans" [*Federal Register* 58(197):53248–53251]. As noted above, the Center for Biologics Evaluation and Research (CBER) has the authority within the FDA to review gene therapy protocols (Table 13.3).

Animal Use and rDNA

rDNA Scope-DNA/RNA Molecules
Nucleic Acids
Gene Therapy
Nuclear Transfer

A. *Concurrent Review of Non-Novel Protocol and NOT Requiring IND*

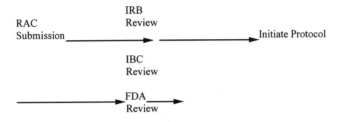

B. *Concurrent Review of Novel Protocol Requiring IND*

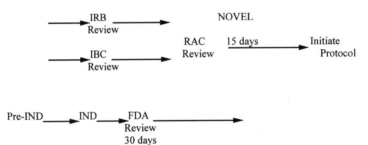

FIGURE 13.2   Proposed non-linear, concurrent review of protocols that involve human/ animal use and rDNA.

TABLE 13.3   Protocols Received by NIH and Pending Review as of 5/99

Protocol	Number
Cancer	
In vitro transduction	2
In vivo transduction	7[a]
Coronary artery disease	1[a]
Gyrate atropy	1[a]
Hemophilia	1

[a] Submission not complete (2 in vitro protocols in cancer).

CBER has produced documents to address somatic gene therapy. They include documents entitled "Points to Consider in Human Somatic Cell and Gene Therapy" as well as an Addendum to this document "Points to Consider on Plasmid DNA Vaccines for Preventive Infectious Disease Indications" and "Guidance for Industry for the Submission of Chemistry, Manufacturing, and Controls Information for a Therapeutic Recombinant DNA-Derived Product or a Monoclonal Antibody Product for In Vivo Use." The original Points to Consider document focused on ex vivo somatic gene therapy, while the addendum provides additional guidance based on current information regarding regulatory concerns for production, testing, and administration of recombinant vector for gene therapy with a particular emphasis on in vivo administration. It also provides a brief outline of information needed for IND applications for gene therapy, regulatory issues related to all classes of vector products for gene therapy, preclinical issues for safety evaluation in animals, and regulatory handling of modifications in vector preparations. Although each new biological requires an independent IND submission, if related vectors are made or "minor" modifications are generated, an argument can be provided for the inclusion of the products as a panel and thus be described in the context of an amendment to the original IND application. Alternatively, the vectors could be described in master files if intended for multiple IND submissions. This simplifies the submission process by not filing redundant materials.

The Points to Consider document for plasmid DNA vaccines was produced in December, 1996, and outlines the CBER approach to regulation of plasmid DNA vaccines. In addition, product consideration for an IND submission is presented as well as considerations for plasmid DNA vaccine modifications, the use of adjuvants and devices for vaccine delivery, and preclinical immunogenicity and safety evaluations. CBER and the Center for Drug Evaluation and Research (CDER) also provide guidance to industry on the context and format of the Chemistry, Manufacturing and Controls (CMC) Section of a Biologics License Application.

A final document of note is the FDA Information Sheets for the Institutional Review Board's (IRB's) and Clinical Investigators. This document provides IRB operations and clinical investigation requirements including the use of drugs, biologicals, and medical devices in clinical investigations. Guidance is provided on cooperative research, foreign clinical studies, study recruitment, payment to research subjects, and informed consent among other topics. These documents provide guidance clinical trial design and development.

MOVING TOO FAST IN A FIELD OF PROMISE

A startling and sobering event occurred in 1999, in the field of gene therapy research, and initiated a series of further events and revelations. That event was the initial death of a patient in an experimental clinical trial involving gene therapy. Investigators at the FDA found numerous violations of federal research regulations and shortcomings in the protection of human subjects in the execution of the clinical research. This resulted in the FDA placing a "clinical hold" or halting all research at the University of Pennsylvania's Institute for Human Gene Therapy. Further government hearings and investigations of the gene therapy research field revealed that a total of 731 adverse events had occurred in human studies involving gene therapy. Of these only 39 had been originally reported as required by law. Six hundred and fifty two adverse events involved studies using adenovirus as the vector while 40 adverse events were belatedly reported for all other vectors. Investigators who receive approval from the FDA to initiate a human gene transfer protocol must report any serious adverse event immediately to the local IRB (see below), Institutional Biosafety Committee, Office of Protection from Research Risks (OPRR), NIH/ORDA, and FDA. The form and information required is presented as Table 13.4. The apparent lack of compliance with this law has resulted in an Executive Order published by President Clinton on February 9, 2000, requiring the FDA and NIH to review all federally regulated gene therapy research for compliance with federal law.

An additional issue arose on February 12, 2000, related to patient safety in gene therapy clinical research. In an apparent oversight in quality control of vector construction (see Appendix), a report surfaced by a clinical researcher of possible viral contamination of a vector preparation used in a cancer pediatric protocol. The investigator reported the possible contamination by pathogenic human immuodeficiency virus (HIV) and/or hepatitis C virus of a viral vector preparation administered to patients in the protocol. The investigator immediately notified the parents of the participants of the possible contamination, halted the clinical trial, and notified government oversight. Further PCR studies are needed to definitively establish contamination. These events show the risks of experimental gene therapy research that must be realized by all and specifically presented to the patient as part of the overall informed consent process.

INSTITUTIONAL REVIEW OF RECOMBINANT DNA RESEARCH

Research involving human subjects and human use, sponsored and not necessarily performed by an institution in the United States, is subject to approval by an IRB and the Institutional Biosafety Committees (IBC). The IRB is a review committee constituted locally and acts on behalf of the sponsoring institution to protect potential research subjects. The IRB balances the competing interests of the medical researcher with loyalty to science and society and the individual patient with a focused interest in personal health. The underlying task of the IRB is to discern the relative risk to a patient of an experimental protocol. Thus, the IRB balances the potential risk to the patient due to adverse effects of the protocol against the potential gain to society through increased medical knowledge.

IRBs grew out of a mandate from the National Commission for the Protection of Human Subjects of Biomedical and Behavioral Research. Statutory authority has come from the Code of Federal Regulations (CFAR) Title 21, CFR part 50, 1981, and (DHHS) Title 45 CFR part 46 1983. The latter Code of Federal Regulations was a component of The Public Health Service Act as amended in 1985 and 1993. Part 46 of the code is entitled "Protection of Human Subjects" and provides the basic federal policy for the protection of human research subjects. The policy defines the code of conduct for research that includes the confidentiality of patient data and identity. Each institution sponsoring research is required to provide the DHHS assurance that the institution does and will continue to comply with terms and conditions of the Public Health Service Act. This assurance includes a statement of principles governing the institution for the protection of rights and welfare of the human subjects enrolled in research protocols conducted at the sponsoring institution, the establishment of the IRB and continued review of the research pro-

TABLE 13.4   Reporting Form for Adverse Events Occurring in Clinical Gene Therapy Research

NIH PROTOCOL NUMBER

(This number consists of a four-digit year/month identifier followed by a three-digit sequence number)

FDA IND NUMBER (This number has four digits)

CLINICAL TRIAL SITE
Name of Institution, Street Address, City, and State

IND Sponsor

IBC Chair
  Name:
  Date Reported:

IRB Chair
  Name:
  Date Reported:

Principal Investigator(s)

Vector Type (e.g., adenovirus)

Vector Subtype (e.g., type 5—also include relevant deletions)

Gene Delivery Method

Route of Administration (e.g., injection + site)

Dosing Schedule and Treatment Group Criteria

Patient Data:

Date of Adverse Event

Complete Description of the Event

Suspected Cause of the Event

Relevant Clinical Observations (For example, there are 24 standard pathophysiological/anatomical categories with defined grades of severity from 0 to 5.) See also Common Toxicity Criteria (CTC) at http://ctep.info.nih.gov/CTC3/Download/ctc gendatacol.doc

Relevant Clinical History

Relevant Tests (That have been conducted to date)
            (That will be conducted)

At this time is the event considered:
RELATED
POSSIBLY RELATED
NOT RELATED
to administration of the gene transfer product?

Any similar observations in other patients treated in this study or a similar study?

In the event of death, has an autopsy been requested? If not, why not?

tocol by the IRB, the names of the IRB members, and the function and reporting of the IRB.

Each IRB must have at least five members by statute. The membership must be diverse with regard to professional expertise, race, gender, and cultural background and must be knowledgeable of community attitudes toward the safeguarding of human rights and welfare.

IRB members must also be knowledgeable about the institutional commitment to human research and applicable legal considerations. Specifically, the committee must not be comprised of individuals of all one gender or profession. It must also be comprised of at least one member with relevant scientific expertise and one lay member and one member not affiliated with the institution or any family member of the institution. No member may have any conflict of interest with the proposed application. The IRB is charged with the authority to approve, require modification, or disapprove an application for human use in research. The IRB must review an approved protocol yearly. The IRB is required to transmit information to the patient through informed consent.

The criteria for IRB approval are that the risks to the patient are minimized through sound research design and procedures and that the risks to the patient are reasonable in relation to the anticipated benefits. The benefits include only those of the research and not standard patient care. The IRB may not consider any long-range benefit beyond the protocol under review regarding knowledge gained in the research. Other approval criteria are that the selection of research subjects is equitable, informed consent is sought and documented, and that data and safety are monitored where applicable through a Data and Safety Management Board (DSMB). Institutions may require additional administrative approval for the research protocol, but no administrative approval can supercede IRB approval or disapproval. The IRB has the statutory authority to suspend or terminate any research study for noncompliance.

Investigators receiving approval from the FDA to initiate a human gene transfer protocol (with or without RAC approval) must report any serious adverse event to a patient, immediately to the local IRB, IBC (below) NIH/Office of Biotechnology Activities, and FDA. This report is to be followed by a written report filed to each group as part of the clinical trial records keeping.

When a protocol involving recombinant DNA molecules is generated, Institutional Biosafety Committee (IBC) approval is required. The IBC acts on behalf of the local community to assure that any and all safety issues involving rDNA in experimental protocols are addressed. The IBC is ultimately responsible to NIH/Office of Biotechnology Activities. The committee must comprise of a botanist or plant pathogen or plant pest containment expert as well as an animal containment expert. The sponsoring institution is also required to train all members of the committee and utilize ad hoc experts as necessary. The sponsoring institution is responsible for the completeness and accuracy of the human gene transfer application in its dual submission to the NIH/Office of Biotechnology Activities (see Fig. 13.1) and the Division of Congressional and Public Affairs, Document Control Center, CBER, FDA.

As a precondition for NIH funding of recombinant DNA research, the institution must ensure that research conducted at the sponsoring institution, irrespective of funding source for a given individual protocol, complies with the NIH Guidelines. In addition, all NIH-funded research involving recombinant DNA techniques must comply with the NIH Guidelines. Failure to comply could lead to the total loss of funding for all research at the institution involving recombinant DNA molecules.

INTERNATIONAL EFFORTS AT REGULATORY OVERSIGHT

Advances in gene therapy research are occurring throughout the world in countries with established biomedical research programs. Significant efforts are being made to generate international guidelines for human gene therapy research. Guidelines can be generated on a country-to-country basis or via a consortium, such as the European Union (EU) or through bilateral agreements.

An example of an individual country establishing its own guidelines for gene therapy research is Australia. In May, 1999, the Australian National Health and Medical Research Council published a consultation draft document entitled "Draft Guidelines for Human Somatic Cell Gene Therapy and Related Technologies." This document contains guidelines for human somatic gene therapy and related technologies and an accompanying background paper. These guidelines should in interpreted in concert with the National Health and Medical Research Council's "statement", Statement on Ethical Conduct in Research Involving Humans. Research protocols are brought to the Genetic Manipulation Advisory Committee that has a mission to oversee the development and use of novel genetic manipulation techniques as well as assess whether these techniques pose a hazard to the community or environment. Such regulatory procedures compare favorably to the process in the United States.

Bilateral agreements include the 4th International Conference on Harmonization (ICH), which focused on virus validation for biotechnology. Joint FDA–ICH documents have been generated on safety, efficacy, and quality. In addition, the FDA has entered into numerous memorandum of understanding (MOU) with individual countries with regard to exports/imports, inspections, laboratory products, and items that represent "dangerous infections to human beings" as covered in the U.S. Public Health Service Act. The FDA has also participated on behalf of the United States and in an Agreement of Mutual Recognition between the United States and the EU in the conformity of assessment procedures. The EU has gone on to harmonize the differing national review processes for clinical trials in individual European countries. Some of these procedures are patterned after the "Appendix M" documents of the United States.

The reader is directed to the executive summary of "Stem Cell Research: Medical Progress with Responsibility" a report from the chief medical officer's expert group reviewing the potential of developments in stem cell research and cell nuclear replacement to benefit human health. This document contains nine specific recommendations for guidelines for medical research in the areas of nuclear transfer and stem cells. It is currently before the Parliament of Great Britain for approval and can be found on the Internet at: www.doh.gov.uk/cegc/stemcellreport.pdf.

SUMMARY

Subsequent to the intense effort to generate an experimental research protocol for human use using recombinant DNA molecules, a labyrinth awaits the investigator in order to initiate the study. Figure 13.1 display a flowchart for the approval needed. At the institutional level, the IRB approval, Institutional Animal Care and Use Committee (IACUC) approval for any animal use, and the Institutional Biosafety Committee approval are needed. Subsequently, institutional administrative approval is needed for submission for approval at the federal level. At the federal level, all novel protocols involving human gene transfer are required to be submitted to the Office of Biotechnology Activities for public notification and discussion by the RAC. IND approval is necessary from the FDA. Then and only then can the protocol be initiated. Investigators who receive approval from the FDA to initiate a human gene transfer protocol must report any serious adverse event immediately to the local Institutional Review Board, Institutional Biosafety Committee, Office of Protection from Research Risks (OPRR), NIH/Office of Biotechnology Activities, and FDA.

The author acknowledges the helpful discussions with Amy P. Patterson, M.D., Director of Office of Biotechnology, NIH.

KEY CONCEPTS

- The RAC, in its original form, constituted a 25-member committee charged with the responsibility of approving all research protocols involving human use and recombinant DNA molecules. Investigators, developing protocols utilizing human gene transfer for marking, nontherapeutic, and therapeutic studies, were required to utilize the NIH Guidelines. The use of the guidelines was mandatory for all recombinant DNA research within the United States and its territories. Human gene transfer protocols were to be submitted in the format described in the appendix documents of the guidelines entitled "Points to Consider in the Design and Submission of Protocols for the Transfer of Recombinant DNA Molecules into the Genome of One or More Human Subjects."

- On July 8, 1996, the director of the NIH proposed amendment to the NIH Guidelines. The amendment would modify the federal oversight of gene therapy research. It was proposed that in order to enhance NIH mechanisms for scientific and ethical oversight of DNA recombinant activities, the RAC would be discontinued and all approval responsibilities for recombinant DNA activities involving human gene transfer would be relinquished to the FDA. Thus, the FDA would retain statutory authority for protocol approval.

- Currently, most gene therapy protocols are phase 1 clinical trials determining toxicity. Investigators who receive approval from the FDA to initiate a human gene transfer protocol must report any serious adverse event immediately to the local Institutional Review Board, Institutional Biosafety Committee, Office of Protection from Research Risks (OPRR), NIH/Office of Biotechnology Activities, and FDA.

- Numerous approaches are being undertaken for the regulation of gene therapy and human cloning on an international basis. However, each country will act and regulate these technologies on an individual basis according to its ethical, religious, and legal traditions. For example, while substantial efforts are being made to outlaw human cloning in the United States, Israel has determined that Rabbinical Law will allow for the cloning of humans.

SUGGESTED READINGS

Advisory Committee on Human Radiation Experiments. Final report of the Advisory Committee on Human Radiation Experiments. Human Radiation Interagency Working Group, Washington, DC, 1995.

Cho MK, Billings P. Conflict of interest and Institutional Review Boards. J Invest Med 45:154, 1997.

Cohen-Haguenauer O. Gene therapy: regulatory issues and international approaches to regulation. Curr Opin Biotechnol 8:361, 1997.

Edgar H, Rothman D. The Institutional Review Board and beyond: Future challenges to the ethics of human experimentation. Milbank Q 73:489, 1995.

Hollon T. Researchers and regulators reflect on first gene therapy death. Nat Med 6:6, 2000.

Jenks S. Gene therapy death—"everyone has to share in the guilt." J Natl Cancer Inst 92:98–100, 2000.

Marchall E. Gene therapy death prompts review of adenovirus vector. Science 286(5448):2244–2245, 1999.

National Bioethics Advisory Commission. Report on cloning by the US Bioethics Advisory Commission: Ethical considerations. Hum Reprod Update 3:629–641, 1997.

U.S. General Accounting Office. Scientific Research. Continued vigilance critical to protecting human subjects. GAO Report, Health, Education and Human Services Division Report No. B259279, Washington, DC, 1996.

CHAPTER 14

Ethical Issues in Molecular Medicine and Gene Therapy

LEONARD M. FLECK, PH.D.

BACKGROUND

The presentation of ethical issues in molecular medicine and gene therapy warrants definition of terms. An *ethic*, as defined by Webster, is a principle of right or good conduct or a body of such principles. *Ethics* is the study of the general nature of morals and the specific moral choices to be made by the individual in his/her relationship with others. A *moral* is defined as of or concerned with the judgment of the goodness or badness of human action and character, that is, pertaining to the discernment of good and evil. Thus, moral decisions or actions require the definition of good and evil. In the context of human gene therapy, the methods or procedure used to provide therapy of a disease is considered a "good." It is in the ethics of gene therapy, when one considers the good choice of gene therapy in the context of others, that controversial issues arise.

INTRODUCTION

In this chapter we will be examining a number of ethical issues related to molecular medicine and gene therapy. In the first part we will be exploring these issues in a clinical context at the level of the doctor–patient relationship. Our focus will be somatic cell gene therapy, work that is now moving quickly into clinical trials. In the second part we will be exploring issues at the level of social and professional policy. In particular, we will give considerable attention to ethical issues connected with germline genetic engineering. This is not a technical capacity we have as yet, but, as many have argued, it would be morally and politically foolish to put off having these discussions until the technology has been actualized. The unexpected announcement of the

An Introduction to Molecular Medicine and Gene Therapy, Edited by Thomas F. Kresina
ISBN 0-471-39188-3   © 2001 Wiley-Liss

cloning of Dolly precipitated moral hysteria and histrionics that do not serve well the goal of more rational democratic deliberation in our society about such issues.

MOLECULAR MEDICINE AND GENE THERAPY: ETHICAL ISSUES IN THE CLINICAL CONTEXT

The Case of David

It is often helpful to begin a discussion like this with a concrete case. A case that is well known in the bioethics literature is that of David, who was usually referred to in the popular press as the "Texas Bubble Boy." He was born in 1971 with severe combined immune deficiency syndrome (SCID). His older brother had died of this disorder; and consequently, physicians were prepared at birth for David, who was expected to be born with the same disorder. Birth took place in an absolutely germ-free environment to protect David from exposure to any and all pathogens. He was immediately transferred to an isolator bubble. Subsequent tests determined that he was in fact afflicted with SCID. His parents were able to modify their living room and create a room-size sterile bubble in which David grew up. At one point he was able to use a space suit of sorts to wander for brief periods in the outside world. He was able to attend school via a speaker phone. He was clearly a very bright child.

At the age of 12 both David and his parents were hoping he would have access to some medical therapy that might free him from the confinement of that bubble. Bone marrow transplant techniques were well developed by then and looked promising. His older sister was a good match, and she agreed to be a donor for him. The hope was that her bone marrow would give David the immune system that he lacked. The surgery itself was successful. Things went well for a few months, but then David began to have a fever whose cause was unknown. Despite the best medical efforts David died about 2 months later. It was later learned that his sister's bone marrow harbored the Epstein–Barr virus. In her own body it was easily kept in check, but in David's body it had the opportunity to run rampant. [For a fuller accessible account of David's story, see Walters and Palmer (1997).]

The case of David is not about gene therapy. But it does raise all the ethical issues that the first uses of gene therapy will certainly generate. Children die of fatal disorders all the time in our society. Such deaths are tragic and regrettable, but we do not usually see them as raising ethical issues. Why does David's death raise ethical issues? Why would David's case have raised ethical issues even if he had survived? And what precisely are the issues that are raised?

We start with a fact that we will stipulate for the sake of discussion, though it does seem like a reasonable stipulation. David might have survived indefinitely, so long as he stayed within that bubble environment. David clearly proved that was possible just by virtue of his surviving for those first 12 years. He and his parents and his physicians took a risk that proved fatal. If an individual is faced with imminent death, and if medicine offers a small chance of a medical miracle, then we see no ethical problem with an individual assuming that risk. He has very little to lose, and perhaps everything to gain. And if that individual is an adult who is capable of rationally assessing the risks and benefits offered by this experimental intervention (to whatever extent they can be known or hypothesized), then too there does not seem to be anything ethically problematic about physicians offering such an intervention to that individual, or that individual choosing to accept that intervention. But neither of these conditions obtained in the case of David.

David was a 12-year old boy who was going to take an irrevocable gamble with the rest of his life by agreeing to an experimental medical intervention. Should David have been allowed to take that chance? Should David's parents have attempted to very strongly dissuade him from taking such a gamble when it was not necessary to take that gamble? Was that one of their moral obligations as parents? There are lots of things that good parents forbid their children to do that are a lot less risky than what David was permitted to do. Should David's physicians have worked very hard to dissuade David from taking this gamble with his life? That is, were his physicians derelict in their duties as physicians to protect David's best medical interests? On the face of it, these questions seem to have obvious answers. And those answers imply harsh moral criticism of both David's parents and his physicians. Is that a fair conclusion?

So far this discussion has been very one-sided. There are morally relevant considerations on the other side of the moral equation. To begin with, one of the cardinal principles of health care ethics today is the principle of respect for patient autonomy. In brief, that principle states that competent patients have a strong moral right to decide for themselves what is in their best medical interests. If such decisions are made freely by patients, if such decisions are a product of careful deliberation (a careful weighing of risks and benefits), if such decisions emerge from a stable set of values and a certain stable understanding of what counts as a life worth living for that patient, then such patient choices ought to be respected. That is, such patients would have a presumptive right to see their choices carried out, and their physicians would be under a presumed moral obligation not to ignore, or worse, overturn those decisions. Among other things, this line of reasoning has led to the conclusion that there are a number of circumstances in which patients have the moral right to refuse life-sustaining care. The case of Dax, a 26-year-old Texas burn victim from 1973, is usually taken as a paradigmatic of just such a right. He saw himself as having been seriously morally wronged because he was forced to undergo 14 painful months of burn therapy in order to save his life. Dax wanted to be allowed to die, something that most physicians at the time were very strongly inclined to resist. David wants to live, not die. But he wants to live in the same large world that he sees his friends and family enjoying, not the very constrained world of a plastic bubble in which he was forever denied, literally, ordinary human touch. If that in fact is what David very strongly desires, then who would have the moral right to deny him the medical interventions that might make that possible?

This latter way of framing the issue does yield a very different moral perspective. But there are still other morally relevant considerations that must be assessed. The principle of respect for patient autonomy does not permit each and every patient to practice medicine as they wish, or to use physicians as mere instruments to achieve whatever health states they see as desirable in the context of their life goals. Physicians have their own moral integrity as physicians, and patients have no moral right to simply violate that integrity as they wish. Thus, a high school senior may very well wish to play football at Notre Dame. This may be the most important goal of his life at this time. He may feel the need to "bulk up" some, and he may see steroids as the key to accomplishing that goal quickly. He might demand that his physician prescribe these drugs, but his physician would have a strong moral obligation not to accede to that request because of the long-term damage that would likely accrue to that young man. This is one of the moral limits on respect for patient autonomy.

Similarly, the argument might be made that David is a young adolescent. He is beginning to rebel against the physical constraints on his life as well as the constraints imposed by his parents and physicians. What he wants is certainly reasonable in a general sense, *but it is not reasonable for him now*. His parents and his physicians need to counsel patience. There will likely be other medical opportunities in the future. Bone marrow transplantation does offer hope of release from his confined environment, but it is a very risky hope. There are too many unknowns. If David is patient, then those unknowns and those risks may be substantially reduced over a period of years through expected medical progress. He should be counseled to let others assume the risks of medical pioneers, others who do not have as much to lose as he does. David is not seeking to misuse a medical therapy, as our hypothetical high school athlete is. So his apparently autonomous choice is not ethically flawed in that respect. But it might be argued that his choice is not autonomous enough that his physicians would be ethically obligated to accept that choice. That is, his choice may be short-sighted, a product of less than adequate deliberation and a less than stable balancing of competing personal values. Someone might argue, for example, that David's decision is in the same moral category as that of any other adolescent who chooses to drive in excess of 100 mph on a dark country road in a race against another adolescent. But then this analogy may also be too stark and too unfair.

Someone might care to see David as a medical hero in a battle against diseases that afflict mankind. After all, the argument goes, adolescents who are only a few years older than David are permitted to fight in wars and risk their lives for what they believe to be noble causes. Might David see himself in the same way? He might, and then a different ethical judgment would be required. However, this raises yet another ethical issue. We are all mindful of the fact that young men can be seduced into participating in wars. Generals cannot prove their military skills by devising clever battle plans in libraries; they need to fight actual wars, and they need to recruit young men as soldiers. Similarly, medical researchers cannot prove their medical ingenuity merely be designing clever lab experiments; they need to recruit patients as participants in these experimental battles against disease. To continue the military analogy, victory does not go to the timid but to the courageous. But there is clearly something ethically troubling about this situation. Rarely do history books record the names of the soldiers who do the courageous things that win battles. Rather, the honors and the social memory are attached to generals. Soldiers take the actual risks; their lives are at stake. But the generals reap the rewards. Medical research, the battle against disease, is very much like that. There are enormous social and professional rewards attached to early medical breakthroughs. We need to emphasize that the rewards are very large *for those who are first*. This can motivate sometimes inappropriate risk taking by medical researchers, or, to be more precise, an inappropriate imposition of risk on those who may be persuaded to participate in medical experiments. Might this have occurred in the case of David? Should the researchers have delayed this experiment until they were more confident that a positive outcome was likely?

In ordinary medicine the practice of informed consent is supposed to protect patients from the risks that are always part of medical practice. Informed consent is aimed at assisting patients to make more autonomous choices about their own medical care. The role of a physician is to present honestly, and in a way that is intelligible to a particular patient, the risks and benefits associated with a proposed therapeutic intervention, along with other reasonable alternatives to that intervention. The physician may make a recommendation for a particular option, but ultimately the patient is supposed to be offered the opportunity to make that choice in the light of their own goals and values as they pertain to this medical encounter. Depending upon the seriousness of their medical problems, patients may be more or less anxious, more or less capable of making a rational assessment of their medical options. Good doctors are ethically obligated to be sensitive to the vulnerabilities of patients in these circumstances, and to assist patients to make decisions that in a meaningful sense are both *their own* and congruent with *their own best interests*. Most certainly, what good doctors are ethically forbidden from doing is advancing their own self-interest, that is, concern about their own economic well being, at the expense of the interests of their patients. Patients trust their doctors, and that trust is violated when doctors use patients and their medical problems to advance their own interests.

We return to the story of David. David was actually cared for by two sets of physicians. There were the physicians who provided his "routine" medical care. (We need to put that word in scare quotes because David was several standard deviations beyond any "routine" or typical patient.) There were also the physicians who proposed and carried out the medical experiment that ultimately proved fatal to David. These were experimental research physicians. Our discussion above of informed consent applied to this first set of physicians. In practice, ethically speaking, our understanding of informed consent needs to be modified when proposed interventions are very experimental.

There are very standard, routine, simplified ways of appropriately eliciting informed consent in routine medical care. This is not true when we are dealing with experimental medicine. The core element of informed consent is *information*, reliable, scientifically and clinically grounded information. This is precisely what is largely lacking in an experimental situation. It is not as if nothing is known. If that were the case, then there would be no ethical justification at all for proceeding with the intervention. On the contrary, enough is known that clinical researchers are morally and medically warranted in believing that this intervention is as likely to yield a net therapeutic benefit as it is likely to yield a net harm. But there are also a lot of unknowns, which represent the potential for serious, perhaps fatal, harm to the patient. In the case of David, for example, we have no reason to believe that the researchers were negligent at that time in failing to detect the Epstein–Barr virus (suppressed) in the bone marrow of his sister. This was a specific danger about which David could not be apprised. In an experimental medical context there are numerous possibilities such as this. This is one feature of experimental medicine that can be a source of ethical pitfalls.

A second feature is that we are usually dealing with patients who are much more anxious (oftentimes desperate) than the average patient. These are patients who are faced with very serious illnesses, often life-threatening, who have failed all conventional medical therapy. That is, they are much more vulnerable than your typical patient; and consequently, physicians caring for them will have a much higher degree of moral responsibility for protecting the best interests of such patients.

But there is a third feature to these experimental circumstances that deepens considerably the ethical pitfall. Specifically, researchers will often have a lot to gain (psychologically and professionally) if their experimental efforts are successful. But this requires that they recruit the patients who will have to assume the risks. And, if the researchers present those risks too candidly or too clearly, they may fail to secure the consent of the patients to participate in the research. We saw earlier that it is easy to denounce as unethical physicians who compromise the best interests of their patients for their own financial gain. This is what might be referred to in colloquial terms as an "ethical no-brainer." But in an experimental context the capacity of medical researchers for clear ethical reflection is obscured by the goals that they are pursuing. Though we have called attention to the professional rewards that await successful researchers, the focus of the public, and the focus of the researchers themselves may be the noble and altruistic motive of defeating some horrific disease that causes premature death and substantial suffering. They may see themselves as taking significant professional risks in pursuing this line of research, and they may have sacrificed much time for relaxation and personal fulfillment to pursue this research; and consequently, their consciences may be numb to the idea that there would be anything ethically problematic about expecting that patients involved in this research would have to take some risks as well.

However, one of the most basic principles in medical ethics is what is referred to as the Kantian principle of respect for persons. In short, that principle says that it is never ethically acceptable to treat persons as if they were mere things, as if one individual could use another individual for purposes that the latter individual did not accept as his or her own. This is the sort of behavior we usually describe as being manipulative. Again, the primary ethical purpose of informed consent is to permit patients to adopt/accept the medical therapies that are being offered by their physician. No matter how genuinely noble the intent of that physician in providing medical care, if that competent patient has not freely consented to those interventions, then the outcome is ethically ignoble.

As noted earlier, eliciting truly informed consent in experimental clinical circumstances is much more difficult and ethically risk-laden than in ordinary medical practice. Patients in these circumstances may desperately want to hear hopeful things from their physicians, which makes it easy for experimental researchers to understate the risks to which such patients might be exposing themselves. Patients are naturally inclined to trust their physicians, which means they are less likely to ask probing questions about competing interests that might motivate that researcher. That makes it all the more ethically imperative that researchers be candid with potential patients about the risks of experimental medicine and the rewards that might accrue to them as researchers. Since researchers themselves may have great difficulty being candid enough in these circumstances, given the genuine mixture of motives that generates such research, the ethically required course of action is that the physician who is primarily responsible for providing therapeutic care to a patient be different from the physician who is responsible for the research. In that way it is expected that the primary care physician will be better able to advise that patient in a suitably neutral fashion about where their best medical interests might lie. Finally, it is assumed that this primary care physician would have no ties at all to that clinical research, which might otherwise potentially compromise his/her ability to protect the best interests of the patient for whom they are caring. If all of this were true in the case of David, then it would be ethically permissible for the bone marrow transfer to go forward.

Case of Donald: Gene Therapy for Cystic Fibrosis

There are a number of other ethical issues that might be raised in connection with the case of David, but we will pass over them. The case of David is not about gene therapy, but it is actually an excellent model for the sorts of clinical ethical issues that are most likely to arise. Having discussed the case of David in some depth, this portion of our discussion can be much more concise. We start by recalling an earlier observation, namely, that patients enrolled in clinical trials for serious medical disorders are often in medically desperate circumstances; they have generally exhausted all other reasonable options. But this was not true with David; and it will generally not be true for many of the early trials with different approaches to gene therapy.

A good example of what we have in mind here would be cystic fibrosis (CF). Some patients with CF will not survive the first decade of life; others, today, may survive for more than four decades. If a 12-year-old CF patient, whom we will call Donald, was aware of the clinical trials now under way with gene therapy for CF, and if disease management seemed to be going reasonably well (so much so that it could be confidently predicted that Donald would likely survive to at least age 30 with reliance on current therapeutic modalities), and if Donald felt that his quality of life were significantly compromised (relative at least to the sorts of experiences and opportunities that were available to other early adolescents), and if Donald were very insistent on having access to one of these trials, then would it be ethically permissible to admit him to one of these clinical trials? As in the case of David, we will assume that Donald is very bright, which is to say he is capable of rationally processing the relevant medical information.

Before commenting directly on this hypothetical case, we will lay out a common ethical framework often used to address cases like this, including a sharper articulation of some of the ethical issues raised by this case. (An excellent source for a more in-depth view of these ethical principles is the fourth edition of Beauchamp/Childress (1994).) We noted earlier that a cardinal principle of health care ethics today is respect for patient autonomy. That prompts the ethical question: Is Donald capable of making an autonomous choice in this matter? Would we (societal representatives) have the moral right to deny all 12-year-old individuals such an option, no matter how bright or mature they were, much as we deny the right to legally consume alcohol to those below age 21?

A second basic principle of health care ethics, probably the oldest of these principles, is what is referred to as the principle of nonmaleficence. It is often interpreted to mean that at the very least physicians should do nothing that will cause unnecessary harm to their patients. Surgeons will cause considerable misery to their patients because of what surgery is, but such surgery does not represent a net harm to the patient because it is confidently believed that surgery will restore the patient's health. Further, the patient has freely agreed to the surgery because he sees this as protecting his best medical interests. So surgery in these circumstances does not represent a violation of this ethical principle. Is this just as obviously true in the case of Donald and the gene therapy clinical trial we are considering for his CF?

In the case of gene therapy there appear to be two obvious possible sources of ethical concern that might make this situation analogous to the case of David. (1) There are various sorts of biological vectors that are used to deliver the genes to some appropriate site in the body. These are not inert substances; they are often modified viruses, which is to suggest that there is some risk of biological modification of those viruses within an individual that could have serious adverse consequences. Again, we have David's actual story as a reminder of the kind of risks that are associated with clinical medicine. (2) The other biological fact is that it is *genes* that are being delivered within the body. It is expected that they will either be destroyed or that they will function in such a way that they produce the proteins with which they are normally associated. Again, it is not expected that these genes will somehow insert themselves into normal cells and disrupt the normal functioning of the genetic machinery. We think we know enough about how things work at that level that it is extremely unlikely that something like that would happen. But this is experimental medicine, and so there can be surprises.

How should we think about this in the context of the case of Donald? For any sort of major surgery patients are assuming significant enough risk of harm. These harms are generally known and quantifiable, though not entirely so. We know in general the risks of anesthesia; and we know in general the risks of infection after surgery. Competent patients generally freely assume those risks of harm. In cases where that surgery is medically necessary (90% occluded coronary arteries), the risk of harm is ethically justified by the confidently expected medical benefit. In the case of Donald, however, we cannot talk about "confidently expected medical benefit." The whole point of describing the intervention as experimental is that we are not at all confident that the hoped-for benefit will materialize. In fact, the degree of success so far with various CF gene therapy trials seems to be minimal. Further, there is some legitimate concern about unknown risks that could be very serious, again a reason why we describe these interventions as experimental. Moreover, Donald is not in medical circumstances where he *must* assume the risks of gene therapy since his CF is being managed well enough as things are now. Facts like that would seem to undermine the ethical warrant for exposing this patient to almost any level of experimental risk. However, that should be taken as nothing more than a tentative conclusion at this point.

There is a third ethical principle that needs to be considered at this point, what is usually referred to as the principle of beneficence. One formulation of this principle would say that physicians always ought to act in such a way as to advance the best medical interests of their patients. We saw in our discussion of the David case how this principle might be violated by allowing third-party interests, or the physician's own self-interests, to compromise inappropriately the patient's interests. Those issues do not need to be revisited in the case of Donald. However, there is one related issue that needs to be mentioned. It is a common practice today to pay physicians a fee for "recruiting" patients for clinical trials. The fee is intended to cover the cost of that physician's time in discussing with a patient the nature of the trial and why he might wish to consider it. It is reasonable to ask whether there is anything ethically suspect about this practice. The short answer is: "It all depends." If the fee is really paying for that physician's time at his normal rate (and nothing more than that), then it seems ethically permissible. If, on the other hand, the fee is very generous, and it is really intended as a strong incentive for that physician to persuade patients to participate in these trials, then it is prima facie ethically suspect.

An easy practical test on this point would be whether a physician who was generously "incentivized" would be willing to reveal to a patient the size and purpose of the incentive. If he were reluctant to do so, that suggests the practice is potentially ethically corruptive. Sound ethical judgments are always capable of standing the light of day, that is, public scrutiny.

That brings us back to the principle of beneficence. There are at least two other construals of that principle that need to be considered in relation to the case of Donald. First, we might take the principle to mean that physicians ought always act in such a way as to maximize the best interests of their patient *from the point of view of their best medical judgment*. This would provide a strong moral warrant for medical paternalism. This would mean in the case of Donald that, if his physician judged that CF gene therapy was too risky (given only a small chance of small benefit), then he would be ethically obligated to deny access to that therapy for Donald. For most medical ethicists today that would be much too broad a warrant for medical paternalism; it would effectively eviscerate the principle of autonomy.

The second construal of the principle of beneficence better protects patient autonomy. It says that patients should have the right to determine what they judge to be in their own best interests (within the constraints of good medical practice) from the point of view of their own stable values and life goals. This means that patients do not have to make medical decisions for themselves that are consistent with what a majority of reasonable persons in similar medical circumstances would choose. From this perspective it might well be ethically permissible for Donald to have access to experimental CF gene therapy. After all, the argument might go, we allow patients to pursue cosmetic surgery, for the sake of nothing more medically urgent than vanity, even though there are some serious medical risks attached to some of those procedures. How could we deny a young man an opportunity to improve markedly the quality of his life (and perhaps length of life) by having access to CF gene therapy, if he judges that the risks are worth the potential benefits? This question has a rhetorical ring to it, as if the answer were entirely obvious, but there is room for argument and judgment. To follow up further on that, we would have to specify a number of empirical facts, largely related to Donald's ability to make decisions that were autonomous enough to warrant moral respect.

The fourth ethical principle that has a bearing on our case is the principle of justice. As with the prior principle, there are several different construals we need to consider. In its formal sense justice requires that like cases be treated alike. Consider the case of a physician who has two patients with the same degree of coronary artery disease. Bypass surgery is indicated for both patients. However, the physician only refers the patient whom he really liked for the surgery. He does not refer the other whom he strongly disliked because the other patient had a history of noncompliance. This would be a clearly recognized case of injustice. How might like cases fail to be treated alike in the case of gene therapy? One response would pertain to ability to pay.

Health insurance will usually not pay for access to experimental interventions. This will mean that individuals who are financially more well off will have greater access to more expensive experimental therapies. Is that unjust? The vast majority of goods and services are distributed in our society on the basis of an individual's ability to pay. Very few will judge this to be ethically objectionable. But many will argue that health care should be regarded as a "morally special" sort of social good, most especially when that care can make the difference between life and death or the difference between permanent injury and the healing of that injury/restoration of function. These sorts of consequences are generally not associated with access to other consumer goods, which is why health care might be seen as being morally special. Still, as we shall see below, very few would be prepared to argue that everyone in our society has a moral right to any or all of the health care they might want or need that would offer them any health benefits. Consequently, the most common justice arguments related to health care are that all in our society ought to have assured access to some basic package of health services, perhaps a fairly thick package of health services as proposed by the Clinton Administration in 1993. But even that more generous package would not have guaranteed anyone access to experimental medical therapies of the sort we are discussing.

We can return now to the case of Donald. If he were denied access to experimental CF gene therapy because his parents did not have the ability to cover those costs, would that represent an injustice? In his case, it would be fair to conclude that this would not be a matter of great moral consequence. After all, his disease process is currently well managed. It is confidently expected that he will likely survive another 20 years. In that period of time there is a reasonable chance these therapeutic interventions will be perfected, which is to say they would become part of the standard medical armamentarium covered by health insurance. Donald would likely have access at that time. In the meantime the quality of his life will continue to be compromised; but he will not have been made worse off by this denial, so this would seem to be a morally tolerable outcome. We have to be careful, however, not to generalize too far from this example.

Imagine another patient, *the case of Edward*. Edward has some sort of cancer that has not been effectively treated by any current therapies. His life expectancy is about one year. We have another gene therapy trial that is aimed at attacking the sort of cancer that Edward has. Edward has no ability to pay for access to that trial. Has he been treated unjustly? The answer to this question is not as obvious as in the case of Donald. There is much more of moral consequence at stake here. Edward will be dead in a year; he will have no opportunity to wait until the therapy is perfected and disseminated. It is true that he, like Donald, is not made worse off by the denial itself. Still, he might argue that he has a just claim to at least a fair chance of access to such a trial. The argument might take this form: Gene therapy is not a product of the private effort and investment of some small group of individuals, as would be true with other consumer products. Rather, enormous public investments (tax dollars and university research facilities and training of the researchers themselves) have made these successes possible. All in our society have contributed to the success of that effort; and consequently, all ought to have at least a fair chance to reap the rewards of that effort. It is unfair that only those who have been especially economically fortunate already should have primary access to such technologies, especially when life itself is at stake.

We will leave the argument at that. We cannot hope to settle the issue in the space of this chapter. For now it is sufficient to note that not all cases of access to experimental gene therapy are morally alike. There are subtle moral considerations that might tilt our judgment one way or another in a given case. Here is one more example of the sort of considerations that might raise issues of justice.

Issues of experimental medicine get put in a distinct ethical compartment because, unlike standard medical practice, there are two distinct objectives being pursued simultaneously. Researchers want a therapeutic outcome for patients in these trials; this is the goal they share with standard medical practice. But they also seek to advance medical knowledge, which needs to be done very systematically. As a result, it will often be the case that patients will be denied access to one of these clinical trials if, in addition to their primary medical problem for which they are seeking entry to the trial, they have a serious co-morbid condition that could potentially compromise the quality of the clinical data they are seeking to gain. Has a rejected patient like this been treated unjustly? Is there any ethical justification for denying such patients an equal chance to be a participant in such a trial, especially if we have in mind an Edward-like patient as opposed to a Donald-like patient? To make certain that the ethical issue here is clear we need to emphasize that no matter what the co-morbid condition is that we imagine, those patients have just as good a chance of a successful therapeutic outcome in that trial as any other patient in the trial.

The standard defense of this experimental practice is utilitarian. That is, these researchers will claim that they are seeking to bring about the greatest good for the greatest number. More precisely, if the clinical data that is gathered in these trials is "tainted" through unknown effects associated with these co-morbid conditions, then there may be some tragic bad outcomes that come about when the results of these trials become part of standard clinical practice. So we have to imagine that there might be at some point in the future a large pool of potential patients who might be seriously harmed as a result of an incorrect clinical conclusion that we draw now. Researchers claim that they are morally obligated to do all in their power to prevent such possible future disasters. But there is another side to this argument.

We have a strong commitment to moral rights. The whole point of both moral and political rights is to protect individuals against the encroachments of large and powerful social organizations, either in the private or public sector. Individuals may not have their deep moral rights violated for the sake of advancing some larger social good. The moral basis for this judgment is the Kantian principle of respect for persons we discussed earlier. Individuals cannot be simply used as mere things to advance even noble social goals without their free consent. A standard example in the ethics literature is the skid row bum who drunkenly stumbles in the street to be hit by a car. He is brought to the emergency room. His head injuries are such that if surgery is done immediately he will likely be restored to something close to his former state of health. However, if surgery is delayed several hours, his condition will worsen and he will end up brain dead. His organs can then be used to save the lives of five other upstanding citizens who very much want to live but who otherwise would die of organ failure. We would certainly judge that any transplant surgeon who took advantage of a situation like this would be doing something grossly unethical. Appeal to a utilitarian justification will not alter that judgment.

Someone might argue in response that there is a major disanalogy between this case and our Edward-like gene therapy case (where Edward is denied access to a clinical trial because of some co-morbid condition), namely, that our skid row bum is negligently harmed by delaying the surgery that will certainly benefit him. Edward is not treated negligently; he is simply being denied the opportunity to benefit from access to the trial. Further, patients have strong moral rights not be harmed by medicine for the sake of some social good, but patients do not have strong moral rights to any and all possible benefits medicine may have to offer. This, however, brings in another line of moral argument.

Edward will be denied access to the clinical trial because of a co-morbid condition on the grounds that the quality of the clinical data might be corrupted by his participation. But the harm that we would hope to avoid there is potential, remote, maybe best described as "speculative" since we cannot assign any specific probability to the likelihood of that occurring. In contrast, Edward will almost certainly be dead from his cancer in a year. He has no other therapeutic options. That harm to him is certain, irreversible, utterly devastating. Even from a utilitarian point of view, the argument goes, should not such devastating harm to an individual be given greater moral weight than remote and speculative possibilities of harm to possible individuals far in the future? We do sometimes make such judgments in medicine, invoking what is referred to as the rule of rescue analogy. *Are these clinical trials for gene therapy one of those circumstances in which this rule is justifiably invoked?*

We move now to our third issue that arises in connection with our principle of justice. To set the stage for that issue we need to make explicit an assumption that has been silently operative in our two prior justice problems. Specifically, we have assumed that these clinical gene therapy trials are more likely than not going to have a therapeutic outcome. Denying individuals access to that therapeutic outcome looks like an injustice. That assumption probably reflects characteristic American optimism about our scientific endeavors. But strictly speaking we are not entitled to that assumption, especially in the earliest stages of clinical trials. And, in some circumstances, it might be more appropriate to have a serious concern about potential harms. That in fact is what motivated the early ethical discussions about medical experimentation. There were the Nazi medical experiments, which are best seen as being maliciously motivated. There are no difficult or complex moral issues to be debated there. But then there were also the Tuskegee experiments that involved African-American men who were allowed to go untreated for their syphilis, even after we had penicillin that would have cured them. The argument given for non-treatment was that we were in the middle of a medical experiment that we had to allow to run its course for the sake of scientific knowledge. This would be another clear case where invoking utilitarian considerations would not be ethically justified. But for now we want to pass over that point. The more serious point is that in the case of Tuskegee, and in the case of Willowbrook (retarded children and orphans), and in the case of the Jewish Hospital in New York (old senile patients), socially disfavored groups were used as experimental material for risky medical interventions. That is, these were individuals who were captives of institutions who were not in a position to give free and informed consent to assume the risks associated with these medical experiments. This violated the principle of respect for persons, but it was also unjust. There was an imposition of risks and burdens on these individuals for the sake of benefits that would go to other individuals. The motivations of the medical researchers may not have been ethically corrupt, as in the Nazi case, but the outcomes were nevertheless strongly morally objectionable. These are concerns that we need to be mindful of in the case of gene therapy as well.

Appropriate Candidates for Gene Therapy

We now turn to another dimension of ethics issues in gene therapy, namely, who the candidates are for therapy. We distinguish four broad types: competent adults, incompetent adults, children, fetuses. We start with competent adults, and we begin by noting that we are using the term competence in its accepted meaning in medical ethics as opposed to law. The courts, of necessity, take a global perspective on competence. If an individual is generally capable of managing the tasks of daily life for himself, then he is competent. If not, then he is incompetent, and a guardian may be appointed to act on that person's behalf. In medical ethics we speak of competence in a "task-specific" sense. The relevant moral question is: Is this patient capable of processing information relevant to the decision at hand in such a way that it would be reasonable to conclude that they are making an autonomous choice? That is, are they capable of giving free and informed consent to this intervention? Can we be morally confident that they have no gross misunderstandings of the risks and benefits associated with this intervention? There are many possible sources of misunderstanding in experimental medicine. Perhaps the two most common would be (1) an excessively optimistic view of what participation in the experimental therapy might yield for them; and (2) in those cases where the trial is structured as a double-blind randomized controlled study, a failure to appreciate that there is a 50% chance that they would not receive the therapy they might expect. The ethical obligation of researchers in these circumstances is to correct these misconceptions so that such patients are making autonomous decisions to participate.

Then there is the case of incompetent patients. In standard medical practice we need to rely upon surrogate decision makers to make medical decisions for incompetent patients. Usually we are talking about close family members, and usually we can be confident that these surrogate decision makers are loyal, caring, and trustworthy, that is, not likely to make deliberately a medical decision for this patient that would be contrary to the best interests of this patient. Again, in ordinary medical practice such surrogates are asked to make substituted judgments. That is, they are asked to make a decision as much as they can from the point of view of the patient, a point of view that best captures the stable goals and values of that patient, as opposed to any decision they might make for themselves were they in the patient's situation. In practice this is not an easy criterion to use or to know with confidence it is being used correctly. The alternate ethical standard is a best-interests test, which means the surrogate is asked to judge whether the benefits of the proposed treatment outweigh the burdens for the patient or vice versa. In experimental medicine both standards can be very difficult to apply with confidence. It will rarely have been the case that patients (now incompetent) had the opportunity to think about the sorts of decisions they would make for themselves if offered the opportunity to be part of a medical trial. It is also more difficult to apply meaningfully the best-interests test because the starting point for such medical interventions is clinical equipoise. Researchers simply do not know whether that intervention will yield a net benefit for that patient, nor do they know more than very imprecisely the range of risks to which that patient may be exposed. A reasonable ethical conclusion to draw from this is that in general incompetent patients should not be included in clinical trials. There will be exceptions.

One general exception might be captured by our Donald and Edward cases. We will modify the cases by reducing the age of each to 8, and attributing to them no more than average intelligence. So from a moral point of view they are clearly thought of as incompetent patients, which means their parents will have to make decisions for them. In the case of our revised Donald patient it would be difficult to ethically justify permitting his inclusion in an experimental gene therapy trial for his CF. The primary reason is that Donald's disease process is well managed; and hence, it would be difficult to justify the risks that this child would be assuming. By waiting several years he will likely have access to a better understood intervention more likely to yield actual benefit. His parents might want him to have "every opportunity for a normal life," but that reasonable desire may not be sufficient to justify their choosing those risks for him. By way of contrast, our revised Edward patient is faced with a terminal prognosis for his cancer. In such circumstances parents may assume for their children a greater level of risk on the grounds that this is the only way to protect the long-term best interests of those children. We should be clear, however, that such tragic circumstances do not warrant parents exposing their children to any level of risk whatsoever. If the failure of the gene therapy is not likely to alter significantly either the quality of life or length of life for that child, then it is justifiable to consider him for the therapy. But if the experimental therapy itself would add to the suffering of that child and yield a worse death, then it is just as clear that it would be morally wrong to consider such a child for this experimental therapy. The sort of case we have in mind would be an extremely aggressive form of chemotherapy, examples of which have drawn media attention in the recent past.

The other sort of patient that deserves separate moral consideration would be fetuses. Such cases are complicated by the fact that the fetus is medically accessible only through the mother, which means specific medical interventions intended for the benefit of the fetus may put her at risk as well. We are all mindful of the fact that there have been several major efforts aimed at fetal therapy in the past few years, often fetal surgery. It may be the case that there will be comparable efforts to employ gene therapy in comparable circumstances. There may be developmental features of fetuses that promise a more optimistic result for such interventions. If so, that would be a feature in favor of such interventions. In general, however, this is another area where great moral caution is warranted. In order to ethically justify fetal gene therapy we would need the moral justifications discussed above in connection with children. But other conditions would have to be met as well. (1) It would have to be the case that a postpartum intervention would very likely prove unsuccessful with death or permanent serious harm as a consequence. (2) It ought to be the case that the basic intervention itself is well understood so far as risks and benefits are concerned, so that the major experimental part of the intervention is associated with its use at the fetal level. In the case of fetal surgery, the techniques and risks of surgery were well understood, but it was recognized that there could be potential problems associated with size and so on that might result in bad outcomes. By way of contrast, gene therapy has been barely introduced into adult medicine. There have been only small successes. There yet remains the potential for serious adverse risks to be discovered. All of this would yield a general ethical counsel against such attempted interventions at this time.

Finally, just to be very clear and explicit, it is absolutely morally imperative that the free and informed consent of the mother be obtained for such interventions. We cannot imagine any circumstances that would permit an exception to that rule. Again, it is common enough in medical practice to treat children against the wishes of their parents when, for example, a Jehovah's Witness parent refuses a blood transfusion for a child who will almost certainly die without it. The ethical justification is that we usually respect the religious beliefs of individuals, no matter what the medical consequences for that individual. But we do not permit parents to make seriously harmful medical decisions for their children on the basis of beliefs that that child does not have the rational capacity to endorse. This would not hold true, however, in the case of fetal therapy. If a woman refused an intervention for her fetus, for religious or other reasons, we would not be warranted in overriding that refusal and imposing therapy upon her. On occasion this rule will yield tragic results; but those tragedies are likely to be so extremely rare that it would be unjustified to take the moral risks associated with permitting breaches of that rule.

GENE THERAPY: ETHICAL ISSUES AT THE POLICY LEVEL

Up to this point we have been discussing ethical issues related to gene therapy within the context of the doctor–patient relationship. But there are ethical issues at other levels as well, most often what we may broadly refer to as "policy issues." These may be either matters of professional policy or social policy. In the remainder of this essay we will identify and address some of those issues.

We should begin by making a large preliminary point. At one level the point will be obvious; at another level its implications may often escape attention. The point is that we live in what political scientists describe as a liberal, pluralistic, tolerant, democratic society. The term "liberal" in this usage is not opposed to the term "conservative." Rather, to say we live in a *liberal* society is to say that we are committed to maximizing the political space in which individuals are free to shape their lives as they wish, so long as they do not use their liberties to violate the comparable rights and liberties of others. Further, our society is *liberal* in the sense that there is no comprehensive value scheme (philosophical or religious belief system) that our government seeks to promote through our public policy choices. Rather, we expect that our public policies will be *neutral* among competing value schemes. This does not mean that our government can be indifferent to all values whatsoever. Our public policies and political practices must be such as to support those values that are central to maintaining a liberal political society. So we will wish to have in place a scheme of basic rights that we wish to protect for all. These include rights we are all familiar with, such as rights of free speech or freedom of religion, and so on. There will be argument about the detailed application and scope of these rights in complex social circumstances (quite unlike the world of our Founding Fathers), but that is to be expected in a democratic society.

Individuals will use their rights and the political space available to them to construct unique lives that will reflect often a personal ordering of values. This is what generates a *pluralistic* society. We are committed to social and political toleration of such ethical and religious diversity. (Historically, America wanted to avoid the destructive hatred that characterized seventeenth and eighteenth century Europe.) But this toleration is not unlimited. We do have to exist *as a society*, not just as a bunch of social atoms. That means that there will be some individual behaviors that will be outlawed as a matter of social practice. To justify such constraints in a liberal society, we have to be able to point to public interests that would be violated or undermined by that behavior.

A public interest is an interest that each and every one of us has, but that *as individuals* we are unable to protect that interest adequately. So we are then authorized to act collectively and coercively (fines/jail) to protect those interests. A simple example would have to do with air or water pollution. All of us have a health interest in having air and water that are "clean enough, safe enough." But there is little we can do as individuals to protect that interest if polluters are very large corporations hundreds of miles distant from us. We also regulate the practice of medicine because, again, we all have an interest in having access to competent, trustworthy medical practitioners, but very few of us have the capacity to assess those skills in individuals who claim the ability to practice medicine. That brings us to our bottom line question: To what extent are there public interests or deep social values that may be justifiably invoked in a liberal society to create social or professional policies that would regulate gene therapy?

The point of that question may be a little obscure at first. There has been no public uproar regarding gene therapy. Do we have any reason for believing that gene therapy should be any more or any less regulated than any other sort of medical intervention? To that question we need to give both a "Yes" and a "No" response. The reason for that will be clear when we make this next distinction. We have been discussing up to this point only somatic cell gene therapy. That is, the effect of somatic cell gene therapy is limited to the body of the person who has the therapy, Donald or Edward. By way of contrast, we need also to discuss germline genetic engineering. That sort of therapy is aimed at addressing genetic disorders, not simply in the genome of the individual who is genetically altered but in the descendants of that individual as well.

We do not presently have the capacity to do germline genetic engineering in any sense, but it is not unreasonable to expect that in the not very distant future we will have that capacity. This is no longer a science fiction scenario we are considering. To be more concrete, the term *germline* suggests we would be altering the genes of gametic material. But there are practical reasons why that is unlikely. The more plausible scenario is that couples will conceive in vitro; embryos will be grown to the eight-cell stage and then genetically analyzed. Genes associated with serious disorders, such as cystic fibrosis or various forms of Alzheimer's or various cancers, will be identified, then deleted from one or more cells of that embryo, then replaced with "normal" copies of the gene that should be there. Alternatively, since cells are totipotent at that developmental stage, a single engineered cell could be used to create a "new embryo."

Before continuing, there is another distinction we need to make. Up to this point we have been discussing what might be termed "negative" gene therapy. That is, our therapeutic efforts, whether somatic or germline, are aimed at correcting what all would agree is a serious medical disorder that threatens to compromise substantially either the length of life or quality of life of an individual. But we can also speak of "positive" gene therapy, which would be aimed at genetically enhancing some specific functioning of that individual or some socially desirable trait in that individual. If there are genes that can mutate from normal and cause serious dysfunction in an individual, then there are likely to be genes that mutate from normal in a way that yields superior functioning. It may be the case that there are genetic reasons why some individuals are taller, will live longer, have better memories, and so on. If this is true, and if we develop the capacity to do negative germline engineering as described above, then we could use the very same tools to accomplish positive germline genetic engineering. With these distinctions in mind we can raise some additional ethics issues regarding gene therapy from a social point of view.

Is there anything intrinsically morally objectionable about gene therapy in any of the four modes that we have identified? That is, are there any essential features of gene therapy that are so morally objectionable that the therapy as a whole ought to be rejected? To put the question into political terms, are there any serious public interests that would be so violated or so substantially undermined that we would be justified in banning these medical interventions? With respect to *somatic cell therapy* the answer to our questions seems clear. There are no deep moral values or serious public interests that are violated. Gene therapy represents a different approach to attacking disease than most standard medical therapies, but there is nothing morally problematic about that. There are risks associated with the development and deployment of this technology, but they seem to be of a kind and an order of magnitude that are comparable to much of accepted contemporary medical practice. Conscientious commitment to informed consent addresses those issues. Germline genetic engineering may be another matter.

There has already arisen a considerable literature attacking germline genetic engineering as unethical. We need to consider some of those objections now.

Violating Individual Autonomy

The first objection against human germline genetic engineering is that it seems to threaten the autonomy of individuals who are engineered. Clearly, eight-cell embryos can have no choice in what happens to them. It is true (we will claim) that embryos are not persons. However, it is equally true that they are *possible* persons, and not in just some remote metaphysical sense. As a practical and moral matter, the parents of these embryos intend them to become persons; and consequently, we would be presumptively obligated to treat them as persons in these circumstances. Still, the obvious response is that no reasonable person would want to be born knowing that they would be afflicted with Tay-Sachs or Huntington's or cystic fibrosis if they had the option of being spared these disorders. If this is true, then this kind of genetic engineering is not obviously violative of autonomy. However, the examples we gave were about negative genetic engineering; parental efforts at positive genetic engineering may not be so obviously respectful of the autonomy of these future individuals.

Parents may want their very young children to excel in music or gymnastics or sports and push their children in these directions quite inappropriately. While this is morally troubling, we (outsiders) take moral comfort in knowing that these children will likely have the opportunity in the future to revolt and make their own choices. In the case of positive genetic engineering, however, it is very unlikely that genetically engineered changes (e.g., for above-average height) could be easily reversed. On the face of it, this seems a much more serious infringement on the autonomy of these future individuals.

We noted earlier that it is a cardinal principle of medical ethics today that the autonomous choices of patients must be respected by physicians. The prior paragraph assumes that the "real" patient is the prospective child, and that parents are merely surrogate decision makers who must act in the best interests of that future child. The concept of best interests works reasonably well as a source of moral guidance in standard clinical circumstances, for example, a child afflicted with life-threatening cancer. We saw that in our discussion of the cases of Donald and Edward. In germline engineering circumstances, however, the concept seems excessively vague and unhelpful at best, conceptually muddled at worst. Hence, the argument can be made that it is the parents who are the "real" patient in this clinical encounter, and that it is their autonomy that must be respected, at least as long as what they request is not clearly violative of basic principles in medical ethics. For now we will merely flag this perspective since the prior question is whether there ought to be a social policy that would prohibit or severely circumscribe the development/deployment of germline genetic engineering.

In a liberal society the reasonable argument is that individual liberty is to be protected, especially the fundamental liberties associated with reproductive decision making, unless there is a compelling public interest that would warrant restriction of that liberty, including a threat to the just liberties of others. Thus, if there were a germline genetic technique for creating very docile children, especially during adolescence, that might be strongly desired by parents, but it would be violative of the fundamental autonomy rights of those future children. The state would be morally and politically obligated to prohibit that bit of genetic engineering. By way of contrast, there are numerous traits (eye color, hair color, predisposition to baldness) that may be genetically alterable and that in themselves do not seem to be of great consequence, morally speaking. Assuming that germline genetic engineering had been perfected to the point where it itself posed minimal risk, a liberal society would have no compelling reason for preventing such parental choices. Choosing the gender of an embryo (for reasons that were not clearly medical) would be morally contentious; there might be public interests that would justifiably foreclose such an option. Choosing the sexual orientation of an embryo (should this prove to be primarily genetically determined) would be even more contentious, morally and politically. Space precludes engaging that issue.

Our first conclusion is that germline genetic engineering is not intrinsically violative of moral autonomy. It could be misused in that way, but that is not an intrinsic feature of the technology.

Human Hubris: The Audacity of Playing God

A second objection often voiced to human germline engineering is that it involves "playing God," or, from the perspective of atheists, interfering with the wisdom of natural evolution. Stripped of the theological or "quasi-scientific" language, what this objection amounts to is the claim that there is something morally privileged about the genetic status quo. This is not a morally self-evident claim. The seventeenth-century philosopher Thomas Hobbes described Nature as being red in tooth and claw. When we consider the level of misery and suffering imposed on humankind by 4000 or so genetically based medical disorders, Hobbes' description does not seem inapt. For us to have the capacity to eliminate many of these disorders, and then to fail to do so because of a worshipful attitude toward the genetic status quo, would evidence a level of moral indifference and moral irrationality that would be wholly unjustified.

Still, there is a deeper issue raised by this objection that is not so easily dismissed. Specifically, genetic changes are fundamental, potentially profound and far-reaching. Traditional medical interventions have effects that for the most part are limited to an individual. Germline genetic changes can profoundly affect multiple generations. From this perspective moral caution regarding the genetic status quo might be morally warranted.

There is a European version of this second objection that deserves moral consideration. It is the "genetic patrimony" argument (Mauron and Thevoz, 1991). In brief, it is the view that the human genome is the common inheritance of humankind as a whole; and consequently, no individual has the moral right to alter or manipulate that common inheritance merely for his or her private benefit, including benefits for familial descendants. This is not an argument that will resonate well with most Americans, since we have a strong commitment to individual rights. However, we also have a strong appreciation for the environmental movement, which we see as justifiably limiting individual property rights. The genetic patrimony argument appeals to an analogous moral and political logic. There are, however, some critical disanalogies.

The environment is actually something shared. Despoiling the environment can have critical consequences for large numbers of people. The human genome is common in more of a metaphorical sense. Only individuals are the actual possessors of genetic information. Genetic engineering is always done to individuals. Obviously those changes can be transmitted to future generations; but if those changes prove deleterious to the first-generation recipient, then the technology that put them in place could presumably be used to reverse or correct them for future generations. [Note: Our intent is not to minimize the risk of harm to that first generation. Our working assumption is that researchers would have taken great care to be as certain as scientifically possible that irrevocable harm would not befall that first generation.] Further, our genetic patrimony is an abstract and idealized notion. Perhaps an argument could be made for preserving the integrity of that patrimony if the genetic changes we contemplated were cosmetic. But our genetic patrimony includes at least 4000 serious medical disorders that shorten life and reduce dramatically the quality of life for hundreds of millions of individuals. If we have the technology and economic resources to prevent that real human misery, then appeal to the preservation of our genetic patrimony is an insipid moral counterargument at best and morally perverse at worst.

Another variant of this argument that is often voiced is that genes that appear to impose serious medical harm on individuals may have temporarily hidden beneficial effects as well, that we might ignorantly lose those evolutionary advantages through short-sighted tampering. The gene associated with sickle cell disease is most often cited in connection with its protective effects for malaria. However, if this line of argument is generalized, it too has consequences that are both irrational and immoral. It is theoretically possible that the Huntington's gene would prove to have beneficial effects in some obscure corner of the world or thousands of years from now. Speculation like that, however, is too remote and uncertain (to say the least) to justify the very real suffering that will be endured by individuals with the Huntington's gene if we have the capacity to remove that gene from their genetic endowment near conception.

Germline Engineering Is Too Risky

A third objection to germline engineering is related to the second: If we have a morally legitimate objective, such as saving human lives or reducing human suffering, and if we have several ways of achieving that objective, then we ought to choose those means that are morally least risky. The assumption here is that human beings do not have the necessary wisdom to engage in germline engineering. Thus, if we have a somatic cell genetic engineering approach to managing cystic fibrosis that is reasonably effective in controlling symptoms and prolonging life, then we would be morally obligated to do that before we engaged in germline genetic engineering. Failing that, however, we would still have the alternative of creating multiple eight-cell embryos in vitro, genetically analyzing them, then implanting only those embryos free of that defect or carrying it only in a recessive form. This is what is referred to today as preimplantation genetic diagnosis and selective elimination.

As a general principle it is true that the morally safer course is to be preferred to morally riskier courses. But this assumes that precisely the same goal is achieved by either course. If this is not the case, then there are judgments involving moral trade-offs that come into play. Thus, let us assume that we have a very effective somatic cell genetic therapy for cystic fibrosis. It is so effective that individuals can be guaranteed something near a normal life expectancy. But this therapy must be taken every 6 months. We can hypothesize whatever cost we want; the bottom line will be that these costs will be substantially greater than the $50,000 for a genetically engineered embryo. Note that this is not simply an economic argument. There are issues of justice: the fair distribution of limited health care resources in the face of virtually unlimited needs. We would need to ask what other health needs will go unmet that make a moral claim on us because we chose to support the more expensive somatic cell intervention for CF rather than a less expensive germline intervention, which would also have the long-run cost-reducing effect of eliminating that CF gene from all future descendants.

The obvious response to the above argument is that we should then adopt the embryo selection strategy. This would achieve the long-term effect we sought at reduced cost. This argument will work so long as we are concerned about only one genetic defect per embryo. But for most embryos, it may turn out, multiple genetic replacements for serious medical disorders would be prudent and cost effective. If this is so, then there is no alternative way of achieving this result other than through germline genetic engineering. Still, the claim will be that this is a risky course, medically, and therefore morally. What right do we (parents, physicians, society) have to expose these future children to these risks? There are, of course, risks that are clearly morally unwarranted. But there seems to be a very large discretionary area of risk taking that is part of current medical practice. In the case of extremely premature infants (600 g, 23 weeks gestation), for example, there is an 80% chance that these infants will die (in spite of aggressive care), and, of the survivors, 70% will be afflicted with moderate to severe mental and physical impairments, mostly as a result of cerebral bleeding. We allow parents to choose aggressive care, in spite of the suffering associated with that and the likelihood of a bad outcome. In fact, some physicians would argue that they and the parents are morally obligated to take these risks. If this is current medical practice, and if it is seen as being morally warranted, then unless germline genetic engineering involves the risk of harm to these future children greater than what we tolerate with respect to very premature infants, we would have no moral justification for forbidding the implementation of this technology.

Embryo Destruction

A fourth argument against germline genetic engineering is what we might refer to as the "excess embryo" argument. It starts with the reasonable assumption that both the development and implementation of germline engineering will require the creation of numerous embryos, only a small portion of which will be successfully engineered and implanted, the rest being rejected and discarded. Strong antiabortion advocates who believe embryonic life makes moral claims on us from the moment of conception will obviously object to this massive destruction of human life. However, this argument does seem rooted in a religious vision not shared by a majority in our society. As a liberal, pluralistic society any moral vision that will govern our shared political life (and potentially employ justifiably the coercive powers of the state) must command something close to unanimous assent.

But there is a variant of the excess embryo argument that does seem to meet this test. Nolan (1991) sees a moral ambiguity in our treatment of these embryos that is difficult to rationally justify. Specifically, we justify germline genetic engineering as an extension of clinical medicine with "the ostensible goal of providing therapy for these 'patients,'" while at the same time we seem "quite comfortable with pursuing germline genetic research that would itself entail substantial destruction of embryonic life" (Nolan, 1991, p. 615). In no other area of clinical medicine or research medicine do we permit the destruction of failed patients.

This objection can be answered by noting an ambiguity in the use of the term "patient," as applied to the eight-cell embryo. Morally speaking, the embryo is not a patient in the same sense that an infant is a patient. The embryo is a patient-whom-we-hope-will-become-a-person-if-all-goes-well. What this signals is a therapeutic attitude toward that embryo, as opposed to a merely experimental attitude. Morally speaking, that is important. However, embryos do not have moral rights; infants do. Two quick examples will illustrate the relevant moral differences. If we have genetically engineered an embryo and implanted it, and then in the third month of gestation some environmental factor causes terrible damage to the fetus, then this mother has the moral right to choose abortion, especially if it is her judgment that this is in the best interests of the fetus who would otherwise face a seriously compromised life of unmitigated suffering. By way of contrast, if this same genetically engineered embryo is born, but some serious medical disorder emerges after birth that is an unexpected consequence of the genetic engineering, then we would have a strong societal obligation to do everything medically reasonable to correct or ameliorate that disorder. Note that we have in mind here a crippling disorder, perhaps one that would be very costly to correct or ameliorate, as opposed to a fatal disorder where heroic medical intervention could only prolong a painful dying process. So long as this moral commitment is in place, we do not see a strong moral objection to germline engineering on the grounds of embryonic destruction.

Threats to Health Care Justice

A fifth objection we need to consider is an argument from justice. This is a complex objection that has many dimensions to it and that is very sensitive to empirical matters of a political and economic sort. The first version of the objection is captured by this question: *Are we morally obligated as a society to develop and implement germline genetic engineering as a therapeutic strategy*? That is, if this were foreclosed as a therapeutic research option by social choice, would we have violated any key moral rights that individuals have, especially future individuals afflicted with specific genetic disorders who could have been spared those disorders if the research had been allowed to go forward? The objector's response to these questions is, in effect, that no one has a just claim to germline genetic engineering. That is, as many have argued, there are an indefinitely large array of therapeutic medical technologies that are possible; but, given limited resources for meeting medical needs in any society, only some fraction of these can be actualized and deployed (Callahan, 1990). If everyone is thereby denied access to these technologies, and if virtually everyone could potentially benefit from them, then all are treated impartially and fairly. That is, it is regrettable and unfortunate that many future individuals will be afflicted with serious genetic disorders they could have been spared, *but, morally speaking, it is not unjust.*

This is a line of argument we have already addressed at great length in another publication (Fleck, 1994). Briefly, if we have only limited resources for meeting virtually unlimited health needs, then those needs must be *fairly* prioritized. This is the problem of health care rationing. One way of thinking about this problem is from the perspective of protecting fair equality of opportunity for all over the course of a life (as opposed to looking at this as a problem of justice at a point in time) (Daniels, 1985). We can imagine that by the year 2003 we will have developed and deployed a totally implantable artificial heart with annual costs of $52 billion ($1997) for 350,000 people per year. By the year 2010 we will imagine we have the capacity to do germline genetic engineering. As a society we cannot afford all of this. [For the sake of establishing a context, we in the United States in (1997) spend more than $1.1 trillion on health care, about 14.5% of gross domestic product (GDP), or about 3 to 5 percentage points more of GDP for health than all other advanced nations in the world. Everyone today is mindful of intense political and business pressures to control health care costs.]

A strong case can be made that the fair equality of opportunity principle would say we are morally obligated to fund germline genetic engineering before we continue funding for artificial hearts. While it is generally true that no one's moral rights are violated if a society chooses not to develop and deploy a new medical technology, *in this case* there is a justice-based argument that would warrant the conclusion that germline genetic resources should command societal resources before ongoing expenditures for artificial hearts. A key element of that argument is that roughly 70% of those artificial hearts would go to individuals over age 65, that is, individuals who would have had the opportunity to lead a full life of reasonable quality. What germline genetic engineering would offer is that same opportunity to other individuals who were at the very beginning of life.

Along these same lines, there is a comparable argument that might be made regarding somatic cell gene therapy. That is, like the artificial heart, somatic cell gene therapy may prove to be another very expensive "half-way" technology. This means that the intervention does not really cure the medical problem; instead, the problem is substantially ameliorated through repeated application of the technology. The need for repeated application of the technology is what adds dramatically to the social costs of the technology. It will be an empirical question as to whether or not this is what happens with the somatic gene therapy work currently being brought to fruition. If this is what happens, however, then we believe a strong case could be made, from the perspective of health care justice, for giving lower funding priority to somatic gene therapy relative to germline gene therapy.

A second version of the justice objection starts with the assumption that it would be very improbable that a technology as promising as germline genetic engineering could fail to be developed in our society. We also assume that this would be a costly technology, perhaps costing $50,000 per genetically engineered and implanted embryo. If this sort of intervention were not covered by health insurance, then only the fairly affluent could afford it. If both positive and negative genetic engineering were options, then the wealthy would be able to buy opportunity-enhancing interventions for their children that would result in a widening of the gap between our professed societal commitment to equality of opportunity and the actual extent to which equality of opportunity was protected in practice (Brock, 1994). Further, the enhanced opportunities thereby purchased would be purchased not simply for their children but for an indefinitely long line of descendants, thereby creating rigidified class structures and diminished opportunities for those already less well off in our society, which is prima facie unjust.

An alternate scenario assumes that private insurance coverage would be available for germline genetic intervention. But this is hardly a more promising scenario if our concern is with protecting justice. The working middle class would then have access to this technology. If two million middle-class embryos were genetically engineered at $50,000 each, that would add $100 billion to the cost of health care each year. This would do nothing to correct the problems of injustice so far as fair equality of opportunity was concerned for those less well off in our society. On the contrary, they would be worsened because a feature of private health insurance now is that it is exempt from federal income tax and the social security tax, which represented a $80 billion subsidy (tax expenditure) for the middle class in 1996. Under this scenario that subsidy would increase by $28 billion per year, real revenue for the federal government, which would translate into program cuts or increased taxes. Medicaid would be politically vulnerable in this scenario. Meanwhile in the private sector businesses faced with increased insurance costs would struggle harder to extract discounts from hospitals, thereby diminishing the resources hospitals now use to underwrite the costs of some health care for the uninsured. In short, the welfare of the poor and uninsured would be most threatened under this scenario. Further, the prospects for more equitable access to health care for the poor and uninsured through national health insurance becomes more remote, politically speaking, because expanding access for them to a constantly improving package of health services would mean higher costs to the middle class and, likely, reduced health benefits for them relative to the generous packages they now have. The conclusion of this argument is that if we want to protect just access to health care for all, then we ought to ban germline genetic engineering as a therapeutic option.

This line of argument has considerable moral bite in the world as it is. But there is an alternate way of concluding the argument that is morally preferable. Specifically, what we are morally obligated to do as a society is to put in place a truly national health care insurance mechanism: universal access to a fairly thick package of health benefits. If, as is arguable, negative germline genetic engineering substantially improved fair equality of opportunity for all in our society, then it ought to have high priority as preventive health care, and it ought to be included in that basic benefit package. As noted earlier, this argument is highly sensitive to medical and economic facts, all of which are hypothetical here. But if there were this strong connection in fact to protecting fair equality of opportunity, then this would be a justice argument supportive of deploying at least negative germline genetic engineering.

Slippery Slope to Eugenics

The last moral objection we need to consider is the "slippery slope to eugenics" argument. We begin with a definitional point. The term *eugenics* is historically associated with the eugenics movement of the 1920s and 1930s as well as with the atrocities of Nazi Germany. A common form of the slippery slope argument is that we would start by permitting negative germline genetic engineering freely chosen by parents for their offspring, which would be viewed as a morally reasonable option, but before long we would have social policies coercively imposed that would mandate genetic enhancement of all embryos to maximize the genetic well-being of society and reduce future health care costs.

The quick response to this objection is that the feared slide toward eugenics can be prevented by putting in place social policies supportive of reproductive freedom and professional practices that restrict such genetic reproductive decision making to the privacy of the doctor–patient relationship. But this may be too facile a response. Diane Paul (1994) and others (Lippman, 1991; Karjala, 1992; Holtzman/Rothstein, 1992) have made the argument that reproductive genetic freedom can have eugenic consequences just as morally objectionable as the most coercive of government policies. What can easily happen is that social pressures and professional "judgment" can conspire with one another to elicit socially correct genetic choices from prospective parents in the privacy of the doctor–patient relationship under the guise of reproductive freedom and informed consent. In addition, private insurance companies can exercise their free market rights and responsibility to protect "actuarial fairness" for all their clients by denying health insurance coverage to children born with preventable genetic disorders. Given this, we must observe that the "virtue" of coercive government eugenic policies is that they are public and visible and open to democratic criticism and change. By way of contrast, social pressures are private, organizationally diffuse, unaccountable but oftentimes morally legitimated (reproductive freedom/best interests of the future child); and therefore, they are extremely difficult to control or change (try to get a court order against social pressure). The conclusion of this line of argument is that we ought to ban germline genetic engineering altogether. This would mean inflicting premature death, chronic disabilities, and considerable suffering on tens of millions of future individuals—all of which would be preventable and, hence, presumptively morally problematic; but the eugenic consequences of the alternative are judged to be even more morally intolerable.

This last line of argument deserves a more subtle and complex response than space permits. We will suggest one possible line of response. There is considerable social pressure on men today not to make sexist comments. Someone might want to argue that this represents effective (but informal) violation of their rights to free speech. But if sexist comments really are bad in terms of their effects on women, as many feminists would argue, then it is not obvious that social pressure to constrain such speech so that it is possible only in the most private of settings is morally or politically objectionable. If there is strong social pressure for negative germline engineering, such that parents who chose to have their children naturally and take the risks associated with the genetic lottery when they had the option of ensuring a healthy genetic endowment for their children would be thought of by the rest of society as being irresponsible, then it is not obvious that this is morally or politically objectionable. Like the Amish, these parents could resist pressures for conformity to contemporary social mores.

Societally available negative germline genetic engineering would have eugenic effects, but it is not obvious that this in itself is morally objectionable. On the contrary, there are numerous moral considerations that would justify seeing this as a morally permissible and morally desirable outcome. Some might claim that this represents an evisceration of reproductive freedom; but the alternate perspective is that this represents social pressure for the responsible use of reproductive freedom. As noted, this does bring about a eugenic effect. But what is most morally objectionable about eugenics is that society would *use* individuals as mere means to eugenic ends, typically employing coercive means rather than methods of rational persuasion. That is, the Kantian principle of respect for persons would be violated. But this is not what is being depicted in the scenario we have in mind. On the contrary, we assume that rational persuasion and rationally well-founded beliefs would be at the core of the social pressure to which we have alluded.

We have deliberated avoided introducing positive genetic engineering in the latter parts of this discussion. The issues become too complex for a brief chapter. What we believe we have succeeded in doing is suggesting moral arguments that would support a presumption in favor of continuing the development of germline genetic engineering. That is, we have argued that germline genetic engineering is not intrinsically morally objectionable. Nor are there obvious and substantial public interests that would necessarily be subverted were the technology to become real. There are potentially harmful social consequences that could come about, but we have the capacity for designing prudent social and professional policies that should minimize that risk. We conclude by emphasizing that we are only at the beginning of this moral dialog. Much more subtle moral and political judgment will be necessary as the technology evolves. The one social imperative that should guide us for now is the imperative to create political forums in which we can engage in rational democratic deliberation about these issues, where "we" includes all strata in our society, not just an educated elite.

KEY CONCEPTS

- One of the cardinal principles of health care ethics today is the principle of respect for patient autonomy. That principle states that competent patients have a strong moral right to decide for themselves what is in their best medical interests.
- In medicine the practice of informed consent protects patients from the risks that are always part of medical practice. Informed consent is aimed at assisting patients to make more autonomous choices about their own medical care. The

role of a physician is to present honestly and in a way that is intelligible to a particular patient, the risks and benefits associated with a proposed therapeutic intervention, along with other reasonable alternatives to that intervention. The core element of informed consent is information, reliable, scientifically and clinically grounded information. The two most common misunderstandings in medical research related to informed consent are (1) an excessively optimistic view of what participation in the experimental therapy might yield for the patient; and (2) in those cases where the trial is structured as a double-blind randomized controlled study, a failure to appreciate that there is a 50% chance that the patient would not receive the therapy they might expect. The ethical obligation of researchers in these circumstances is to correct these misconceptions so that such patients are making autonomous decisions to participate.

· The most basic principle in medical ethics is what is referred to as the Kantian principle of respect for persons. This principle says that it is never ethically acceptable to treat persons as if they were mere things, as if one individual could use another individual for purposes that the latter individual did not accept as their own.

· A second basic principle of health care ethics, probably the oldest of these principles, is what is referred to as the principle of nonmaleficence. It is often interpreted to mean that at the very least physicians should do nothing that will cause unnecessary harm to their patients.

· A third ethical principle that needs to be considered is usually referred to as the principle of beneficence. One formulation of this principle would say that physicians always ought to act in such a way as to advance the best medical interests of their patients. A construal of the principle of beneficence better protects patient autonomy. It says that patients should have the right to determine what they judge to be in their own best interests (within the constraints of good medical practice) from the point of view of their own stable values and life goals.

· The fourth ethical principle that has a bearing is the principle of justice. In its formal sense justice requires that like cases be treated alike.

· In medical research, the ethically required course of action is that the physician who is primarily responsible for providing therapeutic care to a patient be different from the physician who is responsible for the research.

· Health care in our society should be regarded as a "morally special" sort of social good, most especially when that care can make the difference between life and death, or the difference between permanent injury and the healing of that injury/restoration of function.

· The whole point of both moral and political rights is to protect individuals against the encroachments of large and powerful social organizations, either in the private or public sector. Individuals may not have their deep moral rights violated for the sake of advancing some larger social good.

· To ethically justify fetal gene therapy the moral justifications in connection with children need to met as well as the following: (1) It would have to be the case that a postpartum intervention would very likely prove unsuccessful with death or permanent serious harm as a consequence. (2) It ought to be the case that the basic intervention itself is well understood so far as risks and benefits are concerned, so that the major experimental part of the intervention is associated with its use at the fetal level.

· Ethical issues relate to professional and social policies in that our public policies and political practices must be such as to support those values that are central to maintaining a liberal political society. Thus, we need in place a scheme of basic rights that we wish to protect for all. Individuals will use their rights and the political space available to them to construct unique lives that will reflect often a personal ordering of values. This is what generates a pluralistic society. To function in this society, however, there will be some individual behaviors that will be outlawed as a matter of social practice. To justify such constraints in a liberal society we have to be able to point to public interests that would be violated or undermined by that behavior. A public interest is an interest that each and every one of us has, but that as individuals we are unable to protect that interest adequately. Society regulates the practice of medicine because, again, we all have an interest in having access to competent, trustworthy medical practitioners, but very few of us have the capacity to assess those skills in individuals who claim the ability to practice medicine.

SUGGESTED READINGS

Andrews LB, Fullerton JE, Holtzman NA. Assessing genetic risks: Implications for health and social policy. Washington, DC: National Academy Press. 1994.

Beauchamp T, Childress JF. Principles of biomedical ethics, 4th ed. New York: Oxford University Press. 1994.

Brock D. The human genome project and human identity. In Genes and Human Self-Knowledge: Historical and Philosophical Reflections on Modern Genetics, Weir RF, Lawrence SC, Fales E eds. Iowa City: University of Iowa Press. pp18–33, 1994.

Callahan D. What kind of life: The limits of medical progress. New York: Simon and Schuster. 1990.

Daniels N. Just Health Care. Cambridge University Press, Cambridge, 1985.

Davis D. Genetic dilemmas and the child's right to an open future. Rutgers Law J 28(Spring):549–592, 1997.

Fleck L. Just genetics: A problem agenda. In Murphy TF, Lappe MA (Eds.), Justice and the Human Genome Project. University of California Press, Berkeley, 1994, pp. 133–152.

Holtug N. Altering humans—The case for and against human gene therapy. Cambridge Q Healthcare Ethics 6:157–164, 1997.

Holtzman NA, Rothstein MA. Eugenics and genetic discrimination. Am J Hum Genet 50(March):457–459, 1992.

Juengst E. Can enhancement be distinguished from prevention in genetic medicine? J Med Philos 22(April):125–142, 1997.

Kahn J. Genetic harm: Bitten by the body that keeps you? Bioethics 5:291–308, 1991.

Karjala DJ. A legal research agenda for the human genome initiative. Jurimetrics 32(Winter):121–222, 1992.

Kitcher P. The Lives to Come: The Genetic Revolution and Human Possibilities. Simon and Schuster, New York, 1996.

Kolata G. Genetic defects detected in embryos just days old. New York Times 142 (Sept. 24):A1, A12, 1992.

Lappe M. Ethical issues in manipulating the human germ line. J Med Philos 16(Dec.):621–640, 1991.

Lippmann A. Prenatal genetic testing and screening: Constructing needs and reinforcing inequities. Am J Law Med 17:15–50, 1991.

Mauron A, Thevoz JM. Germ-line engineering: A few European voices. J Med Philos 16(Dec.):649–666, 1991.

McGleenan T. Human gene therapy and slippery slope arguments. J Med Ethics 21:350–355, 1995.

Nolan K. Commentary: How do we think about the ethics of human germ-line genetic therapy? J Med Philos 16(Dec.):613–619, 1991.

Paul D. Is human genetics disguised genetics? In Weir RF, Lawrence SC, Fales E (Eds.), Genes and Human Self-Knowledge: Historical and Philosophical Reflections on Modern Genetics. University of Iowa Press, Iowa City, 1994, pp. 67–83.

Peters T. "Playing God" and germline intervention. J Med Philos 20(Aug.):365–386, 1995.

Robertson J. Genetic selection of offspring characteristics. Boston University Law Review 76(June):421–482, 1996.

Walters L, Palmer JG. The Ethics of Human Gene Therapy. Oxford University Press, New York, 1997.

Weatherall DM. Gene therapy in perspective. Nature 349:275–276, 1991

Zohar NJ. Prospects for "genetic therapy"—can a person benefit from being altered? Bioethics 5:275–289, 1991.

CHAPTER 15

Epilogue: Personal Genetic Medicine—The Future Is Now

THOMAS F. KRESINA, PH.D.

INTRODUCTION

A major component of genetic-based research involves the identification of protocols that can be applied to diseases or specific pathogenesis. However, the rapidly advancing technology revolution and the elucidation of the human genome (see the Appendix) have opened other windows for the application of molecular genetics in the practice of medicine. A significant window in the medical management of patients is the use of specific identified human genetic polymorphisms occurring in genes. These polymorphisms are cataloged and used in the context of susceptibility to disease and the pharmacokinetics of prescription drugs. The field has advanced so far that individualized medical care based on the identification of genetic polymorphisms has recently been introduced and is currently offered in Sweden.

DNA DATABASES

Deoxyribonucleic acid (DNA) polymorphisms are changes in nucleotide sequences in genes noted when gene sequences are compared between individuals. Certain nucleotide sequence changes in DNA genes result in overt defective gene function and pathogenesis. They are targeted by gene therapy approaches. However, nucleotide sequence changes can also be "minimal" and result in subtle changes in gene function without inducing pathogenesis. These modifications can be based on the change of a single nucleotide, and they are termed "single nucleotide polymorphisms," or SNPs. SNPs are thought to play a major role in determining the etiology of differing expression between individuals with regard to drug responses and disease susceptibility. Thus, there are major efforts ongoing to establish and utilize SNPs databases. Currently, there are two types of databases, public and private. Public databases, such as HGBASE (Human Geneic Bi-Allelic Sequences) and

An Introduction to Molecular Medicine and Gene Therapy, Edited by Thomas F. Kresina
ISBN 0-471-39188-3 © 2001 Wiley-Liss

the University of Minnesota Biocatalysis/Biodegradation Database (UMBBD), are constructed as public access databases through the Internet. They can be non-specific, such as HGBASE, and function as a repository for all human gene-linked polymorphisms derived from any source. Alternatively, they can specifically categorize integrated genetic and metabolic data as in the case of UMBBD, which provides approximately 100 pathways for microbial catabolic metabolism of organic compounds. On the other hand, bioscience pharmaceutical companies may maintain proprietary databases for use in drug development. A major source of support for the identification of genetic polymorphisms and database generation is the pharmaceutical industry and the Wellcome Trust. It has funded, in 1999, the SNP Consortium, which is a worldwide research initiative ($45 million, 2 years) that plans to identify all of the 300,000 SNPs in the human genome and to map at least 50% of them.

INDIVIDUAL GENETIC PREVENTIVE MEDICINE

The identification and databasing of genetic polymorphisms, including SNPs, combined with data derived from DNA arrays (gene expression levels; see the Appendix) will provide physicians and health care workers a personalized genetic map of a patient. These genetic maps can be used preventively and establish a correlation between genotype and phenotype. These genetic maps will allow health care workers to identify genetic interactions that, along with environmental factors, cumulatively result in disease. Thus, interventions at an environmental, behavioral, pharmaceutical, or genetic level could be introduced to prevent the developing disease at its earliest stage, inception.

The practice of preventive genetic medicine and obstetrics will be of great significance and also be controversial. With the advance of personalized genetic medicine, the concept of genotyping and genetic counseling will become particularly important. Routine genotyping would not only facilitate personalized medicine, it would also permit effective genetic and prenatal counseling. Currently, many individuals first discover that they carry a gene for a recessive disease when an affected descendant is born. Only subsequent pregnancies can be planned with foreknowledge. Routine genotyping would allow couples to know what genetic diseases to anticipate in all of their offspring. This advance knowledge is crucial for genetic counseling to be most effective and will be increasingly important as family size declines. If potential parents knew what genetic diseases to anticipate in their progeny, they could take advantage of preimplantation diagnosis. In 1992, it was shown that in vitro fertilization and "embryo-biopsy" could be used to detect specific genetic lesions in early embryos. Also described was how a couple could experience the birth of a normal baby girl even though both parents carried the F508 mutation of the cystic fibrosis transmembrane regulator (CFTR) gene. This achievement was important. About 4% of Caucasians carry a CFTR mutation. In this segment of the population, about 1 in 2500 babies have lesions in both copies of the gene. Assuming that two-Caucasian couples constitute about 50% of the 4 million couples having a baby in the United States each year, about 1000 babies will be born annually with cystic fibrosis. Because cystic fibrosis is the most common potentially fatal autosomal recessive disease in the white population, a successful intervention has the potential to benefit more individuals than a treatment for almost any other genetic disease in Western culture. Furthermore, since preimplantation diagnosis allows selected embryos to be transferred to the uterus and others to be cryopreserved, this procedure may be an option for couples who find pregnancy termination unacceptable.

INDIVIDUAL GENETIC MEDICINE AND PHARMACOTHERAPY

Genetic differences between individuals strongly alter responses to conventional drugs. Thus, genetic information will be needed to achieve maximum benefit from all types of therapeutic interventions. For the vast majority of individuals who are born without genetic mutations causing disease in childhood, this information could guide adult life-style and therapeutic choices. For example, it would improve methods for calculating the cost/benefit ratio of an intervention such as hormone replacement therapy—where the relative risks and benefits are likely to depend upon a patient's genetic constitution. Genetic information could also be used to select the best drug for an individual patient. The emerging field of pharmacogenetics examines the relationship between gene polymorphisms and drug responses. Because they play a central role in drug metabolism, the six forms of human cytochrome P450 are currently under intense investigation. An emerging field of research is the genetic manipulation of drug metabolism via the cytochrome system.

However, the initial experience with individual genetic-based medicine in Sweden is utilized for cardiovascular disease and drug metabolism. Eurona Medical AB, in Uppsala, Sweden, has made its angiotensin-converting enzyme (ACE) inhibitor responder assay available. The ACE inhibitor responder assay kit predicts an individual patient's response to ACE inhibitors for hypertension. Data from this assay allow the health care worker to prescribe antihypertensive drug treatment based on the patients genetic information, that is, genotype. Drug regimens will be tailored to the pharmacology predicted by the patient's genetic makeup. It is anticipated that additional assays for other drug regimens (β-blockers, antiotensin II antogists) for cardiovascular pathology will be on the market in the near future.

Additional databases are being developed to correlate DNA polymorphisms with clinical trial outcomes. This will establish genetic databases defining the most efficacious therapeutic approach and the correlate as well, a relationship between genotype and drug toxicity.

A final emerging question is: Will genotyping render gene therapies prohibitably expensive by lowering the incidence of genetic diseases? Since 1976, it has been reported that the research and development costs of bringing a new drug to market increased from $54 million to more than $500 million. Allowing for a 5% annual rise due to inflation, this constitutes a fourfold increase in real research and development (R&D) costs over a 20-year period. On average, non-R&D costs add an additional $200 to $300 million. It appears that pharmaceutical companies are deciding what drugs to add to their development portfolio based solely on their financial potential, not their novelty. "Admission fees" for new compounds in terms of sales expectations have become so exorbitant that many indications for innovative drugs previously considered in the past are no longer on the table. Assuming that R&D

costs suddenly cease rising and that development costs for gene therapies are no greater than those of conventional drugs (about $500 million), a gene therapy for a target population of 10,000 (the number of babies with cystic fibrosis expected to be born in the United States during a 10-year period) would need to return $50,000 per patient to break even and perhaps $100,000 to be acceptably profitable. Given these calculations, it is reasonable to consider that declining costs of DNA analysis and increasing options for preimplantation testing may combine to reduce the incidence of certain heritable disorders to the point that gene therapies for these diseases will not be cost-effective to develop. Thus, genotyping or genetic screening coupled with genetic counseling may replace the need for gene therapies. We can currently see that conventional genetic screening has already reduced the incidence of thalassemia and Tay-Sachs disease. This decline shows how powerfully genotype information can influence decision making and modify the cost of disease.

SUGGESTED READINGS

Genomics

Borrebaeck CA. Tapping the potential of molecular libraries in functional genomics. Immunol Today 19:524–527, 1998.

Dutton G. Computational genomics: Medicine of the future. Ann Intern Med 131:801–804, 1999.

Guengerich FP. The environmental genome project: Functional analysis of polymorphisms. Environ Health Perspect 106:365–368, 1998.

Leipe DD. Biodiversity, genomes and DNA sequence databases. Curr Opin Gen Dev 6:686–691, 1996.

The Human Genome Project. Br Med J (BMJ) 7220:1282–1286, 1999.

Single Nucleotide Polymorphisms: SNPs

Marth GT, Korf I, Yandell MD, Yeh RT, Gu Z, Zakeri H, Stitziel NO, Hillier L, Kwok PY, Gish WR. A general approach to single-nucleotide polymorphism discovery. Nat Genet 23:452–456, 1999.

Databases and Infomatics

Brookes AJ, Lehvaslaiho H, Siegfried M, Boehm JG, Yuan YP, Sarkar CM, Bork P, Ortigao F. HGBASE: A database of SNPs and other various variations in and around human genes. Nucleic Acids Res 28:356–360, 2000.

Dicks J, Anderson M, Cardle L, Cartinhour S, Couchman M, Davenport G, Dickson J, Gale M, Marshall D, May S, McWilliam H, Omalia A, Oughham H, Trick M, Walsh S, Waugh R. UK CropNET: A collection of databses and bioinformatics resources for crop plant genomics. Nucleic Acids Res 28:104–107, 2000.

Ellis LB, Hershberger CD, Wackett LP. The University of Minnesota Biocatalysis/ Biodegradtion Database: Microorganisms, genomics and prediction. Nucleic Acids Res 28:377–379, 2000.

Overbeek R, Larsen N, Pusch GD, D'Souza M, Kyripides N, Fonstein M, Maltev N, Selkov

E. WIT: Integrated system for high-throughput genome sequence analysis and metabolic reconstruction. Nucleic Acids Res 28:123–125, 2000.

Smigielski EM, Sirotkin K, Ward M, Sherry ST. dbSNP: A database of single nucleotide polymorphisms. Nucleic Acids Res 28:352–355, 2000.

Thompson C. The impact of infomatics. Br Med J (BMJ) 7220:1294–1297, 1999.

Wheeler DL, Chappey C, Lash AE, Leipe DD, Madden TL, Schuler GD, Tatusova TA, Rapp BA. Database resources of the National Center for Biotechnology Information. Nucleic Acids Res 28:10–14, 2000.

Technology and Medicine

Berger A. Changing the doctor-patient relationship. Br Med J (BMJ) 7220:1279–1281, 1999.

Wilson CB. Hospital of the Future. Br Med J (BMJ) 7220:1287–1290, 1999.

Stevens A, Milne R, Gabbay J. How do new technologies get into practice. Br Med J (BMJ) 7220:1291–1293, 1999.

Commercial Implications: Large-Scale DNA Production and Quality Control; Technology Advancement and Elucidating the Human Genome

CHARLES P. LOLLO, PH.D., ROY MUSIL, DEBORAH Y. KWOH, PH.D., and ANDREA D. BRANCH, PH.D.

BACKGROUND

The extensive and diverse worldwide research activity of academic and industrial laboratories is currently focused in the nascent field known as gene therapy. While a wide variety of approaches and methodologies are being applied to a broad spectrum of diseases, most of these efforts are based upon the concept of introducing genes into patients to correct disease conditions. As such, gene therapy is an endeavor where new basic research insights are being rapidly translated into preclinical and clinical applications.

INTRODUCTION

Commercial companies, both large and small, are major players in the rapid development and use of genetic manipulation for therapeutic application as well as in technology development and application to elucidating the human genome. It is the nature of commercial business ventures to identify and exploit a competitive advantage in the marketplace. Even in the rarefied atmosphere of scientific research and development, some regard must be given to the eventual applications and utility of inventions. In the commercial environment, basic research efforts are evaluated for both scientific merit and economic potential. The economic potential usually is defined by a clinical application. Large companies will typically focus on disease states with large markets. Smaller companies tend to be open to any opportunity that fits their strategic intent. There is a continued emergence of small, start-up

An Introduction to Molecular Medicine and Gene Therapy, Edited by Thomas F. Kresina
ISBN 0-471-39188-3 © 2001 Wiley-Liss

biotechnology venture companies in the commercial field of gene therapy. Thus, a viewpoint of the commercial application of gene therapy from a small company or contract research laboratory working to develop gene therapies to meet existing clinical needs is presented.

PROPRIETARY TECHNOLOGY

There are primarily two points of entry into gene therapy research for the commercial venture defined by two equally important components of gene therapy. In this context, antisense applications and oligonucleotides are included because their chemistries and uses overlap significantly with strict gene replacement therapy. The first component for point of entry is the active ingredient. This can be a gene of interest that encodes information relevant to a treatment protocol. The information may be employed in a wide variety of applications including gene replacement, suicide gene therapy, gene therapy immunostimulation, and gene inhibition by antisense. The second component is equally important and much more challenging. It is the delivery system. Gene delivery systems that have been assayed in vivo range from naked (or formulated) deoxyribonucleic acid (DNA) to naturally occurring viruses that have been modified to circumvent some of their undesirable characteristics (see also Chapter 4). Between these two extremes is the arena of nonviral gene delivery systems. These can be described as artificial viruses, synthetic viruses, or self-assembling systems for gene delivery. A wide variety of reagents have been investigated for these delivery vehicles. They can be assigned to two main categories: lipids and cationic polymers. Both categories have preferred attributes, but it is accurate to indicate that neither has yet achieved the potency of viruses.

Large pharmaceutical companies may have both gene discovery efforts and gene delivery programs in place and can integrate them to create a proprietary pharmaceutical. In contrast, most small companies or budding entrepreneurs will only have one of the two main components in hand. For example, numerous researchers around the world are discovering new genes and their biological roles. The Human Genome Project and other developments in molecular biology accelerate these efforts. Newly discovered genes are investigated to determine their mode of action, genetic interactions, as well as what biological functions are modified upon deletion or mutation. In short, it is common to have access to knowledge about these genes as to their therapeutic utility. This information regarding a proprietary gene can be used to create interest and investment in a licensing arrangement or a start-up company. For example, recombinant proteins such as recombinant factor VIII or recombinant interferon-α have been used to treat a variety of medical conditions. It seems as an obvious application, the delivery of a gene that expresses the recombinant protein(s) endogenously for current medical conditions. Here the recombinant protein could be delivered by intravenous injection. If the gene is expressed for weeks or months at a sufficient therapeutic level, then the treatment would require far fewer administrations to the patient. Ramifications would then occur for the medical management of the patient. Gene therapy approaches, in this case, may consequently lead to better patient compliance. Also, the regulated secretion of the expressed protein would provide physiological serum levels and thus be advantageous in many disease states including hemophilia.

In addition to the introduction and expression of a gene for a missing or non-functional protein, the term gene therapy is also used to describe the introduction of genes into cells that do not normally express the products of these genes. Gene therapy can describe the introduction of foreign genes into cells to change the biology or drug sensitivity of cells (see Chapter 10). Besides approaches that "add function" to target cells, the term gene therapy also includes strategies for reducing expression of specific genes in tissues. Such reductions have been pursued by the introduction and expression of antisense ribonucleic acid (RNA), ribozymes, and dominant negative gene products (see Chapter 11). Treatment with antisense oligonucleotides is also directed at modulating gene expression and shares a number of challenges facing gene therapy approaches using larger DNA molecules. For these approaches, as for "functional" gene therapy, a proprietary position can be created through demonstration of novel or improved effects.

An analogous situation arises in the gene delivery arena with regard to developing a proprietary position. It is only necessary to show an improved system for gene delivery in animal models or even in vitro. Developmental work is typically performed on reporter genes such as luciferase or β-galactosidase. It is important to measure the efficiency of new delivery systems or to demonstrate a biological effect arising from the treatment.

Observing a consistent biological effect from treatment is perilous because of biological diversity. Thus, consequences may arise from unexpected mechanisms in individual situations. For example, toxicity inherent to a delivery system may stimulate endogenous gene up-regulation and cell proliferation. If the gene being delivered encodes for a protein that is being endogenously up-regulated, the results may be misleading. Using a reporter gene in initial studies avoids this pitfall. Reporter genes lead to expression of proteins that are not endogenous to the test system. These proteins are then easily quantitated and suitable for comparison to results achieved with other gene delivery technologies. Comparative improvements in delivery can be assessed on a variety of attributes including manufacturability, stability, toxicity, efficacy, and cost.

The research and development needed to advance a proprietary technology is largely defined by the expected clinical applications. Once more, it is appropriate to split the discussion to focus upon the active ingredient and, in turn, the delivery system. For formulated DNA or nonviral DNA delivery systems, manufacturing concerns regarding the components are not different from those developed for protein and drug pharmaceuticals. Formulated DNA is merely the active ingredient (DNA) dissolved or admixed into a carrier solution. The carrier solution can be quality controlled by a defined standard. Nonviral delivery systems involve components that interact electrostatically with the plasmid DNA to create an interpolyelectrolyte complex as the final pharmaceutical. The conjugates or lipids used to condense or coat the plasmid DNA can also be well-characterized by *U.S. Pharmacopeia* (USP) methods to meet standards of purity, identity, potency, and stability. These methods are the typical methods used to characterize protein reagents. None of these methods or standards is unique to gene therapy applications. The active component in gene therapy pharmaceuticals is the plasmid DNA. Standardization has evolved as DNA characterization and quantitation methods advance. The presently recognized preparation and purification methods and specifications are discussed subsequently in this chapter.

CONSIDERATIONS IN CHOOSING A TARGET DISEASE FOR GENE THERAPY

A variety of approaches have been utilized for the introduction of nucleic acids (principally DNA) into cells. These include the use of viral and nonviral methods for gene delivery. Modified retroviruses, adenovirus, adenoassociated virus (AAV), and herpes virus have been investigated for virally based delivery (see Chapter 4). Naked DNA, cationic lipids, liposomes, and cationic polypeptides are being pursued as nonviral approaches for gene therapy (see Chapters 4 and 5). Matching a gene therapy methodology to a target disease involves a number of factors. The technical issues that must be considered include determining the tissue and cell specificity needed for expression of the therapeutic gene, the number of cells that need to be targeted, and therapeutic level and duration of transgene expression. The delivery vehicle identifies the tissues and cell types that the therapeutic DNA can be delivered. If the choice of delivery vehicles is limited, then target diseases or genes will also be limited. The delivery vehicle will dictate the number of cells targeted and the duration of expression of the transgene (therapeutic gene). Therapies that require high levels of gene expression or require targeting a large percentage of cells likely require viral delivery vectors rather than nonviral delivery vectors. This is because, at present, viral vectors are more efficient at delivery. The delivered gene may be integrated into the host chromosome using AAV or retrovirus vectors (see Chapter 4). These may give longer duration of expression of the transgene than would be expected with adenovirus or nonviral delivery vectors. However, if the gene is to be delivered multiple times during the course of treatment, nonviral vectors may avoid the development of immune responses that can occur with viral delivery systems.

Regulation of the therapeutic gene is another factor to consider when choosing a target gene. How gene expression is regulated may determine which and how many cells need to be targeted. At present, gene expression regulated at the level of transcription is less problematic than gene expression regulated posttranscriptionally. Posttranscriptional gene regulation is, in most cases, less well understood. The consideration of posttranscriptional regulatory mechanisms could complicate or slow the development of gene therapy. As discussed later, the levels of transcription of a gene can be manipulated by modification of the plasmid or viral vector DNA.

Unusual requirements for gene product processing needed for activity of the expressed gene must be considered when choosing a target disease. Many genes can be expressed in cell types other than the normally expressing cell types and still be therapeutic. However, other gene products require special processing in a particular cell type or in a particular organelle. Thus, such genes would not be effectively expressed in other cell types. Still other proteins may have cofactors (proteins) that are essential for activity and must be made in close proximity (same cell or organelle) as the cofactor.

Another key factor in choosing a target gene is the availability of the gene. Questions that should be asked are:

· Is the gene sequenced and cloned?
· Is it a cDNA clone or full-length gene (containing introns)?

· Does the cloned gene also contain the native promoter and regulatory sequences at the 5' and possibly 3' ends?

The commercial development process is faster when maximal information is known about a targeted gene (regulation, sequence, etc.). The overall size of the gene to be delivered is also an important consideration since many viral vectors are limited in the size of DNA that can be packaged. The nonviral delivery systems are less restricted in the size of DNA that can be delivered.

The development of a commercial gene therapy product is also facilitated by the availability of an animal model of the genetic disease being targeted. Although not all human genetic diseases currently have animal models of disease, the number of transgenic and knock-out mouse strains (see Chapter 3), as well as larger animal models, has increased exponentially in the last few years. These animal models prove valuable in developing effective gene therapy treatment approaches for many single-factor genetic disorders and possibly some multifactor diseases as well.

As for any commercial venture, patent and licensing issues for a particular gene will necessarily be important factors in choosing a target. The size of the potential patient population and the accessibility of patients for a particular product are also crucial. There are numerous genes that could be targeted for gene therapy, however, many of the single-factor genetic diseases are relatively rare (see Chapter 1). Diseases currently treated with recombinant proteins (severe immune deficiency, hemophilia A and B) provide larger markets where gene therapy could have an impact. As with any new therapy, gene therapy approach for a disease state would need to have advantages over treatments currently in use.

DNA PRODUCTION AND QUALITY CONTROL

Introduction

The large-scale, commercial production and quality control of DNA and viral vectors to be used in clinical research protocols is critically important. Assurance of purity must be provided to investigators who purchase or contract for reagents to be used in basic or clinical research. As can be seen from the recent events, poor quality control of reagents can lead to the cessation of clinical trails of gene therapy protocols (see Chapter 13).

Laboratory Scale Purification

As the clinical aspects of gene therapy continue to grow, one of the challenges facing industry is the large-scale purification of plasmid DNA. Within the typical research laboratory, plasmids continue to be routinely obtained by the standard method of CsCl–ethidium bromide density gradient ultracentrifugation. CsCl–ethidium bromide gradients are popular since large numbers of different plasmid preparations can be processed simultaneously. The approach applies to both plasmid and viral DNA of varying sizes; and a single band in the density gradient contains the monomeric, supercoiled form of the DNA partially resolved from the intrinsic host cell contaminants (protein, DNA, and RNA). But there are numerous drawbacks and limitations to this process. For the researcher at the lab bench, it is time consuming, labor intensive, and expensive. For the biotechnology company, however, this method is completely unacceptable for the production of clinical-grade materials because of its use of mutagenic reagents and its inherent inability to be a process of scale.

Recently, a number of companies have initiated market-adapted micropreparative methods for the production of larger quantities of plasmid DNA. These modified "mini-prep" kits, make use of the alkaline lysis method for cell disruption followed by a chromatographic cartridge purification. The composition of the stationary phase used in these kits varies. Some kits use a silica-based stationary phase, while others are based on an agarose stationary phase. In most cases, the mechanism of binding is anion exchange. These kits are aimed at a particular market niche: the production of small quantities (milligram or less) of research-grade material for molecular biology applications. They do not meet the rigorous requirements for the development of a highly controlled drug manufacturing process and most do not have a Drug Master File (DMF). Purity in these applications is usually evaluated by agarose gel electrophoresis. Trace impurities such as endotoxin and host DNA are not as thoroughly investigated as needed for human clinical use.

The Food and Drug Administration's (FDA's) "Points to Consider" on plasmid DNA was drafted in October, 1996, and provides the U.S. approach to regulation of plasmid preventative vaccines (see Chapter 13). The same general criteria that guide the manufacture of recombinant protein pharmaceuticals apply to the development of processes for the production of plasmid DNA for human clinical investigations. The common thread linking these processes is the basis of well-documented research. This basis allows for the final product to meet defined quality standards supported by validated analytical methods and controlled unit operations. All components of the process must be generally recognized as safe and must meet all applicable regulatory standards. It is precisely because of these reasons that plasmid DNAs used for clinical investigations are not produced using kits intended for laboratory research.

LARGE-SCALE PRODUCTION: AN OVERVIEW

To proceed to advanced clinical trials and ultimately gain regulatory approval, the pharmaceutical development of gene therapy products need to meet the requirements for cGMP (current good manufacturing practices) production. While the "c" ostensibly stands for "current," when actually following the spirit and intent of the FDA guidelines the "c" represents control of the process and characterization of the product. GMP is defined as "the part of quality assurance that medicinal products are consistently produced and controlled to the quality standards appropriate to their intended use and as required by the Marketing Authorization or product specification."

There are two main components of cGMP, comprising both production and quality controls. Production control is concerned with manufacturing. This includes the suitability of facility and staff for the manufacture of product, development of standard operating procedures (SOPs), and record keeping. Quality control is concerned with sampling, specifications, testing, and with documentation and release procedures ensuring satisfactory quality of the final product.

For a typical production conducted under the principles of cGMP, the major points to consider in the manufacture of plasmid DNA are:

- SOPs are in place to ensure the control and consistency of the entire production cycle starting from the initial receipt of raw materials to the final formulated drug product.
- All raw materials used in the manufacturing process are put on a testing program based on the *U.S. Pharmacopeia*.
- A master cell bank (MCB) and manufacturer's working cell bank (MWCB) has been prepared under conditions of quarantine to ensure the purity and identity of the fermentation seed pools. Thorough vector characterization has been carried out, including a detailed history on the construction of the vector, complete nucleic acid sequence determination, and plasmid stability within the host strain. MCBs should be shown to be free of adventitious agents.
- Details of the fermentation process must be elucidated and consistency data generated. Several commercial media have been designed for plasmid production, but a defined medium that has been empirically developed for a specific strain plasmid is preferable. This should assist in achieving a reproducible well-controlled process. Bacterial strains should be compatible with high copy number plasmids, high biomass fermentations, and the selection system cannot be ampicillin based.
- Purification processes must be developed to meet the challenges inherent with a high cell density fermentation process. Plasmid DNA purification kits routinely fail when challenged with high cell density starting feed streams. Recovery and purification must be controlled and validated. Special attention must be paid to the removal of host cell proteins, DNA, and endotoxin. Documented reproducible removal of key host-cell-derived impurities is essential for setting accurate limits and specifications on the bulk drug product.
- Appropriate analytical assays must be developed for both the monitoring of the production cycle as well as for final quality control release criteria. The FDA's "Points to Consider" lists some of the tests needed to confirm purity, identity, safety, and potency of plasmid DNA. A functional in vivo or in vitro bioassay that measures the biological activity of the expressed gene product, not merely its presence, should be developed. Measuring the relative purity and concentration of plasmid DNA by agarose gel electrophoresis or by high-pressure liquid chromatography (HPLC) is only a small part of the battery of analytical measurements necessary to confirm product quality. All assays must be fully validated.
- Ongoing stability and efficacy testing must be conducted on the product in support of the ongoing clinical trials. This data is critical in eventually determining product shelf life for the approved drug.

LARGE-SCALE PRODUCTION: THE PURIFICATION PROCESS

The basic unit operations for the manufacture of plasmid DNA are basically the same as those for the production of any recombinant biopharmaceutical (see Fig. A.1). Typical process steps for the production of plasmid DNA include initial vector

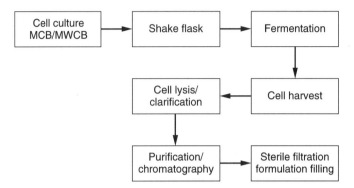

FIGURE A.1 Steps in a typical large-scale biotechnology process.

design, fermentation, cell harvesting, alkaline lysis precipitation, chromatographic purification, formulation, and filling. The process cannot rely on the use of animal-derived enzymes such as lysozyme, proteinase K, and RNAase. Use of these reagents in any manufacturing process for a drug substance raises regulatory concerns about residuals in the final product. Disregarding such purity issues would increase the difficulty in process validation and ultimately putting final regulatory approval at risk. The process should also not include toxic organic extractions. The various forms of plasmid DNA including supercoiled, relaxed, and concatamers should be separated. The final product must be free of contaminating nucleic acids, endotoxins, and host-derived proteins.

Fermentation is generally considered the starting point in designing the purification process. By careful selection and control of the variables associated with the fermentation process, the subsequent purification may be greatly simplified. Various fermentation feed strategies (batch, fed-batch, continuous) should be explored. While somewhat more difficult to optimize, as well as document, continuous fermentations may offer several advantages in terms of production cycle times. Normally, fed-batch fermentations allow quicker process development times, simpler process control and sufficiently high biomass. The growth stage at which the fermentation is harvested must also be tightly controlled since it will greatly impact on the final yield of purified plasmid. Harvesting too early will result in suboptimal final yields. Harvesting too late in the fermentation cycle will not only result in low yields but also plasmid of poor quality. The optimal stage of harvest is late log phase.

The monitoring of fermentation process parameters including temperature, glucose addition, dissolved oxygen, and carbon dioxide evolution are critical for the development of a reproducible process. By manipulation of these parameters or through the use of an inducible plasmid system, the growth characteristics of a strain can be effectively changed, resulting in an increase in the plasmid-to-biomass ratio. Any increase in this ratio will aid in the design of the purification process. As well, it can result in higher final yields of plasmid. Chloramphenicol has been traditionally used just for this purpose.

The host cell and plasmid are the most important starting materials in the production fermentation. The key parameters in choosing a host strain are a low endogenous endotoxin, the capability of growing to high biomass, and relevant genotypic markers. These markers could be *rec*A1, *end*A1, and *deo*R: *rec*A1 prevents recombination and improves stability of plasmid inserts; *deo*R allows for the uptake of large plasmids; *end*A1 improves plasmid quality. The plasmid should be structurally as well as segregationally stable and have a high copy number origin of replication. Typically it is pUC derived.

Scale-up for the purification of plasmid DNA is a definite issue. Chromatography is the tool that has enabled the biotechnology industry to achieve the purity levels required for today's biotherapeutics, diagnostics, and other biologicals. These include enzymes and plasma products. Chromatographic purification of DNA presents a novel set of problems. These are based on the physical characteristics of the biomolecule as well as the intrinsic impurities derived from the host cell of choice, *Escherichia coli*. The chief culprits that hinder the purification of plasmid DNA are the large amounts of polymers of similar structure (chromosomal DNA and RNA) and high levels of endotoxin.

Plasmid DNA is a highly anionic polymer that is sensitive to shear and to degradation by nucleases. Plasmids are as large or larger than the pores of almost all chromatographic resins. Several chromatographic procedures for the purification of biologically active plasmid DNA (without the use of CsCl–ethidium bromide ultracentrifugation) have been developed, at least at laboratory scale. They include gel filtration chromatography, hydroxyapaptite chromatography, acridine yellow affinity chromatography, anion exchange chromatography, reversed phase chromatography, silica membrane binding, and binding to glass powder. Unfortunately, many of these methods are not well suited to the purification of large quantities of DNA. In choosing the method of purification for large-scale production of plasmid DNA, there is a most important physical characteristic of the biomolecule to consider. It is that DNA is a highly anionic polymer that is sensitive to shear and to degradation by nucleases. Any large-scale manufacturing process must address all of these characteristics. Currently, the most successful methods of extraction and purification involve large-scale alkaline lysis in sodium deodecyl-sulfate (SDS). This step efficiently removes chromosomal DNA, nuclease enzymes, and other contaminants. Therefore, cell lysis conditions must be carefully optimized. Low shear mixing must also be used during this step. Large-scale tangential flow systems, which are routinely used for the processing of recombinant proteins, can easily nick the supercoiled form of the plasmid. Cross flow rates, pump design, as well as the mixer's impeller design must all be carefully scrutinized. Plasmid extracts are primarily contaminated with low-molecular-weight cell components, process chemicals, and RNA. These contaminants and trace host protein contamination may be removed by a combination of selective precipitation, anion exchange chromatography, and a final polishing step.

A major drawback of using anion exchange chromatography as the sole high-resolution purification step in the purification of plasmid DNA is that a portion of endotoxin and pyrogen contaminants will co-purify with the plasmid. Given the limitations of currently available commercial matrices and the similar structure and charge profile of biomolecule species passing over the column, anion exchange chromatography is best used as a primary capture and initial purification step. A second polishing step, which is orthogonal to the principles of anion exchange, is prudent and ensures rigorous process control.

Historically, gel filtration has been used in the biotechnology industry as a polishing step. Plasmid DNA, host cell DNA, and endotoxin resolve using gel filtration chromatography. This is a simple and reproducible method that also offers the advantage of simultaneously incorporating a buffer exchange step within the chromatographic process. Contaminating salts and/or residual metals can thus be removed allowing for the careful control of the counter ion in the final drug product. However, the main drawback in using gel filtration is that it is a very slow and volume-dependent method. It is not a high throughput method and often becomes the bottleneck within a given process.

Reversed phase chromatography (RPC), on the other hand, can also offer excellent separation and resolution of trace contaminants as well as the removal of endotoxin. It is commonly the method of choice for the purification of small pharmaceutical compounds. When purifying biologically active molecules, care must be taken so that biological activity is retained. Through its use of volatile solvents, RPC can also serve the function of a buffer exchange step. But it is precisely this point that contributes to reversed phase's own set of unique problems. The use of combustible organic solvents (acetonitrile or ethanol) requires explosion-proof facilities. This safety factor can dramatically increase the cost of waste management. With the heightened awareness of environmental issues in today's industrial nations, the cost and feasibility of waste disposal are major considerations when designing or deciding on a purification process. Ion-pair RPC, while again providing excellent separation, resolution, and endotoxin removal, introduces ion-pair reagents that must be assayed for in the final product. Their removal must be assured by validated methods.

The final crucial aspect in deciding on a chromatographic support is the necessity of cleaning in place and sanitization by cycles of caustic washing. The ability to withstand repeated cycles of regeneration, sterilization, and sanitization with $0.5 N$ NaOH while maintaining run-to-run reproducibility of the column profile is an important consideration in manufacturing pharmaceutical-grade plasmid DNA in accordance with cGMP manufacturing guidelines.

LARGE-SCALE PRODUCTION: QUALITY CONTROL

Recombinant proteins and plasmid DNA are both derived from *E.-coli*-based expression systems. This results in a fair degree of similarity in their contaminant profiles. The FDA has presented a general list of contaminants that should be quantified in all biopharmaceutical products. They include pyrogen, nucleic acid, antigen, and microbial and residual contamination. Most assays that have been developed for the quality control of recombinant protein drug substances need only slight adaptation for the quality control of plasmid DNA production. The most challenging assays in terms of unique or specific analytical tests for plasmid DNA bulk products are the measurements of protein (antigen), nonplasmid DNA, and RNA trace contaminants. Documentation and validation of all assays must adhere to cGMP guidelines.

Fermentation cultures need to be routinely monitored for microbial contamination. Sterility checks should be performed on inoculation flasks, the fermentor, and the fermentation media. The presence of contaminating organisms will alter the production levels of plasmid produced and thereby invalidate data on the levels of contaminating impurities within the final DNA product.

The most common and routine analysis of plasmid DNA is through the use of ethidium-bromide-stained agarose gels. In research settings, this assay is usually used as a standalone technique for determining RNA contamination, residual genomic DNA, as well as quantifying the relative amounts of supercoiled plasmid in relation to the relaxed or nicked form. It is well known, though, that ethidium differentially stains linear, nicked, and supercoiled plasmid DNA as well as host cell RNA. Thus, care must be used when using this assay as the sole tool for judging relative amounts of DNA or in determining residual RNA levels. To accurately characterize the purified product (and monitor in-process samples) an array of electrophoretic, chromatographic, and spectrophotometric assays should be employed. In particular, the use of analytical high-resolution HPLC can avoid the detection and quantitation problems associated with ethidium bromide staining of plasmid DNA since detection is based on ultraviolet absorption.

Another common quality control test for plasmid DNA used in most research laboratories is the A_{260}/A_{280} absorbance ratio assay. It highlights the discrepancies between true cGMP production and laboratory-scale purification. The test was originally designed to measure enzyme concentrations in the presence of low levels of nucleic acid contamination. The original usage has been corrupted, however, and now it is routinely used in molecular biology laboratories to assess DNA purity. An A_{260}/A_{280} absorbance ratio of 1.8 to 2.0 is generally considered "pure." In fact, when one does the actual calculation using the true extinction coefficients of nucleic acids and proteins (nucleic acids have extinction coefficients on the order of 50 times higher than proteins), it becomes obvious that an $A_{260}/A_{280} = 1.8$ can contain as much as 60% protein contamination. Therefore, this method can only be used as a functional test and cannot in itself be used to determine DNA purity.

Equally critical for achieving pharmaceutical-grade plasmid DNA is the monitoring of any chemical reagents introduced into the manufacturing process. If alcohol is used in a precipitation step in the process, an assay must be included to determine the residual trace levels of alcohol that remain in the final product. If antifoam (a common fermentation additive) has been used, an analytical assay must be in place for its determination as well as a final release specification for its concentration. Choosing the appropriate analyses in this area requires careful control and sourcing of all raw materials. One of the hallmarks of a fully FDA-compliant production process is the use of well-characterized reference standards. These are necessary for the completion of analytical assay assessment and for use in ongoing validation studies. The most critical reference standard is the plasmid DNA. Ideally, the plasmid should be fully characterized and be derived from a manufacturing batch that has been clinically evaluated. Having a well-characterized reference standard greatly aids in the successful evaluation of product stability testing.

With the proper appropriate supporting data, background information and supplementary studies, the development of a minimal panel of characterizing assays can be put in place. These would provide the necessary level of confidence to reliably determine identity, purity, potency, and stability of the manufactured plasmid DNA (Table A.1). The foundation that makes this possible is rooted in the compliance to cGMP throughout the entire plasmid production cycle. While there may be differences in the specific physiochemical assays for the determination of identity and purity, pharmaceutical-grade plasmid DNA and recombinant-protein-based therapeutics share a similar quality control characterization strategy.

TABLE A.1 Sample Plasmid Specifications and Test Methods

Assay	Method	Specification
DNA homogeneity	1% agarose gel electrophoresis Anion exchange HPLC	>95% supercoiled plasmid
E. coli chromosomal DNA	Slot blot hybridization	<1%
RNA	Slot blot hybridization	<1%
Endotoxin	LAL kinetic	<5 EU/mg plasmid
Identity	Restriction digest followed by 1% agarose gel electrophoresis	Conforms to map
Sterility	USP membrane	No colonies at 14 days
Purity	A_{260}/A_{280}	1.75–2.0
	A_{260}/A_{230}	>2.2
Protein contamination	Optical density scan	$\lambda_{min} = 230$ nm
	BCA microtiter	Below limit of detection
Potency	Transfection assay	"X"ng expressed gene/μg reporter gene
Residual ethanol	Gas chromatography	≤250 ppm

TECHNOLOGY ADVANCEMENT

Techniques for Profiling Proteins and mRNAs

Two general approaches can be used to take a census of the proteins in a cell or tissue: direct analysis of the proteins and indirect analysis of cDNAs reversely transcribed from mRNAs. The advantages of the former are straightforwardness and the ability to detect protein modifications. The advantages of the latter are sensitivity and the ability to tap into the awesome power of molecular genetics through DNA databases. At the moment, nucleic-acid-based techniques are more widely used. However, techniques for analyzing proteins are rapidly advancing.

Galactic-Scale cDNA Techniques Methods for analyzing cDNA populations reversely transcribed from mRNAs are expanding in number and variety. They are roughly quantitative because cDNAs are synthesized in proportion to the amounts of individual mRNAs present in the population. A partial list of cDNA methods includes differential display, direct sequencing and counting of "tags," and hybridization to DNA arrays. Selected methods are briefly described below to illustrate the choices available. The latest techniques allow the expression patterns of thousands of genes to be monitored simultaneously, generating a rough outline of the "transcriptome"—the complete set of genes expressed in a particular cell. These powerful techniques are generating a tsunami of information. They promise to revolutionize the way biology is studied and the way drug development is carried out.

Differential Display In old-fashioned differential display methods, cDNAs are primed at the 3′ end of mRNA, with one of three oligo-dT (A, C, G) oligonucleotides, and are anchored at the 5′ end by a specific (but arbitrary) primer. In modified versions of that method, other types of primers, such as oligonucleotides optimized to detect coding sequences, are used. In either case, about 400 cDNA bands can be resolved by DNA gel electrophoresis in ultrathin polyacrylamide gels. Typically, cDNAs from two populations of mRNAs are compared to each other. To identify the differentially expressed genes, cDNAs are recovered from gels, cloned, and sequenced.

SAGE (Serial Analysis of Gene Expression)   SAGE is based on two principles: (1) that a gene can be identified by a short sequence tag (9 to 11 bases long) provided that a second piece of information about the tag sequence is known, such as its position in the mRNA relative to that of another short sequence, for example, a restriction site and (2) that many tags can be concatenated into a single molecule and sequenced to determine the abundance of each tag and to identify the gene (if it is present in a database, such as GenBank).

Using this method, more than 300,000 cDNA tags representing a minimum of 45,000 different genes have been examined and compared with mRNA populations in human intestinal tumor cells and control cells. Contrary to expectations, two widely studied oncogenes, c-*fos* and c-*erb3*, were expressed at much higher levels in normal colon epithelium than in colorectal cancers. Such surprising results emphasize the value of studies carried out on clinical samples (patient specimens). This approach provides insight into the gene expression patterns of human malignancy and helps to identify genes that may be useful targets for gene therapy.

Expressed Sequence Tags (ESTs) and DNA Arrays   ESTs provide gene "identifiers." They are made by copying mRNA populations into cDNA clones, which are then partially sequenced and entered into databases. About 800,000 ESTs of human genes are currently available in public databases and at various Web sites (see also Chapter 15). These represent 40,000 to 50,000 of the estimated total of 70,000 to 100,000 human genes. ESTs can be expressed as DNA molecules and hybridized to cDNAs under a variety of conditions. Recently, ESTs have been combined into high-density DNA microarrays (Figure A.2). DNA microarrays consist of thousands of individual gene sequences attached to a surface in a precise and reproducible pattern. Because these arrays are minute, they can be used with tiny quantities of cDNA.

In a tour de force that paints a scintillating and detailed picture of a eukaryotic cell's inner workings, microarrays were used to study changes in yeast gene expression during the shift from anaerobic to aerobic metabolism (Fig. A.3). This study showed the feasibility of a small group of researchers to PCR amplify more than 6000 open reading frames (ORFs), representing all the genes of *Saccharomyces cerevisiae*, in about 4 months. In two days, 110 microarrays, each containing 6400 elements, were produced and were ready for hybridization to fluorescently labeled cDNA copied from mRNAs extracted from cells at various time points. According to the investigators, preparation of fluorescently labeled cDNA probes, hybridization, and imagine analysis proceeded quickly. When analyzed, the data revealed that during the metabolic shift, mRNA levels for approximately 710 genes were induced by at least a factor of 2, and the mRNA levels for approximately 1030 genes declined by a factor of at least 2. Since the equipment for DNA microarrays was chosen for its rel-

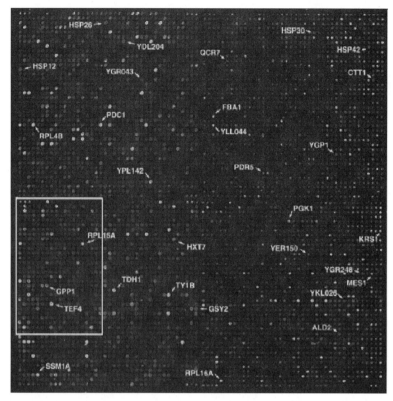

FIGURE A.2 Molecular analysis of the shift from anaerobic to aerobic metabolism using a yeast genome microarray. Fluorescently labeled cDNAs were prepared from mRNA from anaerobic cells by reverse transcription in the presence of Cy3-dUTP and from aerobic cells in the presence of Cy5-dUTP. Hybridization of the Cy3-dUTP-labeled cDNA appears in green and that of the Cy5-dUTP-labeled cDNA appears in red. Genes up-regulated after the metabolic shift appear in red, while those down-regulated appear in green. Genes expressed at roughly equal levels appear in yellow. The actual size of the microarray is 18 by 18 mm. The image was obtained with a scanning confocal microscope. (Reproduced with kind permission from *Science*, 278, 1997.)

atively modest cost, it may be feasible for small academic groups strongly committed to comprehensive expression analysis to establish this technique for local use.

Gene Expression Microarray (GEM) This technique is useful for probing expression patterns in human cells. Their microarrays contain PCR products over 100 bases in length, which permits the use of stringent hybridization conditions. GEM's two-color competitive hybridization process is reported to detect twofold changes in the level of expression. The first commercial product, a chip that contains tags from 10,000 human genes, is being replaced by a microarray containing tags for 55,000 genes. Although the 3-year subscriptions are expensive—from $300,000 to $9,000,000 depending upon usage—the establishment of academic collaborations will reduce costs. In addition, several additional commercial hybridization formats are available for "expression profiling."

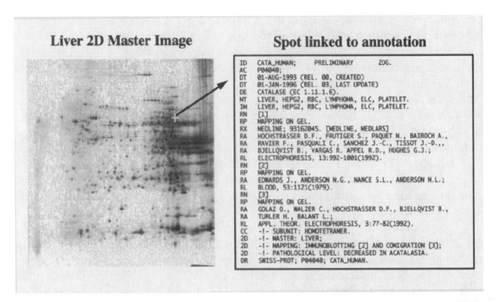

FIGURE A.3   Two-dimensional gel master image of liver proteins. The image is available from the URL http://expasy.hcu.ge.ch/cgi-bin/map1. Two-dimensional spots can be selected and the annotation linked to the spot displayed. The spot marked with the white cross is catalase. The annotation indicates how it was identified (by immunoblotting) and how it alters in disease states (decreased in acatalasia) and provides a futher link to the entry in the SwissProt database. (Reproduced with kind permission from Biochemical Biophysical Research Communications, 231, 1997.)

High-Resolution Two-Dimensional (2D) Gel Electrophoresis of Proteins, and ProteinChip Microarrays   Efforts to define the "transcriptome" are paralleled by efforts to define the human "proteosome," the total set of proteins within a particular cell. Direct information about cellular proteins is needed for two reasons. First, protein function is often altered by posttranslational modifications, which cannot be discerned from mRNA analyses. Second, mRNA concentrations and protein concentrations do not have a strict one-to-one relationship. To investigate the magnitude of the disparity, investigators compared mRNA and protein abundance in extracts of human liver. They found that of the 50 most abundant liver mRNAs, 29 encoded secretory proteins; none of the 50 most abundant proteins appeared to be secretory. The correlation coefficient of RNA and protein abundance was only 0.48. These results underscore the need to ensure that cDNA-based methods sample both high and low abundance mRNAs and demonstrate the value of protein analyses.

High-Resolution 2D Gels   High-resolution 2D gels separate proteins according to charge (in the first dimension) and size (in the second). Scanned images can be analyzed by computer programs. Small changes in the concentration of individual proteins ($\pm15\%$) can be quantified. This sensitivity may be important when measuring the effects of medical interventions. In addition, these gels can detect posttranslational protein processing events, including proteolytic cleavage and phosphorylation. The introduction of immobilized pH gradient electrophoresis has extended the pH range and improved reproducibility. Progress is being made toward the development of fully automated 2D systems. A World-Wide-Web (www) federation is being established to facilitate the exchange of 2D gel images. A common interface for data accession already allows 2D gel databases from all www federation sites to be searched and may soon allow gel images to be matched over the network. The SWISS-2DPAGE database (Fig. A.4) contains 2D master gel images of cells representing the "normal" state, while the complementary SWISS-2Disease database consists of annotated gel images from cells and tissues of various disease states, such as renal failure and myeloma. Both are available on www. Typical 2D gel patterns stained with silver contain about 1000 to 2000 spots, about 75% of which contain less than 500 femtomoles of protein. Large-format gels allow up to 10,000 spots to be detected.

High-resolution 2D protein gels are used in combination with various techniques for identifying the proteins comprising the spots, such as microsequencing and antibody binding. Furthermore, new "soft" ionization techniques in mass spectrometry are creating a revolution in spot identification. Matrix-assisted laser desorption and ionization (MALDI) and electrospray ionization allow minute quantities of protein to be analyzed. The powerful new techniques for obtaining reproducible gel patterns, for analyzing the gel images, and for identifying individual proteins are yielding detailed snapshots of cellular protein populations.

Microarrays (Chips) This technology for protein analysis is being developed. Chip technology will examine protein expression and structure. The ProteinChip microarrays are comprised of molecules that bind proteins, such as antibodies, receptors, or ligands. Cellular proteins are incubated with the array, and then laser pulses are used to probe each site. Proteins can be released from the surface of the chip and analyzed by mass spectrometry.

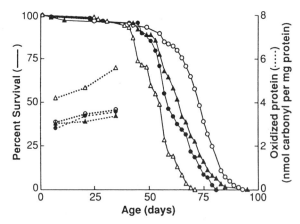

FIGURE A.4 Effect of overexpression of antioxidative enzymes on life span and protein oxidative damage in *D. melanogaster*. Survival curves (solid) and protein carbonyl content (dashed curves) at different ages for a control group (blue triangles) and three different lines (remaining symbols) of transgenic *D. melanogaster* overexpressing both Cu,Zn-superoxide dismutase and catalase. (Reproduced with kind permission from Science, 273, 1996.)

Public-Access, Systematic Database of Functional Genomics Information about gene expression patterns is accumulating at a blistering pace. In the study of yeast gene expression, it is perhaps the greatest challenge to develop efficient methods for organizing, distributing, and interpreting the large volume of data generated with microarrays and related technologies. Adequate methods for storing and analyzing these data are essential. Something akin to a four-dimensional GenBank needs to be constructed: (1D) primary sequence of the mRNA or protein, (2D) post-translational modifications, (3D) concentration, and (4D) temporal changes. The gene expression database will need to be densely annotated with information about the physiological state of the organism, tissue, or cell and about experimental treatments. Listing of publicly available bioinformatics resources, which give a useful starting point, are currently available. In the long-run, the information about expression patterns will need to be correlated with information about the genetic heterogeneity of the human population.

Elucidating the Human Genome Through DNA Analysis

The human genome is the nuclear composition of genetic material. It is estimated that 50,000 to 100,000 genes encompass the human genome. Elucidation of the composition of the human genome would result in information and spin off technologies that would revolutionize the study of disease. A map or descriptive diagram of each human chromosome would involve dividing each chromosome into smaller fragments that can be propagated and characterized as well as ordering the fragments to correspond to their location on the chromosome. After mapping, the next step is to determine the nucleotide sequence of each chromosome fragment. The ultimate goal would be to locate and assign a biological function to all the genes in the DNA sequence.

A research effort, entitled the Human Genome Project, is an international effort designed to construct detailed genetic and physical maps of the human genome, to determine the nucleotide sequence of human DNA, to localize all the genes of the human genome, and to perform a similar analysis on the genomes of several other organisms as model systems. Ninety percent of the human genome nucleotide sequencing project has been completed. Most recently, a highly public announcement was made that the human genome has been sequenced. Chromosome 22 has been fully characterized and the data put in the public domain. The elucidation of the entire genome and public disclosure of the data would almost certainly identify most if not all of the major genes involved in common diseases. Already, as noted above, correlations between genetic mutations and disease susceptibility are being established with the hope that such information will lead to novel therapies targeted at defined patient populations. Genetic based diagnosis and treatment of disease have the potential to radically improve the practice of medicine (see Chapter 15).

Human Genome Project The Human Genome Project began in the 1980 as part of a national scientific research effort supported by the Department of Energy. Recently, an institute, The National Human Genome Research Institute (NHGRI), has been created at the National Institutes of Health to lead this research effort. The scientific priorities of the NHGRI can be broken down into the areas of (1) genetic mapping of the human genome, (2) physical mapping of the human genome, (3) DNA sequencing of the human genome, (4) new technologies for interpreting human genome sequence, and (5) analysis of genomes of model systems.

The Human Genome Project is driven by technology. As mentioned previously, new techniques are constantly being developed and introduced into the research environment. A long-term objective of the Human Genome Project is to identify all coding sequences, including genes and regulatory elements in the human genome. This once unimaginable goal is now feasible through new methods of DNA analysis.

Superfast DNA Sequencing   Faster methods for DNA sequencing are being developed. Speed is increasing as a result of both incremental improvements in current methods and from the introduction of entirely new approaches, such as "sequencing by hybridizing" on microarrays or chips. In July, 1997, the first of such systems was launched and called GeneChip (p53) assay for research applications. This assay is capable of analyzing the full-length coding sequence of the human p53 tumor suppresser gene, frequently mutated in human cancers. The chip contains a matrix of more than 50,000 DNA molecules and is designed to detect more than 400 distinct mutations in the p53 gene. Success of the p53 chip is certain to spawn similar chips for heritable diseases, such as cystic fibrosis, which can result from any of a number of mutations within the *CFTR* gene.

In addition to sequencing by hybridization, entirely different approaches are being explored. For example, the possibility of sequencing DNA molecules by measuring effects on conductance as DNA passes through ion channels in membranes is being explored. Many additional initiatives are underway with a major breakthrough in this field likely. It is unclear the form that superfast DNA sequencing will take, but most would agree that it will depart from the current techniques as dramatically as the current techniques depart from the depurination fingerprinting method that started the field.

Making use of human genetic information will be challenging. Genetic epidemiologists and statisticians will be needed to validate genetic linkages in conditions with polygenic inheritance patterns. Information needs to be gathered about the factors affecting expression of genes and gene mutations, so that predictions can be made about the outcome in particular individuals.

Model Systems   The mapping and sequencing of the genomes of other organisms as model systems is fundamental to the elucidation of the human genome. Model systems are also useful in the testing of new technologies to be applied to the human genome. Currently, five organisms are being used as model systems: *E. coli* (bacterium), *S. cervisiae* (yeast), *C. elegans* (round worm), *D. melanogaster* (fruit fly) and *M. musculus* (the laboratory mouse). The physical map and genomic sequence of *E. coli*, *S. cervisiae*, and *C. elegans* are completed. The current goal of the project is to complete a genetic map, an STS content map of 300 kilobase resolution and sequence regions of the mouse genome in a side-by-side comparison with human genomic sequences. Given the conservation of genetic information and the use of the mouse in animal models of disease, these data are anticipated to be highly informative.

SUMMARY

Commercial development of pharmaceuticals for gene therapy is a burgeoning field. Discovery and clinical applications of novel genes is expected to continue at an accelerating pace. Gene manipulations to increase expression levels and to provide cellular specificity and control mechanisms will lead to added safety and efficacy. There does not appear to be many limitations to the accomplishments of molecular biologists with regard to gene discovery and engineering. Methods and inventions to deliver these genes in vivo to specific cell types is an area needing improvement, and the variety of approaches presently being investigated bodes well for future breakthroughs. In particular, present delivery systems are often lacking in both specificity and efficiency. Since the active ingredient, DNA, is by far the most expensive, improvements in delivery will be beneficial on the bases of both cost and potency. It is expected that both components of gene therapy (plasmid DNA and delivery vehicles) will see large improvements in the near future. Large-scale manufacturing methods for production and purification will fall into place as the utility of gene therapies is demonstrated in many of the ongoing clinical trials.

KEY CONCEPTS

- In the commercial environment, basic research efforts are evaluated for both scientific merit and economic potential. The economic potential usually is defined by a clinical application. Large companies will typically focus on disease states with large markets. Smaller companies tend to be open to any opportunity that fits their strategic intent.

- Large pharmaceuticals companies may have both gene discovery efforts and gene delivery programs in place and can integrate them to create a proprietary pharmaceutical. In contrast, most small companies or budding entrepreneurs will only have one of the two main components in hand.

- The research and development needed to advance a proprietary technology is largely defined by the expected clinical applications. For formulated DNA or nonviral DNA delivery systems, manufacturing concerns about the components are not different from what has been developed for protein and drug pharmaceuticals.

- As for any commercial venture, patent and licensing issues for a particular gene will necessarily be important factors in choosing a target. The size of the potential patient population and the accessibility of patients for a particular product is crucial. In addition, as with any new therapy, gene therapy approach for a disease state would need to have advantages over treatments currently in use.

- One of the challenges facing industry is the large-scale purification of plasmid DNA. A number of companies have begun to market adapted micropreparative methods for the production of larger quantities of plasmid DNA. These modified mini-prep kits generally make use of the alkaline lysis method for cell disruption followed by a chromatographic cartridge purification. These kits have been aimed at one particular market niche: the production of small quantities (milligram or less) of research-grade material for molecular biology applications. They do not meet the rigorous requirements for the development of a highly controlled drug manufacturing process and most do not have a Drug Master File (DMF).

· FDA's "Points to Consider" on plasmid DNA was drafted in October, 1996, and provides the U.S. approach to regulation of plasmid preventative vaccines. The same general criteria that guide the manufacture of recombinant protein pharmaceuticals apply to the development of processes for the production of plasmid DNA for human clinical investigations.

· To proceed to advanced clinical trials and ultimately gain regulatory approval, the pharmaceutical development of gene therapy products will have to meet the requirements for cGMP (current good manufacturing practices) production. There are two main components of cGMP, comprising both the production and quality controls. This includes the suitability of facility and staff for the manufacture of product, development of standard operating procedures (SOPs), and record keeping. Quality control is concerned with sampling, specifications, testing, and with documentation and release procedures ensuring satisfactory quality of the final product.

· A group of promising new tools is emerging that will allow patterns of gene expression to be compared in healthy and diseased tissue. On the one hand, these gene profiling techniques will detect gene therapy targets—genes whose products contribute to disease. On the other hand, they will identify genes whose products may be useful when delivered as replacement genes.

SUGGESTED READINGS

Commercial Implications

Chang PL (Ed.). Somatic Gene Therapy. CRC Press, Boca Raton, FL, 1995.

FDA (CBER). Points to consider on plasmid DNA vaccines for preventive infectious disease indications, Docket No. 96N-0400. Food and Drug Administration, Washington, DC. 1996.

Friedman T. The future for gene therapy—a reevaluation. Ann NY Acad. Scc 265:141–152, 1976.

Gene therapy therapeutic strategies and commercial prospects. TIBTECH 11 (5, Special Issue), May 1993.

Hines RN, O'Conner KC, Vella G, Warren W. Large scale purification of plasmid DNA by anion exchange high performance liquid chromatography. Biotechniques 12:430–433, 1992.

Horn NA, Meek JA, Budahazi G, Marquet M. Cancer gene therapy using plasmid DNA: Purification of DNA for human clinical trials. Hum Gene Therapy 6:565–573, 1995.

Ledley FD. Pharmaceutical approach to somatic gene therapy. Pharmaceut Res 13:1595–1614, 1996.

Mahato RI, Smith LC, Rolland A. Pharmaceutical perspectives on nonviral gene therapy. Adv Genet 41:95–156, 1999.

Maniatis T, Fritsch EF, Sambrook J. Molecular Cloning: A Laboratory Manual, 2nd ed. Cold Spring Harbor Laboratory, Cold Spring Harbor, NY, 1989.

Mrsny RJ. Special feature: A survey of the recent patent literature on the delivery of genes and oligonucleotides. J Drug Target 7:1–10, 1999.

Richter J. Gene transfer to hematopoietic cells—the clinical experience. Eur J Haematol 59:67–75, 1997.

Smith J, Zhang Y, Niven R. Toward development of a non-viral gene therapeutic. Adv Drug Delivery Rev 26:135–150, 1997.

Technology Advancement

Anderson NG, Anderson NL. Twenty years of two-dimensional electrophoresis: Past, present and future. Electrophoresis 17:443–453, 1996.

Drews J, Ryser S. The role of innovation in drug development. Nat Biotech 15:1318–1319, 1997.

Kasianowicz JJ, Brandin E, Branton D, Deamer DW. Characterization of individual polynucleotide molecules using a membrane channel. Proc Natl Acad Sci USA 93:13770–13773, 1996.

Kozal MJ, Shah N, Shen N, Yang R, Fucini R, Merigan TC, Richman DD, Morris D, Hubbell E, Chee M, Gingeras TR. Extensive polymorphisms observed in HIV-1 clade B protease gene using high-density oligonucleotide arrays. Nat Med 2:753–759, 1996.

Krieg AM, Yi AK, Matson S, Waldschmidt TJ, Bishop GA, Teasdale R, Koretzky GA, Klinman DM. CpG motifs in bacterial DNA trigger direct B-cell activation. Nature 374:546–549, 1995.

Liang P, Pardee AB. Differential display of eukaryotic messenger RNA by means of the polymerase chain reaction. Science 257:967–971, 1992.

Schneider SW, Sritharan KC, Giebel JP, Oberleithner H, Jena BP. Surface dynamics in living acinar cells imaged by atomic force microscopy: Identification of plasma membrane structures involved in exocytosis. Proc Natl Acad Sci USA 94:316–321, 1997.

The impact of new technologies in medicine. Br Med J (BMJ) 7220 (entire issue), 1999.

Web site:
Biotechnology & Pharmaceutical Universe
www.navicyte.com/biolink.html